# FINE RESERVOIR DESCRIPTION

# FINE RESERVOIR DESCRIPTION

## Techniques, Current Status, Challenges, and Solutions

**Huanqing Chen**
Research Institute of Petroleum Exploration & Development, PetroChina, China

ELSEVIER

Gulf Professional Publishing
An imprint of Elsevier

Gulf Professional Publishing is an imprint of Elsevier
50 Hampshire Street, 5th Floor, Cambridge, MA 02139, United States
The Boulevard, Langford Lane, Kidlington, Oxford, OX5 1GB, United Kingdom



ISBN: 978-0-323-95401-3



*Publisher:* Gwen Jones
*Acquisitions Editor:* Glyn Jones
*Editorial Project Manager:* Naomi Robertson
*Production Project Manager:* Prem Kumar Kaliamoorthi
*Cover Designer:* Miles Hitchen

Typeset by MPS Limited, Chennai, India

# Contents

# Preface

Fine reservoir description is one of the basic activities in oilfield development. This increasingly accurate activity has dramatically accelerated the improvement of oilfield development. Its results are widely and successfully integrated into the design and implementation of the plans for progressive extension, development adjustment, and comprehensive management of mature oilfields, the foundation year of oilfield development and waterflooding treatment, secondary development, major development tests, large-scale application of horizontal wells, etc.

We started to work on fine reservoir description for the Yangchang Formation in the Ordos Basin in 2003. Over 19 years, great progress has been made in this discipline in the aspects of content description and its methods/techniques. Thus a systemic summarization is necessary to adapt fine reservoir description to its practical application and to provide a reference for oil and gas field development. Since 2009, we have explored and tested fine reservoir description techniques/techniques. This has been done firstly through a series of research projects, such as the "Technologies for Safe Development of $CO_2$-bearing Gas Reservoir and Utilization of $CO_2$" (the No. 16 project under the National Major Science and Technology Project during the 11th Five-Year Plan "Development of Large Oil and Gas Fields and Coalbed Methane"). This general program has been supported by the China Postdoctoral Science Foundation, while the project was sponsored by the Innovation Funds for Youth and Middle-aged of PetroChina Research Institute of Petroleum Exploration Development. Secondly, it's been done through several production projects, such as the volcanic rocks in the Daqing Oilfield, glutenite in the Xinjiang Oilfield, and heavy oil in the Liaohe Oilfield. Since 2014, we have been engaged in the fine reservoir description projects of PetroChina Company Limited ("PetroChina"). In this regard, we reviewed and analyzed much literature around the world, and have understood the current status and problems of fine reservoir description by site visits to oilfields. The results thereof comprise a part of this book. Fine reservoir description is a systematic discipline, which covers almost all aspects of development geology, and the fine reservoir description techniques/techniques are diverse depending on available data and objects. It can hardly be expounded in just one publication. Based on the literature review, this book illustrates the fine reservoir description from aspects of its current status, content, key methods/techniques, and trends, taking the volcanic reservoirs of the Yingcheng Formation in the Xudong area of the Songliao Basin, the glutenite reservoirs of the Lower Karamay Formation at the northwestern margin of the Junggar Basin, and the sandstone reservoirs of the first member of the Shahejie Formation ("Sha 1 Member") in the western Liaohe Basin, as examples.

In this book, the key problems and the current status of fine reservoir description for three types of reservoirs (i.e., high water-cut, low-permeability, and complex lithology) are summarized, and key contents of fine reservoir description are highlighted. High-resolution sequence stratigraphy is applied to fine stratigraphic classification and correlation, with eight principles for layer classification proposed for the first time, and the correlation between stratigraphic boundaries defined by high-resolution sequence stratigraphy and traditional techniques is clarified. The existence of fractures in volcanic reservoirs is delicately depicted by synthesizing various data to provide guidance for development planning. The sedimentary facies-based architecture characterization of the glutenite reservoirs of alluvial fan refines the description of a single sand body. The corresponding classification system for different classes of architecture is established, which corresponds to the traditional sedimentology-based system and highlights the role of lithology analysis in reservoir architecture characterization. The pore structure of the reservoir is microscopically classified and evaluated according to its origin, and its influence on oilfield development is analyzed. The research methods of reservoir heterogeneity and the permeability heterogeneity characteristics are systematically introduced. Based on geological genesis analysis, qualitative analysis is integrated with quantitative classification for comprehensive reservoir evaluation. Through cluster analysis and discriminant analysis, the flow units of the reservoir are studied, thereby providing technical support for the conversion of the exploitation method from steam stimulation to steam flooding for better heavy oil recovery. Focusing on multipoint geostatistics, we have introduced geological modeling technology for fine reservoir description using examples. The current status of the remaining oil is summarized systematically, and the methods for the remaining oil characterization are presented, and their advantages and disadvantages elucidated. Finally, the challenges for fine reservoir description are outlined, with recommendations supplied for their corresponding solutions.

We are particularly grateful to the guidance and assistance provided by Prof. Yongle Hu, Prof. Chengfang Shi, Prof. Jiuqiang Jin, Prof. Changbing Tian, Prof. Baozhu Li, Prof. Yixiang Zhu, Prof. Xingjun Gao, Prof. Jigen Ye, Prof. Yingcheng Zhao, Prof. Qiquan Ran, Prof. Hujun Zhang, and other experts from the PetroChina Research Institute of Petroleum Exploration Development! We also appreciate the guidance and help of Dr. Lin Yan, Dr. Min Tong, Dr. Yongjun Wang, and Dr. Jing Zhang of the Daqing Volcanic Rocks Research Team; Dr. Shunming Li and Dr. Ping Jiang of the Xinjiang Glutenite Research Team; Engineer Jue Wang, Engineer Yijing Du, and Engineer Yao Yao of the Liaohe Heavy Oil Research Team; and Yao Hong and Yuhao Sui of the Fine Reservoir Description Research Team.

We would like to express our heartfelt thanks to the directors and experts from the Daqing, Xinjiang, Liaohe, Changqing, Jilin, Dagang, Jidong, Huabei, Qinghai, Yumen, Tuha, Tarim, and other oilfields of PetroChina for their lavish support and assistance to this work. Our special thanks also go to Songquan Li (the Chief Geologist of Changqing Oilfield), and Haiyan Hu, Hongbiao Wu, and Chen Cao (the Deputy Chief Geologist, the Deputy Director, and the Senior Executive, respectively, of the PetroChina Exploration and Production Company).

We thank Prof. Zhihao Qu, Prof. Yushuang Zhu, Prof. Wenhou Li, and Prof. Zhichao Mei from the Northwest University for their guidance and help.

We also thank Prof. Yongle Hu, Prof. Chengfang Shi, and other experts for reviewing the manuscript and putting forward many valuable suggestions.

Due to the limited level of knowledge and experience, we could not avoid inappropriate statements in this book. Your comments and criticism are thereby warmly welcomed.

# CHAPTER 1

# Introduction to fine reservoir description

Reservoir description is a new technology used for the exploration and development of oil and gas fields. It cropped up as a discipline in the 1930s and grew in the 1970s. In China, since being initially introduced during the period of the 7th Five-Year Plan, this technology has spread quickly and gained significant attention of researchers in oil and gas exploration and development. At present, it is widely applied in production, with remarkable economic and social benefits (Yiwei et al., 1997; Huanqing and Xiaomin, 2008; Ailin, 2010).

## 1.1 Overview

Qiu and Chen (1996) defined reservoir description as the development geological characteristics description of an oil (gas) reservoir after discovery, and its main purpose is to provide necessary and reliable geological basis for the development strategy and technical measures of this oil (gas) reservoir; in short, reservoir description is comprehensive research into and evaluation of a reservoir. Yongmin et al. (2004) indicated that the fine reservoir description in the middle-late development stage of an oilfield is the fine geological study of the oilfield in the middle-high or extra-high water-cut stage to clarify the distribution and controlling factors of remaining oil in the oilfield through constantly improving the geological model of the reservoir and quantifying the distribution of remaining oil, with the deepening of reservoir exploitation and the availability of more production performance data, eventually aiming to make the oilfield develop economically and effectively with enhanced oil recovery (EOR). In this book, fine reservoir description refers to the discipline of fine geological research and remaining oil description, together with the improvement of existing geological models and quantification of remaining oil distribution, carried out with the exploitation of the reservoir and the increase of dynamic and static

*Fine Reservoir Description*
DOI: https://doi.org/10.1016/B978-0-323-95401-3.00001-9

data after an oilfield is put into production. For mature oilfields in the middle-late development stage, the fine reservoir description is helpful to understand the geological characteristics of the reservoir comprehensively, and it is of great practical significance for enhancing the oil recovery and tapping the potential of the remaining oil.

Internationally, Lake and Carroll (1986) compiled the *Reservoir Characterization*, which collects the papers on the latest progress in reservoir characterization. Stoudt and Paul (1995) published the *Oil and Gas Reservoir Characterization—Modeling of Geological Framework and Flow Unit*, which involves the reservoir description and characterization using sequence stratigraphy, comprehensive petrology, and engineering data, the improved understanding of reservoir properties using outcrop data, the flow characterization of fluids in dolomized carbonate slope reservoir, the three-dimensional modeling of shallow carbonate slope based on geostatistics, and the influence of reservoir geological characterization on flow unit modeling. Richard and Jordan (1999) edited the AAPG album "Reservoir Characterization—Recent Progress." Rajesh et al. (2001) characterized the oil-bearing reservoirs under complex geological conditions in mature oilfields based on geostatistics, with an example from Carpinteria, California, and he used a large amount of data for statistical analysis to minimize the uncertainty of reservoir prediction. Masoud and Aminzadeh (2001) made a comprehensive analysis of the current status and trend of intelligent reservoir characterization technology; he concluded that this intelligent technology includes expert system, artificial intelligence, neural network, fuzzy logic, genetic algorithm, probabilistic reasoning, and parallel processing and proposed the process of intelligent reservoir characterization (Fig. 1.1). Lars et al. (2003) made a

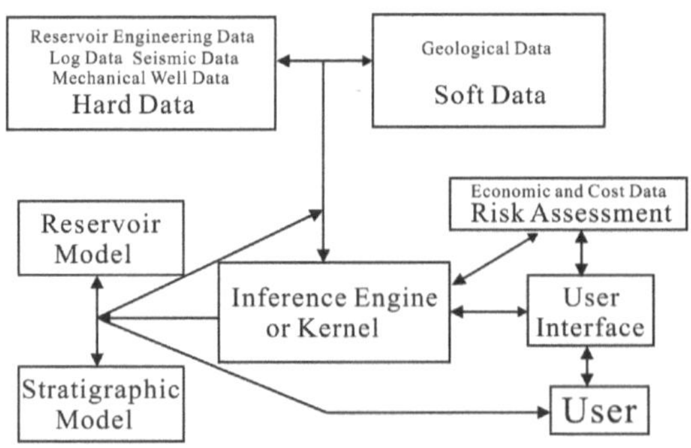


**Figure 1.1** Process of intelligent reservoir characterization (Masoud and Aminzadeh, 2001).

detailed analysis and comparison of reservoir stochastic structure modeling and expounded the modeling of fault and horizon consistency; he indicated that, in addition to seismic data, borehole data are available for determining the petrologic features and the plane development characteristics of strata and faults. Ouahed et al. (2005) used the artificial intelligence method to study the natural fault belts in Hassi Messaoud Oilfield, Algeria, with the two-dimensional map of fracture intensity obtained from the well to depict the development trend and location of major faults. Isha and Roland (2005) improved the reservoir description to great effect by using the multiresolution wavelet analysis technique. Hisafumi et al. (2006) analyzed the application of magnetometric resistivity model in three-dimensional reservoir characterization, using the geothermal reservoir at the western margin of Mt. Aso Volcano in southwestern Japan as an example, and proved that this method is efficient for ascertaining the deep reservoir structure. Zafari and Reynolds (2007) analyzed the uncertainty in reservoir description and reservoir properties prediction using Ensemble Kalman Filter, which can be used easily in combination with reservoir simulation. Jerry (2007) published the *Carbonate Reservoir Characterization—An Integrated Approach (Second Edition)*, which elaborates on the petrophysical rock properties, rock-fabric classification, wireline logs, depositional textures and petrophysics, reservoir models for input into flow simulators, limestone reservoirs, and dolostone reservoirs. Remeysen and Swennen (2008) analyzed the possibilities in and limitations of fine reservoir description to understand what happens when the micro-computed tomography (Micro-CT) is used to characterize carbonate reservoirs, and they pointed out that Micro-CT can acquire three-dimensional images of minerals and pore structure of sedimentary reservoirs without damage to the reservoir, which is extremely competitive to the traditional two-dimensional thin section analysis. Pyrcz et al. (2009) proposed the event-based stochastic modeling of the alluvial fan depositional system. Khidir and Catuneanu (2010) took the fluvial sandstone of Scollard sequence in the precontinent of Alberta, Canada, as an example to conduct detailed reservoir characterization, revealing that diagenesis has a great influence on reservoir properties and that the hydrocarbon storage capability of the reservoir is closely related to the rock-forming history. Darabi et al. (2010) used artificial intelligence tools to construct the three-dimensional model of the naturally fractured reservoir in the Parsi Oilfield, where with the support of two neural networks (multilayer induction and radial basis function) he showed that the radial basis function plays a better role in the depiction of fracture index. Ramstad et al. (2010) used the lattice Boltzmann method to simulate two-phase flow in reservoir rocks. Artun and Mohaghegh (2011) analyzed the workflow of intelligent seismic inversion in high-resolution reservoir characterization and completed the inversion using well point data to constrain seismic data. Leite and Vidal (2011) used seismic inversion and neural network technology to predict the three-dimensional pore structure of the reservoir, where seismic reflection data were firstly used to establish the petrophysical model of the reservoir. Qazvini Firouz et al. (2012) investigated the relationship between productivity index and diffusion coefficient and analyzed their application in reservoir description, with the use of the Asmari reservoir of the Iranian Persian Gulf coast as an example. Chekani and Kharrat (2012) made a comprehensive characterization of carbonate reservoirs in an oilfield in Iran; in this study, rocks were classified according to flow zone indicator and initial water saturation, and the results were verified by scanning electron microscopy (SEM) photographs, pore-throat radius, grain size analysis data, and thin sections, which showed a good agreement. Gupta et al. (2012) analyzed the petrophysical model using log data and seismic data applied to lithology and fluid classification, with the use of the Cambay Basin, India, as an example and concluded that the petrophysical model is a very effective tool for identifying oil and gas anomalies in undrilled areas and for assisting in the prediction and evaluation of reservoirs based on seismic data. Kadkhodaie-Ilkhchi et al. (2013) analyzed the logging facies in different flow units in the Willespie tight sandstone reservoir in Whicher Range, the Perth Basin, Western Australia, and proposed a method that makes it possible to analyze the reservoir flow units by using the response characteristics of logs.

Through analysis, it is found that the foci in reservoir characterization/description research efforts abroad have changed from conventional outcrop analysis, experimental analysis, and numerical simulation on the one hand, to seismic inversion-based reservoir prediction, microseismic reservoir characterization, artificial intelligence neural network, and other innovative techniques and methods, on the other, and the objects have changed from traditional sandstone and carbonate rocks to tight oil and gas reservoirs. In these researches, comprehensive seismic technology and computer technology are further strengthened. Heterogeneity has always been the emphasis in reservoir characterization. At present, the research on it has expanded from the influence of heterogeneity on porosity, permeability, and other reservoir properties to the influence of reservoir heterogeneity on the development and production of oil and gas fields. The seismic data were applied to simple reservoir inversion forecast originally, and now three-dimensional or four-dimensional seismic data are used to monitor and analyze the development process of oil and gas reservoirs. Initially, reservoir characterization depended on Kriging and various geostatistical methods; nowadays, more emphasis is laid on the establishment of various forms of models for getting an in-depth understanding of reservoir properties through simulation. For reservoirs that are obvious in geological characteristics but difficult to investigate, some specific researches (naturally fractured reservoir characterization, for example) are available.

In China, the research efforts on fine reservoir description began with learning from relevant advanced experiences abroad. Yinan et al. (1993) translated and published the *Foreign Reservoir Description Technologies*, which presents the latest progress in reservoir description internationally from the perspective of geostatistical technology, seismic technology, and logging technology, thereby generating positive impacts on the research efforts in China. Yiwei et al. (1997) published the *Nonmarine Reservoir Description*, which comprehensively introduces the description technologies for unique nonmarine reservoirs within China in respect of oil accumulation model and reservoir model, reservoir description in the exploration stage, reservoir description in the early development stage, and reservoir description in the middle-late development stage. Longxin and Yinan (1999) published the *Reservoir Description in Different Development Stages*, which elaborates the main characteristics, technical requirements, and key contents of reservoir description in different development stages and, also, presents the main techniques of reservoir description. Zhongran et al. (2004) used the logging-constrained inversion to describe low-permeability reservoirs and confirmed that this method can achieve tracking prediction of a reservoir, guide drilling operation reasonably, adjust the well location in time, and improve the implementation effect of the development plan. Lixin (2006) expounded on the important role of reservoir geological modeling in reservoir description using the Nanpu Oilfield as an example. Huanqing and Xiaomin (2008) summarized the progress of sedimentary microfacies modeling in fine reservoir description and introduced the widely used object- or pixel-based stochastic modeling techniques to model reservoirs using the geologic, geophysical, and oilfield development performance data and some new modeling techniques like architecture analysis and interwell seismic. Anna et al. (2009) gave a detailed introduction of the reservoir description and its evaluation method integrating seismic, logging, and geologic data, using the first member of Dongying Formation in the Nanpu 1 Structure as an example. Ailin (2010) established a program of digital fine reservoir description. Shujuan et al. (2011) developed a set of unique methods/techniques for fine reservoir description applicable to the late development stage of the carbonate buried-hill reservoir in Renqiu Oilfield. Lideng et al. (2012) summarized five key techniques of seismic reservoir description, including seismic petrophysical analysis, well-controlled seismic data processing, well-controlled fine structure interpretation, logging-seismic joint inversion, and seismic-constrained reservoir modeling and numerical simulation. Xiantai et al. (2013) made a fine description of the reservoirs with low-order faults, together with four effective techniques, that is, coherence cube, multiscale frequency division, curvature and seismic attribute fusion, and RGB display. To sum up, in China, a set of mature fine reservoir description procedures and techniques has been established, instead of the simple imitation of reservoir description techniques found in other countries initially. The reservoir description research is relatively comprehensive and perfect, covering almost all the aspects of reservoir geology research. Moreover, works related to the fine reservoir description of glutenite reservoir, reservoir with heavy oil thermal recovery, and low-permeability reservoir have been published (Shanwen et al., 2003; Qinglong et al., 2010; Lin et al., 2013). Although the mentioned reservoirs do not differ too much from conventional reservoirs in description techniques and contents, it is indicated at least that they have become attractive and concerned to the researches.

As to the content, fine reservoir description involves two major concerns: fine structure interpretation and accurate reservoir prediction. As to the purpose, fine reservoir description mainly deals with the depiction of the remaining oil distribution. Based on available literatures (Lake and Carroll, 1986; Lake et al., 1989; Zhihao et al., 1994; Longxin and Yinan, 1999; Masoud and Aminzadeh, 2001; Pingping et al., 2003; Ouahed et al., 2005; Shouyu, 2005; Remeysen and Swennen, 2008; Huanqing and Xiaomin, 2008; Anna et al., 2009; Adeniran et al., 2010; Qigu and Gongyang, 2010; Qazvini Firouz et al., 2012; Ailin and Lihua, 2012; Fic and Pedersen, 2013; Daqing et al., 2013; Huanqing et al., 2015a,b,c,d), we compared the advantages and disadvantages of fine reservoir description in China and abroad (Table 1.1). Although the research of fine reservoir description in China is much more systematic, the extent of research abroad is worthy of being learned. Through literature review (Chengyan, 2000; Ping et al., 2004; Wu, 2004; Narr et al., 2006; Zafari and Reynolds, 2007; Khidir and Catuneanu, 2010; Ramstad et al., 2010; Shenghe, 2010; Leite and Vidal, 2011; Huanqing et al., 2011a,b,c; Gupta et al., 2012;

**Table 1.1** Fine reservoir description researches in China and other countries.

| Division | Advantages | Disadvantages |
|---|---|---|
| Other countries | 1. Experimental study on reservoir porosity, permeability, and water saturation is relatively thorough and mature<br>2. Study on various mathematical and geostatistical algorithms in fine reservoir description is thorough<br>3. Artificial neural network and other new methods/techniques that can fully reflect computer technology are frequently engaged<br>4. Application of 4D seismic technology in dynamic monitoring of reservoirs in the middle-late development stages is frequently discussed<br>5. Application of nuclear magnetic resonance (NMR) logging and other new techniques to understand the reservoir properties is frequently and successfully put to use<br>6. Significant achievements have been made in the study of unconventional reservoirs like tight oil and heavy oil reservoirs | 1. Inadequate attention is paid to the fine stratigraphic classification and correlation in fine reservoir description<br>2. Insufficient emphasis is given to the programmability and systematization of fine reservoir description |
| China | 1. Study on reservoir geological properties is thorough<br>2. Establishment of high precision stratigraphic framework based on high-resolution sequence stratigraphy is mature<br>3. Activity and content of fine reservoir description research are systematic and programmatic<br>4. Adequate attention is paid to the remaining oil distribution | 1. Experimental study concerning fine reservoir description is weak<br>2. Mathematical statistics analysis is not deep enough<br>3. New methods/techniques like artificial neural networks, 4D seismic, and NMR logging are rarely applied<br>4. Research of unconventional reservoirs, including tight oil and heavy oil reservoirs, is still in the initial stage, with imperfect methods/techniques thereof in use<br>5. The features of research on fine description of low-permeability and naturally fractured reservoirs with distinct characteristics are not highlighted |

*Source:* Based on Huanqing, C., 2021. Progress in the fine description of reservoirs in China and its prospect. Geology in China 48 (2), 424–446.

Kadkhodaie-Ilkhchi et al., 2013; Huanqing et al., 2013a,b,c; Huanqing et al., 2014a,b; Zhilin, 2014; Huanqing et al., 2015a,b; Huanqing et al., 2016a,b,c,d), combined with our practical researches, we put forth the thesis that the fine reservoir description researches are vastly different between China and other countries (Table 1.1). To fully understand the current status of research on fine reservoir description around the world, and to be updated on how far it has progressed, this book presents a summary from three aspects: high water-cut oilfield, low-permeability oilfield, and complex lithologic reservoir.

## 1.2 Fine reservoir description of high water-cut oilfield

Fine reservoir description of high water-cut oilfields, different from that in the early and middle development stages, is mainly based on logging and reservoir dynamic data when the oilfield with a relatively perfect development well pattern enters the stage with high water-cut (more than 80%). Such reservoir description requires a high accuracy; for this purpose, a sound three-dimensional reservoir geological model should be built, and the spatial distribution of remaining oil should be determined, so that specific measures to control water and stabilize oil production can be figured out to ultimately enhance the oil recovery (Wanchao, 2003). The viscosity of crude oil produced from continental sedimentary reservoirs in China is generally high. For example, the crude oil viscosity in the majority of oilfields in major oil-producing areas in eastern China is higher than 5 mPa/s. In this circumstance, a large number of recoverable reserves in these oilfields should be produced at a high water-cut stage (Naiju et al., 1999). Qinlin (1999) also pointed out that 40%–60% of the recoverable reserves in high water-cut oilfields in China should be recovered at the stage with a water-cut of 80%–98%. Therefore, the fine reservoir description of a high water-cut oilfield is greatly significant to the production practice. This book comprehensively reviews the progress of studies on fine reservoir description of high water-cut oilfields to provide a reference point centering which future research directions can grow out promoting the technology of fine reservoir description and improving the development level of water-cut oilfields.

Many scholars have worked a lot on fine reservoir description of high water-cut oilfields. Raju et al. (2010) studied the scale inhibitor treatment in horizontal wells in high water-cut carbonate oilfields based on the mathematical model, indicating that addition of phosphonate can effectively prevent calcium carbonate scaling. Manichand et al. (2010) made a preliminary evaluation on the polymer flooding pilot area in the Tambaredjo heavy oilfield and selected a well cluster with inverse five-spot flooding in an onshore block of Suriname for testing, which demonstrated a certain performance with increasing oil production and declining water-cut after one year of continuous injection. Taware et al. (2011) developed a practical numerical simulation approach for history matching using grid coarsening and streamline-based inversion, taking a giant carbonate reservoir with high water-cut as an example. Ghosh et al. (2012) studied the optimization of formation tester sampling and perforation positions using multidimensional nuclear magnetic resonance (NMR) technique in high water-cut mature oilfields, with the use of the Nilan area in the southeast of Great Mumbai Oilfield in the West Coast Basin, India, as an example; he indicated that the identification of formation fluids was crucial and the multidimensional NMR technique could effectively optimize the sampling and perforation positions in carbonate reservoirs through observing the change of reservoir physical properties taking place with the change in vertical depth. In China, many researchers have conducted in-depth research for Daqing, Shengli, Xinjiang, Dagang, and Jianghan oilfields. Wanchao (2003) presented, in his work "Development Techniques and Methods for High Water-Cut Oilfields," three types of development techniques, that is, fine reservoir description, enhanced water flooding recovery, and chemical flooding, with the use of the Jiyang Depression of Shengli Oilfield as an example. Gao et al. (2013) took the Well Yi11 area in Bonan Oilfield as an example and proposed the techniques for developing the low-permeability oilfield in ultrahigh water-cut stage, including the techniques of unstable water injection, optimized single-well liquid production intensity, improved injection-production well pattern, and enhanced producing degree of reserves in nonmajor pay zones. Zongbao et al. (2014) took the Putaohua reservoir in the north fault block of Xingnan Oilfield in the Songliao Basin as an example to investigate the enrichment and potential tapping of remaining oil at the edge of the fault in a high water-cut stage. Yupu et al. (2014) divided the development of continental sandstone oilfields in China into four stages according to water-cut, as shown in Table 1.2. Chao et al. (2015) figured out the methods for identifying the communication between injection wells and production wells and those for optimizing the injection volume in high water-cut mature oilfields, and he successfully applied such methods in the Gao 5 fault block in Jidong Oilfield. Lihong et al. (2015) analyzed the three major contradictions in the development of large continental multilayer sandstone reservoirs in the ultrahigh water-cut stage, with

**Table 1.2** Classification of oilfield development stages by water-cut (Qinlin, 1999; Wanchao, 2003; Yupu et al., 2014).

| Stage | Water-cut/% | Oilfields | |
|---|---|---|---|
| | | China | Other countries |
| Low water-cut stage | 0–20 | | |
| Middle water-cut stage | 20–60 | Changqing | |
| High water-cut stage | 60–90 | Yumen, Tuha, Xinjiang, Dagang, Liaohe, Jilin, Yanchang, Henan, and Jianghan | Great Mumbai Oilfield in the West Coast Basin of India, Romashkinskoye Oilfield, Dumaz Oilfield, Alan Oilfield, Samotlol Oilfield, East Texas Oilfield |
| Extra-high water-cut stage | >90 | Daqing, Shengli, Zhongyuan | Marmul Oilfield in Oman, Rapdan Oilfield in Canada, Yates Oilfield in the United States, Hankesbuetted Oilfield in Germany |

the use of Lasaxing Oilfield in Daqing as an example, and they put forward corresponding countermeasures, which were successfully applied in six demonstrative zones of potential tapping by water flooding in Lasaxing Oilfield. Xiaojie et al. (2015) carried out a comprehensive study on remaining oil by fine numerical simulation for the lower oil formation in Triassic in Tahe 1 zone. At present, the studies on fine reservoir description of high water-cut oilfields in China focus on remaining oil characterization and potential tapping, flooded layer logging interpretation, and dominant flow pathway, by means of geologic, experiential, numerical, and physical simulations, exclusively aiming to stabilize oil production, control water-cut, and tap the potential of remaining oil. The fine reservoir description of a high water-cut oilfield can provide evidence for implementing the tertiary recovery program and measures of tapping the potential of the remaining oil, including those of well infilling, horizontal well deployment design, the progressive extension of mature oilfields, and polymer flooding.

Domestically, many oilfields entered the high water-cut stage in the 1980s; thus, the research efforts of high water-cut oilfields have been going on for decades, with some satisfactory progress made. Some scholars summarized the problems in developing such oilfields (Yu, 2016). This book holds that the fine reservoir description of high water-cut oilfields is mainly challenged by the issues in five aspects. First, the fine structure interpretation and reservoir prediction based on the increasing availability of data are defective. The development of most high water-cut oilfields for over three decades has contributed a large number of dynamic and static data, which facilitate and also impede the relevant research efforts. Taking the interpretation of fault system as an example, only the fault interpreted as fourth- or fifth-order in the structural interpretation in the high water-cut stage is considered feasible for development. However, limited to the seismic data dominated by 2D or old 3D data, the structural interpretation results cannot match the dense well pattern data given accuracy and precision, indicating a serious problem. The inconsistency between well logging and seismic data in reservoir prediction also exists. Influenced by the accuracy of reservoir prediction, the optimization design of horizontal wells is greatly restricted. Second, the underground oil and water movement law is complex, and a dominant flow path exists. After a long-term water flooding development, the distribution and movement of underground oil, gas, and water become complex, and thus they are difficult to identify accurately. In addition, due to the strong heterogeneity of continental sedimentary reservoirs, the dominant flow path leads to an ineffective circulation of injected water, which increases the difficulty in effective water flooding development. Third, in the development process, reservoir properties change as influenced by fracturing, compatibility of injected water, and other factors. The resulted reservoir damage brings big challenges to subsequent development. Fourth, rapid variation of continental sedimentary facies, strong reservoir heterogeneity, development of fault system, diagenesis combined with the early development measures lead to the complex distribution of remaining oil in reservoirs and which present a difficulty in its accurate characterization. Fifth, after a long period of water flooding, abundant static and dynamic data of high water-cut oilfield are collected. Most of these data are paper-based. It is very difficult to establish a unified database to realize information management and application of these data and, also, to build a platform for fine reservoir description data and results management and application for improving the efficiency and research level of fine reservoir description.

### 1.2.1 Key problems

Fine reservoir description is a systematic discipline, covering all aspects of, especially, those of high water-cut oilfields. In the current context of low oil prices, all oilfields try to cut costs and increase efficiency. Given the significantly reduced investment, it is necessary to sort out and analyze the key problems in fine reservoir description of high water-cut oilfields. We conclude such issues in five aspects.

#### *1.2.1.1 Fine stratigraphic classification and correlation*

Fine stratigraphic classification and correlation are fundamental for all matters in fine reservoir description, and especially so for high water-cut oilfields. The reservoirs in high water-cut oilfields are diverse in genetic types and involve various rocks like clastic rocks and carbonates (Hongwen et al., 1997; Zhigui et al., 1998; Hongwen et al., 2002; Hanqing, 2005; Weidong et al., 2007; Rongcai, 2010; Youliang et al., 2012). For the high water-cut oilfields in the middle-late development stage, the layer-level stratigraphic classification is no longer valid and feasible for oilfield development and production. A more detailed stratigraphic classification is required, while the traditional stratigraphic classification based upon "cyclic correlation, hierarchical control" is far beyond this requirement. The high-resolution sequence stratigraphy, based on a single sand body, is considered ideal for current research on fine reservoir description of high water-cut oilfields for its geologic genesis analysis and spatial classification and correlation of stratigraphic base-level cycles at different orders (Fig. 1.2). Theoretically, it realizes the stratigraphic classification and correlation in the sense of geologic genesis, agreeing more with the underground geology; technically, it refines the stratigraphic classification and correlation to the level of the single layer (Huanqing et al., 2014a,b). In practice, however, the theory and method of high-resolution sequence stratigraphy have not really played an effective role in fine reservoir description. On one hand, a weak database, work continuity and habits, insufficient efforts to research, and inadequate financial support impose constraints. On the other hand, the high-resolution sequence stratigraphy itself is defective, for example, in the correspondence between the stratigraphic classification by high-resolution sequence stratigraphy and the stratigraphic classification by traditional systems. Nevertheless, we should envisage more the high performance of the high-resolution sequence stratigraphy when it is applied in the fine classification and

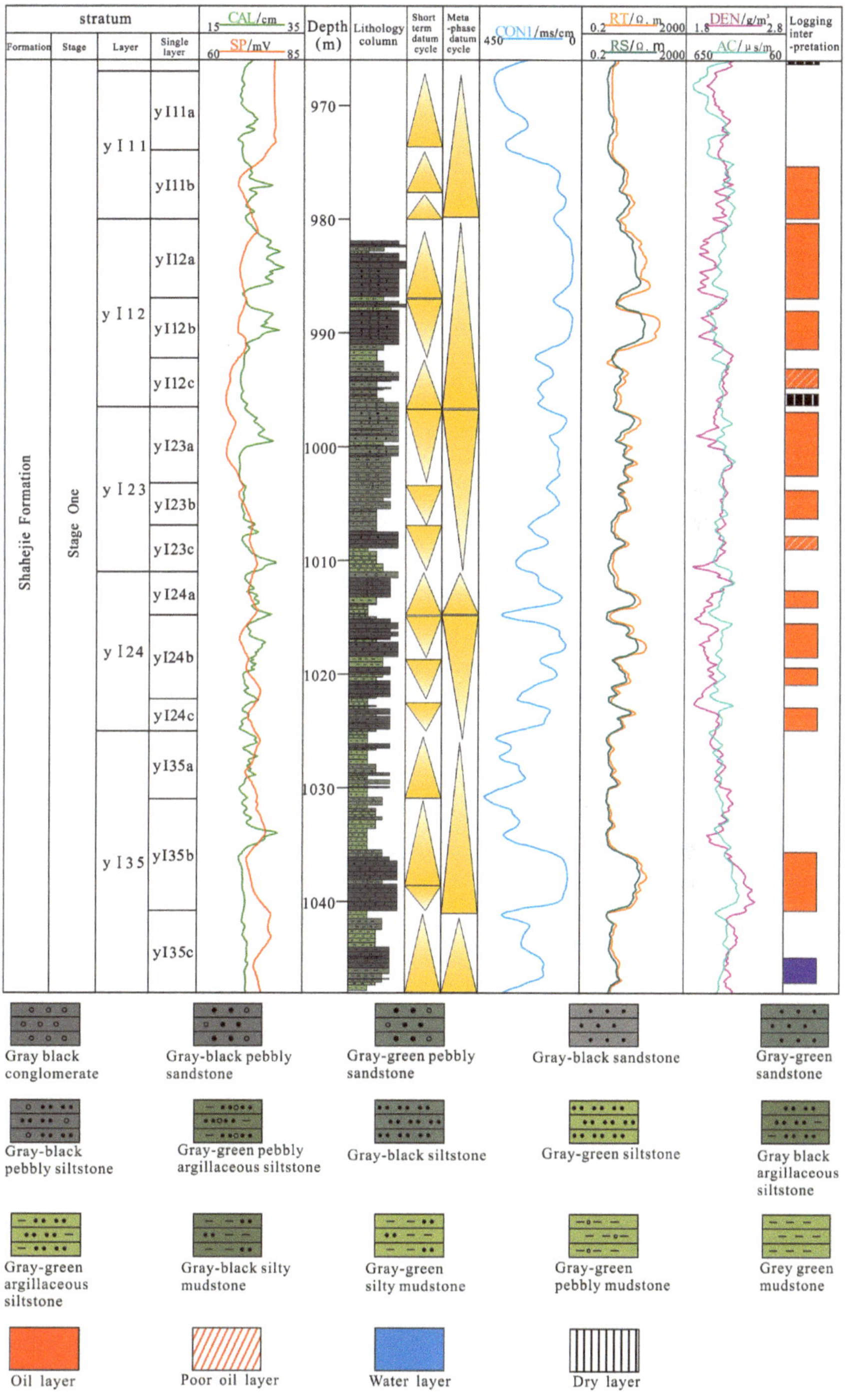


**Figure 1.2** Response model of the short-term base-level cycle for Well W3 in Yulou reservoir in Western Sag, the Liaohe Basin.

correlation of continental sedimentary strata in China, despite the problems mentioned earlier. We believe that this technique, after improvements and innovations are applied constantly, can play its due role in the fine reservoir description of high water-cut oilfields and, thus, lay a solid foundation for the research advancement thereof.

#### 1.2.1.2 Reservoir heterogeneity

Reservoir heterogeneity is always a core area of reservoir geologic study; thus, it is also a key subject in the research of fine reservoir description. Reservoir heterogeneity involves abundant contents, typically including the study of flow field heterogeneity and fluid heterogeneity (Yongsheng, 1993). First, the flow field heterogeneity includes interlayer heterogeneity, intralayer heterogeneity, plane heterogeneity, and pore heterogeneity (Shenghe and Qihua, 1998). Then, fluid heterogeneity refers to the differences in properties of oil, gas, and water stored in reservoirs. The former is a focus area in most studies, while the latter has been rarely reported. As the uppermost matter deserving of concern in the study on reservoir flow field heterogeneity, the permeability derived from fine logging interpretation is often used to calculate the variation coefficient, breakthrough factor, and differential, which are then analyzed for their spatial variation to quantitatively characterize the reservoir flow field heterogeneity. Although log data are quantitative and accessible, the accuracy of the above parameters is somewhat in question as the parameters are constrained by the accuracy of logging interpretation of reservoir physical properties.

In high water-cut oilfields, with the progress of development and production, interlayer and intralayer contradictions are becoming increasingly prominent, which means fine reservoir description requires much more elaborate efforts at researching to resolve these contradictions. The practice of simply using the calculated permeability heterogeneity to describe the reservoir flow field heterogeneity is considered far from the need of production. Emphasis should be placed on the barriers or baffles between or inside layers or beds (Fig. 1.3). Many baffles in a pay zone become barriers between layers or beds under a more detailed classification at layer or bed level; such barriers are smaller in spatial scale and vertical thickness, and the baffles also become more scattered. This book proposes two aspects that the research of intercalation (barriers and baffles collectively) in high water-cut oilfields should focus on. One is evaluating the effectiveness of intercalation, especially the intercalation that can effectively

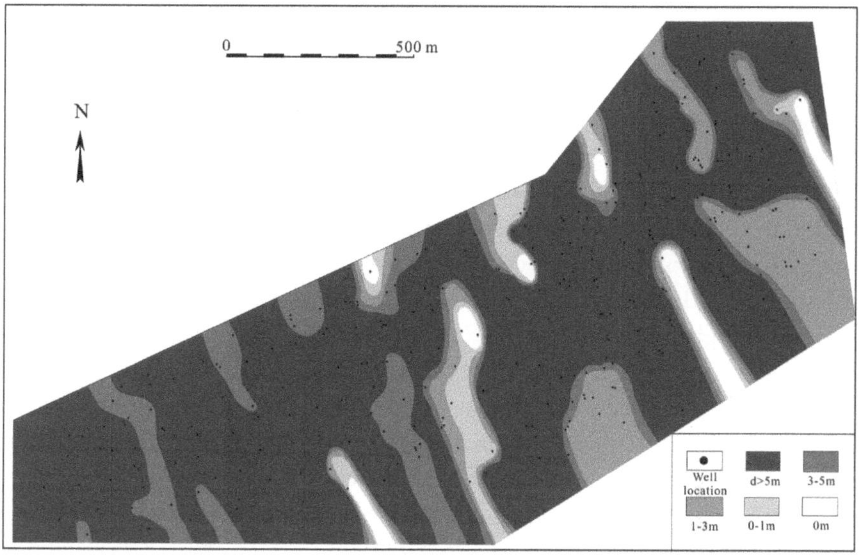


**Figure 1.3** Barriers between $yl_3^{6c}$ and $yll_1^{1a}$ of Yulou Reservoir in Western Sag, the Liaohe Basin.

isolate oil, gas, water, and other fluids in space. Intercalations of different lithologies are significantly different in sealing capability, for which the evaluation standards may be different depending on oilfields or oil zones. At present, the main methods for evaluating the effectiveness of intercalation include laboratory testing, physical simulation and numerical simulation, and a simple and effective test is of course the summary of experience based on production practices. The other is the analysis of the influence of intercalation on the movement of underground oil, water, and gas. Since the researches have been further refined, the workload of intercalation characterization increases dramatically. More attention should be paid to the intercalation with significant influence on development techniques like water, steam, or polymer flooding, based on the evaluation of intercalation effectiveness. For small ones which only complicate the motion trace of underground fluids and have no substantial impact on production, devoting too much time and energy on these should be avoided.

The research of fluid heterogeneity is still languishing where making significant achievements is concerned. In the study of fine reservoir description of high water-cut oilfields, various test data (e.g., water test data) and geochemical methods should be considered to describe the spatial, especially vertical variation and heterogeneity, of reservoir fluids such as oil, gas, and water, to provide the basis for tapping the potential of remaining oil and enhancing oil recovery in the late development stage. This book holds that fluid heterogeneity will be an essential object and orientation in reservoir heterogeneity analysis.

#### 1.2.1.3 Change law of reservoirs in the process of development

In the development process of an oilfield, with the implementation of water injection or steam injection measures, the reservoir, including pore structure and clay mineral properties, exhibits a series of changes, which correspondingly result in the change of reservoir permeability and reservoir fluid properties. In the fine reservoir description of high water-cut oilfields, such changes should be fully identified to reduce the damage to reservoirs caused by development, thereby facilitating the potential tapping of remaining oil and thus enhancement of oil recovery. Zhizhang et al. (1999a,b) took the water flooding reservoirs in Shuanghe Oilfield and steam-flooding reservoirs in Block 9 of Karamay Oilfield as examples to give an elaborate account of the variation regularity and mechanism of reservoir parameters in the middle-late development stage. Dario Grana and Mukerji (2015) used the actual seismic data of an area in the Norwegian Sea after Bayesian transformation to predict static and dynamic changes of reservoir properties, and they found that the change of hydrocarbons in reservoirs can be reflected by seismic data. This book analyzes the change of reservoirs before and after steam flooding in an area of Western Sag, the Liaohe Basin, with SEM data (Fig. 1.4). The results reveal a significant increase in clay minerals, especially kaolinite, after steam flooding. The reservoir after steam flooding contains a relatively stable type of clay mineral, while the reservoir before steam flooding is mostly composed of unstable clay minerals like illite/smectite (I/S). The dramatic increase of clay minerals after steam flooding leads to pore and throat blockage, so that the reservoir porosity and permeability reduce significantly.

#### 1.2.1.4 Multiinformation remaining oil characterization

Determination of the remaining oil distribution is a core study area in fine reservoir description of high water-cut oilfields and, also, a hot and difficult topic in current research efforts. Certainly, it will be a key research orientation in the future (Shouyu, 2005). Dakuang (2010) believed that the remaining oil in high water-cut oilfields shows a generally highly scattered and locally relatively enriched distribution pattern. Chengyan et al. (2013)

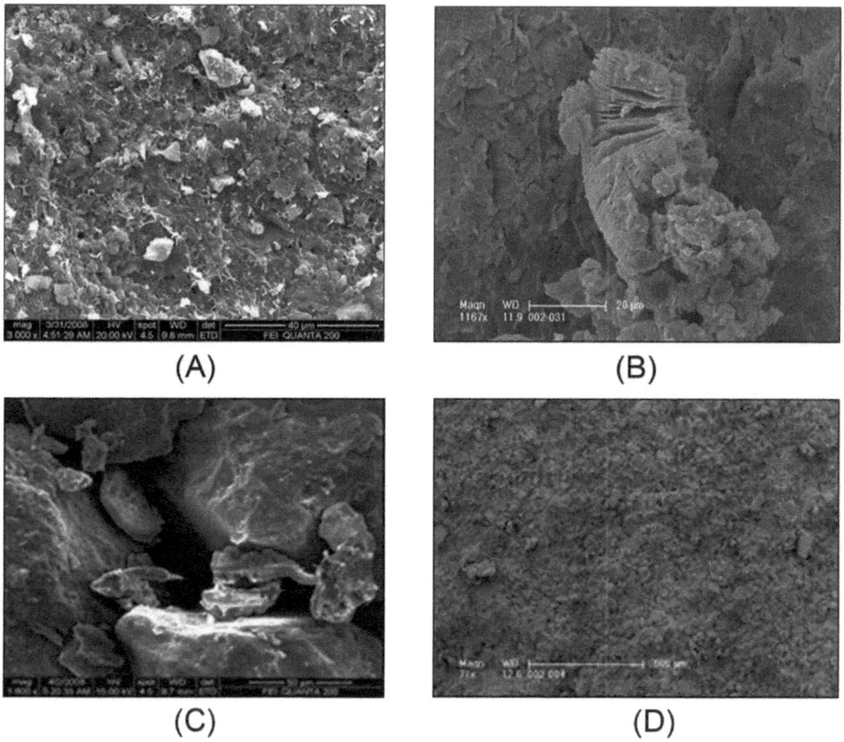

Figure 1.4 Reservoir change before and after steam flooding in Yulou reservoir in an area of Western Sag, the Liaohe Basin. (A) Well W2, before steam flooding, sandstone, I/S, 957.69 m; (B) Well W41, after steam flooding, sandstone, kaolinite, 758.4 m; (C) Well W2, before steam flooding, sandstone, with pores, 1000.76 m; (D) Well W41, after steam flooding, sandstone, with moderate pores, 739.53 m.

proposed the distribution law of remaining oil, potential tapping unit, and corresponding techniques based on single sand body, with the use of the Pu-I reservoir with thin and narrow sand bodies developing by water flooding in high water-cut stage in Pubei Oilfield as an example. Junlong et al. (2013) summarized the logging evaluation technology of remaining oil in middle to the high water-cut stage, divided the oil saturation logging into open-hole logging and cased-hole logging, and introduced different logging techniques and their applicable conditions (Tables 1.3 and 1.4). Yan (2014) took the delta front reservoir in Layer 81 in the second member of Shahejie Formation in Shengtuo Oilfield, Dongying Depression as an example, and he systematically brought to light the original oil-bearing property of the reservoir and the distribution of remaining oil in the late extra-high water-cut stage with the data of cores and remaining saturation

**Table 1.3** Applicable conditions and characteristics of open-hole logging methods for remaining oil saturation (Reed et al., 1992; Yujiao et al., 2000; Ming and Haining, 2002; Peihua, 2003b; Yan et al., 2005; Yuhong et al., 2006; Yingli, 2008; Junlong, etc., 2013; Bintao et al., 2014; Jun et al., 2016).

| Logging method | Applicable conditions and characteristics | Deficiencies | Application examples |
|---|---|---|---|
| Resistivity logging | It is widely applicable, as the main means of reservoir oil-bearing property evaluation | It is influenced by the properties of injected water and water flooding degree | Cyro Oilfield of Denver Basin, Western Sag of the Liaohe Basin |
| Dielectric logging | It is applicable to reservoirs with low-salinity formation water, where dielectric constant is less affected by the change of salinity of formation water. When the formation porosity is greater than 8%, it can distinguish oil and water. The larger the porosity, the higher the identification accuracy of oil and water layers | The detection depth is shallow | Baorao structural belt in Jirgalangtu sag of Erlian Basin, Well Zhong11–016 in Shengli Gudao Oilfield, Wells Xi2–6–3 and Xi3–7–1 in Dagang Oilfield |
| Induced polarization + spontaneous potential logging | It can calculate the salinity and resistivity of formation water point by point and can also eliminate the influence of clay on the saturation in the calculation of water saturation | It is only applicable to sandstone and mudstone sections with freshwater mud and low-salinity formation water (less than 30,000 mg/L). It is poorly applied in case of severe heterogeneity, great permeability change, and high salinity | Jidong Oilfield, Block Qi40 in Huanxiling Oilfield of Liaohe Basin |

| | | | |
|---|---|---|---|
| Chlorine energy spectrum logging | It is applicable to both cased hole and open hole. This simple, fast, and low cost method can overcome the deficiency of the open-hole logging which may generate unreliable interpretation results in the case that mud penetrates deeper (than the tool's detection depth) in low-resistivity reservoirs with high mudstone content and high-permeability formations | If a reservoir contains Ca, the measuring result will be affected. Moreover, when the method is applied to high salinity oilfields with $Cl^-$ concentration >40,000 mg/L, the formation porosity must be more than 10% | Weicheng Oilfield of Zhongyuan Oilfield, Well Zhong6−15 in Jianghan Oilfield, Well La8-B in Lamadian Oilfield of Daqing Oilfield |
| Electromagnetic propagation logging | As insensitive to formation water salinity, it is applicable to cases with unknown or anomalous salinity. This method with small detection range can couple with the resistivity logging to achieve better performance | The detection range is small | Dongfang block in western South China Sea |
| Nuclear magnetic resonance (NMR) logging | The measurement is not related to lithology. The T2 relaxation time reflects the size distribution of oil- or water-bearing pores and the fluid contents in pores with different sizes. The analysis of T2 relaxation time distribution can intuitively show the microscopic distribution of remaining oil in pores with different sizes and can accurately calculate oil content | Since the nuclear magnetic properties of hydrogen in formations are determined by the nature of fluids and their interaction with solid phase, the nuclear magnetic properties of fluids in rocks should be known before studying NMR of rocks | Adjustment Well Jing67−541 in Block Shen84-An12 of Liaohe Oilfield, Well L3−8 in Junggar Basin |

**Table 1.4** Applicable conditions and characteristics of cased-hole logging methods for remaining oil saturation (Wei et al., 2007; Chunhua et al., 2008; Xiaoqin et al., 2008; Feng and Xinguang, 2009; Jianjiang and Jianguo, 2013; Junlong et al., 2013).

| Logging method | Applicable conditions and characteristics | Deficiencies | Application examples |
|---|---|---|---|
| Neutron lifetime logging | It is especially applicable to high salinity formations. Neutron gamma logging shows its superiority in these aspects, and it is easy to be popularized and applied | Two factors, $\tau$ and $\Sigma$, need to be calculated in log data interpretation, which makes the method problematic. The logging cost is high | Well XiII3–11 in Jianghan Oilfield |
| Boron neutron lifetime logging | It is applicable to low-salinity formations. It has large capture cross-section and highly accurate interpretation | It has high requirements for well killing and flushing operations | Wells He143-Xie52, He68–24, He51-Xie100, Shi8-Xie110, He146–45, He2-Xie4 and He146–55 in Xianhezhuang Oilfield of Shengli Oilfield |
| Gadolinium neutron lifetime logging | It has large capture cross-section and highly accurate interpretation. The less-dosage and low-cost operation can help minimize the reservoir damage. Gadolinium has better solubility than boron, and it can be disposed off at low temperature | Leakage layer and seriously swept layer are difficult to distinguish and nonpermeable gadolinium layers are difficult to identify | Wells 9285 and 1427 in the Well Bai 21 area in Baikouquan Oilfield of Karamay Oilfield |
| Pulsed neutron–neutron logging | It can capture the cross-section of low porosity (>5%) and low-salinity (about 5,000 ppm) reservoir water | Logging interpretation results are greatly influenced by formation water salinity when determining reservoir water saturation | Qinghai Oilfield |

| | | | |
|---|---|---|---|
| C/O logging | It is not affected by the change of formation water salinity, especially. This method is uniquely advantageous when the salinity of injected water is different from that of formation water. Good results can be achieved in high-porosity (> 15%) formations | The detection depth is small. The results are influenced by invaded mud filtrate, borehole size, borehole fluid salinity, capture background value, neutron pulse period, and neutron tube | Well 19–123 in Liaohe Oilfield, Lamadian Oilfield of Daqing Oilfield |
| C/O spectral logging | It is economical, effective, fast, intuitive, and accurate | It is only applicable to formations with porosity greater than 20%. The vertical resolution difference is about 0.8 m. Response of thin and poor layers is poor | Well 1–2–153 in Shengtuo Oilfield, Lamadian Oilfield of Daqing Oilfield |
| Pulsed neutron decay spectrum logging | It is less influenced by lithology. It is applicable to formations with the porosity more than 10%. It can distinguish oil layer from low-salinity water layer. Requirement for wellbore is not high; logging through tubing is acceptable | For reservoirs in middle-late development stage, the water saturation measured is higher than that derived from open-hole logging due to water flooding. Logging tools are costly | Well Gudong7–36–346 |
| RMT logging | It is widely applicable with multiple measuring modes and is not affected by the change in salinity of formation water | It is not applicable to low porosity (<8%) formations | Well Lu2025 in Luliang Oilfield of Karamay Oilfield, Well Jian552 in the northwestern margin of the Junggar Basin |

(*Continued*)

**Table 1.4** (Continued)

| Logging method | Applicable conditions and characteristics | Deficiencies | Application examples |
|---|---|---|---|
| Through-casing resistivity logging | Well flushing and scraping are not required under the push-the-bit mode, so that the well-occupying time and operation costs are reduced greatly. The detection depth is large. Multiple measurements can be made at each point. The dynamic range of measurement is large and the resistivity range of detectable formation is 0–300 Ωm. The stratigraphic resolution is strong, and the influence of surrounding rocks is relatively small | It is difficult to record signals in DC mode. Cementing quality has a great influence on logging quality. Negative differences may occur in nonperforated sections. Salt mud and injected water salinity changes have great impacts on interpretation and evaluation conclusions. Thin layer resolution is low. Length of logging interval is generally less than 400 m | Wells C2285 and C1103 in Cainan Oilfield of Karamay Oilfield |

logging. Hao et al. (2015) took Layer 42 in the third member of Hetaoyuan Formation in Anpeng district, Zhaowa Oilfield as an example, and he quantitatively evaluated the distribution law of remaining oil in high water-cut stage. There are many methods to study remaining oil, including geological, reservoir engineering, well testing and numerical simulation methods, laboratory experimental techniques, and various dynamic monitoring methods. This book states that, for understanding the remaining oil distribution in high water-cut oilfields, the data of sealed coring wells, relevant laboratory experiments, dynamic monitoring, and production are extremely important, in addition to numerical simulation. Moreover, the identification process of the reservoir genetic model and microstructure (especially small faults) should be strengthened. In Daqing Oilfield, infill wells have been drilled near the faults to successfully tap the potential of remaining oil. This is a good inspiration for remaining oil characterization as a part of fine reservoir description of high water-cut oilfields.

#### 1.2.1.5 Tertiary oil recovery in high water-cut reservoirs

As mature oilfields enter the middle-late development stage when the water-cut is high or extra-high, it is more and more difficult to enhance oil recovery by relying on the overall adjustment of the oilfield and to tap the potential using conventional techniques. In order to substantially enhance oil recovery for the purpose of oilfield sustainability, it is necessary to increase recoverable reserves through a variety of ways and to explore new development methods to improve the development level of oilfields. The tertiary oil recovery has been proved, theoretically and practically, as an economic and feasible EOR technique, which can expand the swept volume and improve the oil displacement efficiency (Wanchao, 2003). For an oilfield in the high water-cut stage, knowing the development geological characteristics of the oilfield is essential for its tertiary oil recovery to happen. Therefore, the study on development geology for tertiary oil recovery becomes a critical orientation of fine reservoir description of high water-cut oilfields. Jiandong et al. (2013) took the H3IV1−3 layer series in the north block of Shuanghe Oilfield in Henan Oil Province as an example and proposed the process control technology to improve the effect of polymer flooding. Water flooding should be improved, and polymer flooding, steam flooding, thermal recovery, microbial flooding, and other tertiary oil recovery techniques should be performed to enhance oil recovery (Xiang'an et al., 2007). This book puts forward two foci in fine reservoir description. One is the study of

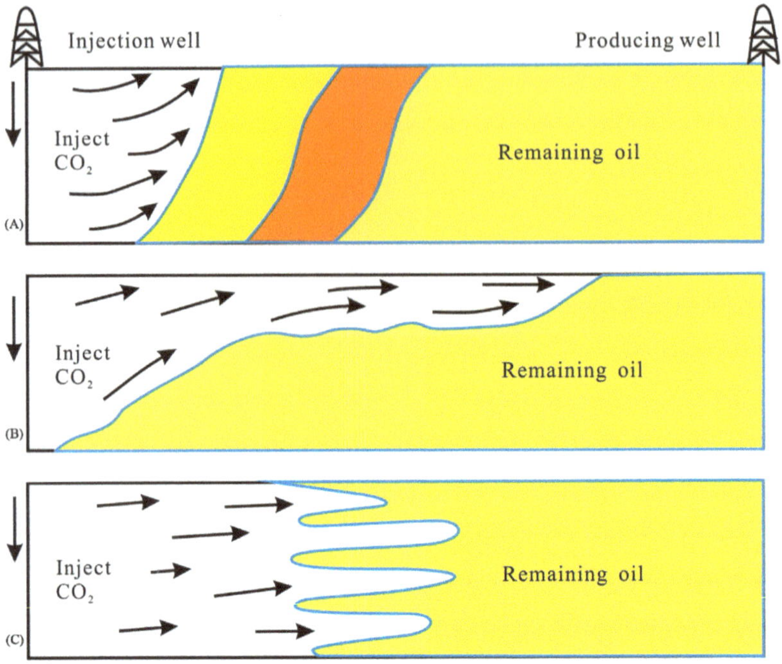


**Figure 1.5** Mechanism of $CO_2$-EOR (Huaiyou et al., 2008; Pingping and Xinwei, 2009; Huanqing et al., 2012a,b,c). (A) In the case of very low $R_{vg}$, the ratio of viscous force and gravity, the oil displacement is characterized by gravity tonguing of gas overlap. (B) In the case of higher $R_{vg}$, the displacement is still characterized by the gravity tonguing of gas, but the vertical sweep does not depend on the specific $R_{vg}$ until it reaches its limit. (C) The sweep efficiency increases rapidly with the increase of $R_{vg}$, until $R_{vg}$ reaches a specific value where the displacement is completely controlled by multiple fingering on cross-section and the lateral sweep efficiency does not depend on the specific $R_{vg}$.

reservoir pore structure and oil displacement mechanism related to polymer flooding, as the tertiary oil recovery techniques like polymer flooding and steam injection are greatly different from water flooding in mechanism; the study is expected to provide a basis for the selection of tertiary oil recovery technique and oil displacement agent (Fig. 1.5). Further, theoretical analysis and laboratory experimental study on reservoir damage caused by tertiary oil recovery should also be carried out. The other focus in fine reservoir description is the study of distribution and movement law of oil, gas, and water. An oilfield in the high water-cut stage has the distribution and movement of underground oil, gas, and water after a long period of development and production as significantly different from what an oilfield in the initial stage of development has. For the purpose of reasonable planning and implementation of tertiary oil recovery measures, the distribution and movement law of underground oil, gas, and water must be accurately understood. The basic study related to tertiary oil recovery, which further broadens the coverage of fine reservoir description of high water-cut oilfields and improves the corresponding research level, lays a solid foundation for further enhancing oil recovery.

## 1.2.2 Trend

### *1.2.2.1 Fine interpretation of fault system*

Fault systems are developed widely in major oilfields of China, such as the Daqing, Dagang, and Shengli oilfields in East China, and the Xinjiang and Tarim oilfields in West China. In the high water-cut oilfields especially, the existence of a complex fault system complicates the corresponding relationship of oil and water distribution between injection and production wells, thereby posing huge challenges to the depiction of remaining oil distribution. Naiju et al. (1999) pointed out that the faults affecting the water flooding performance of reservoirs are mainly the fourth- or lower-order faults. The occurrence of these faults is controlled greatly by the presence of large regional fault systems as well as the occurrence of local structural factors. The lower order the fault, the greater the influence of the local structural factors. Therefore, the fine characterization of the fault system is of great significance to the characterization of the remaining oil and the enhancement of oil recovery in high water-cut oilfields.

According to this book, the following three aspects need to be specially considered in the fine reservoir description of high water-cut oilfields.

1. Sectional fine interpretation and plane combination of fault systems based on the logging-seismic data (Fig. 1.6).

   Depending on the accuracy of data and the requirements of development and production, fifth-order faults with throw >5 m should be identified for medium-shallow reservoirs; fourth-order faults with throw >10 m should be identified for deep reservoirs; and third- or fourth-order faults should be identified for ultradeep reservoirs. Accurate interpretation of higher-than-third-order faults is the basis for fine interpretation of the fault system. Currently, the results obtained during the exploration or evaluation stage are often instrumental in the interpretation of the fault system in practical oilfield production. However, given such results may contain errors due to the limitations in data availability, and research

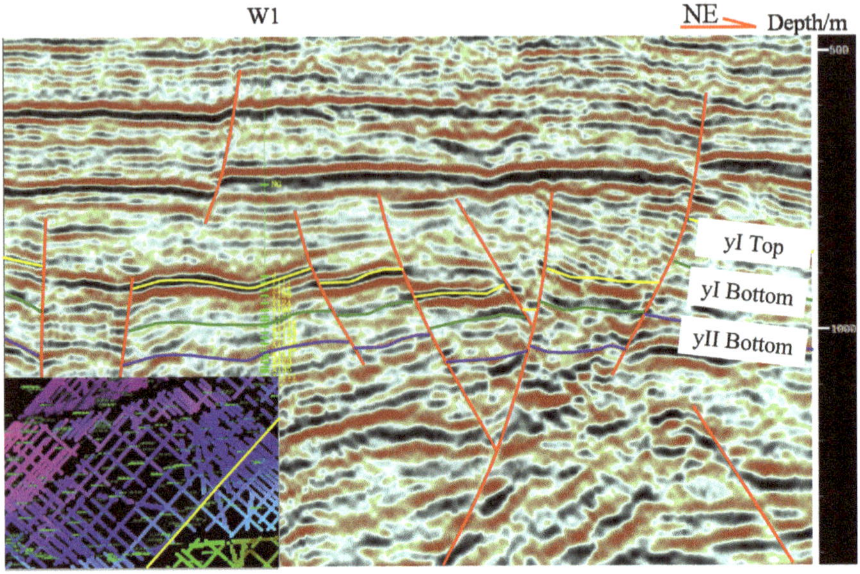


**Figure 1.6** Interpretation results of fault system section of Yulou reservoir in an area of Western Sag, the Liaohe Basin.

level and researchers' knowledge in these stages, it thus follows that the conclusions of fault system interpretation in fine reservoir description vary every year, and sometimes even the general structural pattern may change. This phenomenon of the variability of conclusions of fault system interpretation in fine reservoir description largely restricts the improvement of research on fine reservoir description. This book holds that, in the interpretation of the fault system, the geological genesis of the fault system should be analyzed firstly by synthesizing all data, if possible; the existing fault system interpretation results should be verified and analyzed to make sure the interpretation results of large-scale fault systems (second-order and third-order) are accurate, before the fourth- and fifth-order faults are interpreted. Yang and Jianmin (2007) provided the classification and identification marks of the fourth- and fifth-order faults in the monograph "Reservoir Development Geology."

**2.** Identification and characterization of microstructures, especially small faults.

For oilfields in the middle-late development stage, the characterization of remaining oil is one of the key objectives of fine reservoir description, while the distribution of remaining oil is controlled by small faults and microstructures. Thus, the fine characterization of small faults and microstructures should be treated specifically. Xuesong et al. (2015)

made the innovation in the logging data-based and seismic data-based fine structural description using technology: the three-dimensional velocity field time-depth conversion is used constrained by dense well pattern and small fault interpretation that is guided by well breakpoint, is based on fine description of fault using well logging and seismic data, and takes the Daqing Oilfield as an example. Li's technology can accurately characterize the spatial distribution of faults; so, his study provides a solid basis for the optimal design of high-efficiency wells near faults in mature oilfields with extra-high water-cut.

**3.** Analysis of fault sealing

After a long period of development, an oilfield enters the high water-cut stage. In this stage, the oil-water relationship is complex, and the injector-producer correspondence becomes one of the foci in practical development and production. The presence of the fault system is a very important determinant of the injector-producer correspondence. Therefore, it is necessary to strengthen the study of fault sealing. Huanqing et al. (2015a,b) used the typical well water analysis data to investigate the sealing capacity of faults as flow barriers in an area of Western Sag, the Liaohe Basin. As shown in Fig. 1.7, there are four major faults, F1, F2, F3, and F4. Water analysis data of 10 wells in five clusters were selected for comparison. According to the development level of faults (Table 1.5), F1 and F2 are large third-order sag-controlled faults, while F3 and F4 are fourth-order

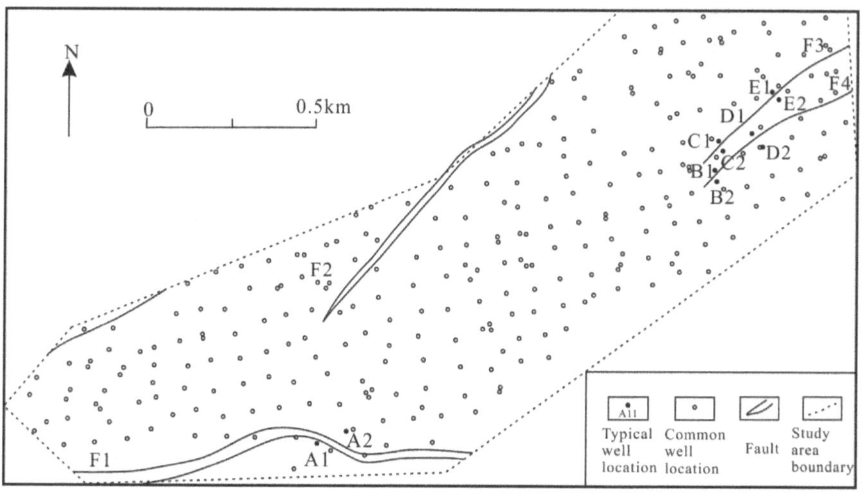


**Figure 1.7** Location of an area of Western Sag, the Liaohe Basin (Huanqing et al., 2015a,b).

**Table 1.5** Water analysis data of some typical wells in Yulou reservoir in Western Sag, the Liaohe Basin (Huanqing et al., 2015a,b).

| Well | Sampling date | Test date | Na + Ka | Mg | Ca | Cl | $SO_4$ | $CO_3$ | $HCO_3$ | Total salinity mg/L | Total hardness | Total alkalinity | Water type | pH |
|---|---|---|---|---|---|---|---|---|---|---|---|---|---|---|
| A1 | June 20, 2005 | June 21, 2005 | 519.8 | 4.86 | 38.1 | 266 | 19.21 | 90 | 854.28 | 1792.44 | 115.1 | 850.8 | $NaHCO_3$ | 7 |
| A2 | June 16, 2005 | June 17, 2005 | 437 | 7.3 | 20 | 212.8 | 4.8 | 0 | 884.79 | 1566.69 | 80.1 | 725.7 | $NaHCO_3$ | 6 |
| B1 | Mar. 24, 2010 | Mar. 25, 2010 | 740.6 | 3.65 | 10 | 248.2 | 24.02 | 0 | 1556.01 | 2582.52 | 40 | 1276.2 | $NaHCO_3$ | 6 |
| B2 | Sept. 27, 2010 | Sept. 28, 2010 | 579.6 | 6.08 | 12 | 230.5 | 14.41 | 60 | 1067.85 | 1970.45 | 55.1 | 975.9 | $NaHCO_3$ | 7 |
| C1 | Oct. 11, 2010 | Oct. 12, 2010 | 545.1 | 9.73 | 18 | 230.5 | 67.24 | 0 | 1067.85 | 1938.45 | 85.1 | 875.8 | $NaHCO_3$ | 6 |
| C2 | July 14, 2010 | July 14, 2010 | 503.7 | 7.3 | 18 | 159.6 | 91.26 | 0 | 1037.34 | 1817.21 | 75.1 | 850.8 | $NaHCO_3$ | 7 |
| D1 | Aug. 3, 2001 | Aug. 4, 2001 | 363.4 | 3.65 | 6.01 | 141.8 | 19.21 | 150 | 427.14 | 1111.25 | 30 | 600.6 | $NaHCO_3$ | 8 |
| D2 | Aug. 3, 2001 | Aug. 4, 2001 | 542.8 | 3.65 | 12 | 106.4 | 24.02 | 240 | 793.26 | 1722.13 | 45.1 | 1051 | $NaHCO_3$ | 8 |
| E1 | Nov. 2, 2003 | Nov. 3, 2003 | 446.2 | 7.3 | 16 | 177.3 | 14.41 | 90 | 762.75 | 1513.99 | 70.1 | 775.7 | $NaHCO_3$ | 6 |
| E2 | Nov. 12, 2003 | Nov. 13, 2003 | 407.1 | 4.86 | 20 | 177.3 | 4.8 | 90 | 671.22 | 1375.32 | 70.1 | 700.6 | $NaHCO_3$ | 7 |

faults. The analysis data of formation water in upper and lower walls of the faults were compared to evaluate the fault sealing and characterize the sealing fault flow barrier. The great difference in analysis results between the upper and lower walls indicates good sealing, namely, the sealing fault flow barrier can be formed; the similarity of analysis results indicates poor sealing, namely, the sealing fault flow barrier cannot be formed. For F1, the water analysis results of Wells A1 and A2, which are drilled on the upper and lower walls, respectively, are quite different, especially in the contents of Mg, Ca, $SO_4$, and $CO_3$, indicating that the fault is a sealing fault. By comparing Wells C1 and C2 with Wells E1 and E2 as two groups, it is found that the indexes of Na + K, Ca, $CO_3$, $HCO_3$, and total hardness are identical or similar; thus, it can be concluded that the fluids in the upper and lower walls are communicated, and the fault is not a sealing fault to serve as the flow barrier. By comparing Wells B1 and B2 with Wells D1 and D2 as two groups, it is found that Mg, $SO_4$, $CO_3$, total salinity, and total alkalinity on both walls are quite different; thus, it is concluded that the fluids in the upper and lower walls are not communicated, and the fault is a sealing fault to form a flow barrier. Among the four faults in the study area, F1 and F4 are sealing fault flow barriers.

#### *1.2.2.2 Fine characterization of reservoir architecture*

Reservoir architecture refers to the pattern, scale, direction, and overlapping relationship of different orders of reservoir architectural elements (Shenghe, 2010). Fine characterization of reservoir architecture plays an important role in the fine characterization of reservoir sand bodies and intercalations, making it increasingly attractive to researchers (Huanqing et al., 2014a,b, 2015a,b). Consequentially, the fine characterization of reservoir architecture is considered an essential part in the fine reservoir description of high water-cut oilfields. We studied the architecture of an alluvial fan conglomerate reservoir in an area of the Junggar Basin, using the field outcrop, modern sedimentary, core, logging, seismic, and production performance data (Fig. 1.8). The research of reservoir architecture should be conducted with the priority to: (1) the determination of the reservoir architecture classification scheme and hierarchy, (2) the identification of the interface between different reservoir architectural elements and the reservoir heterogeneity, and (3) the analysis of the quantitative spatial scales and spatial combination mode of different types of reservoir architectural elements.

(A) (B) (C) (D)

**Figure 1.8** Field outcrop and modern sedimentary characteristics for architecture characterization of alluvial fan reservoir at the northwest margin of the Junggar Basin. (A) and (B) are the outcrop characteristics of alluvial fan deposition. (C) and (D) are modern sedimentary features of alluvial fan deposits.

There are many reservoir architecture classifications. The commonest one is the eight-order classification proposed by Miall et al. Log data, which are accessible and quantitative in the production process of oil and gas fields, are widely used. Limited by the data accuracy, the practice is for the four-order and five-order reservoir architecture classifications to be adopted, generally. It should be emphasized that a finer classification is not necessarily a better one. The classification hierarchy mainly depends on the data accuracy and the production demand. Since the availability of core data is very limited, the reservoir architecture classification into bedding or laminae is only meaningful in scientific research but cannot be popularized to solve the problems in practical production. Moreover, a classification which pursues the hierarchy blindly but neglects the data accuracy can only produce wrong results.

For a long time, researchers in fine reservoir description have focused on analyzing reservoir architectural elements themselves, but which rarely involved the interface between them. Actually, reservoir architectural elements of different orders and types play nonnegligible roles in fine reservoir description. The interfaces between these elements with different origins and scales are the main part of intercalations in reservoirs, and they are very important in the study of reservoir heterogeneity. Especially, for the high water-cut oilfields in the middle-late development stage, a correct understanding of these interfaces allows the researchers to more accurately clarify the interwell reservoir connectivity and other bottlenecks in development.

Identifying the quantitative spatial scale of reservoir architectural elements is the most direct purpose of reservoir architecture characterization. The statistical analysis of these quantitative data can provide a geological basis for the design and implementation of well infilling, well spacing reducing, and other adjustment measures in the middle-late development stage. Moreover, the summary and analysis of the spatial combination mode of reservoir architectural elements with different origins can trigger a profound understanding of the spatial distribution and superimposition law of different sand bodies and thereby provide a reference for the smooth implementation of EOR techniques in the middle-late development stage. This can also effectively contribute to a more quantitative and scientific fine reservoir description.

#### *1.2.2.3 Interpretation of swept layer logging*

In China, the interpretation of swept layer logging started in the 1950s in Yumen Oilfield. Great progress has been made so far. For example, an experimental study was carried out systematically on physical properties of the swept layer; consequent to oilfields having established their specific swept layer logging techniques, the swept layer logging interpretation model and method were studied systematically; and the swept layer logging had been successfully applied to oilfield development (Peihua, 2003b). Peihua (2003b) reported that the vast majority of oilfields in China have entered the high water-cut stage, talking from the perspective of the composition of water producing from oil-producing wells or the composition of crude oil production. Underground oil and water are characterized by highly dispersed remaining oil and disordered distribution of high water-cut and low water-cut zones. To identify the distribution of underground oil and water and determine the remaining oil-enriched zones, the most important and effective method is swept layer logging interpretation. Many scholars in China are engaged in the swept layer logging interpretation for high water-cut oilfields. Jiang et al. (2013) summarized the difficulties and orientation of swept layer logging interpretation for oilfields in the late development stage with high water-cut, and he

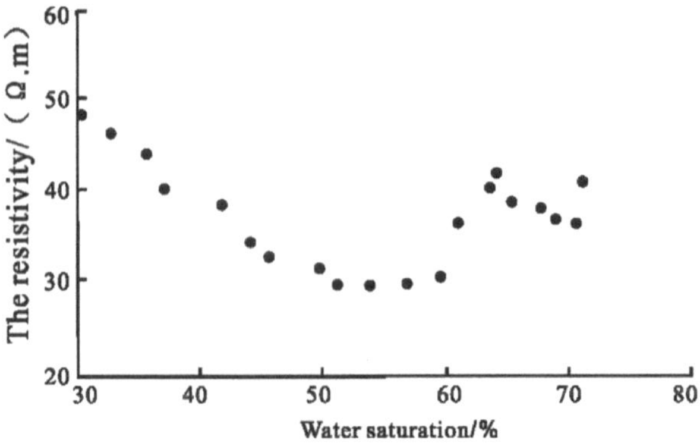


**Figure 1.9** Rock resistivity in fresh water—clear water + polymer—fresh water flooding (Jiang et al., 2013).

analyzed how the resistivity changes with the properties of injected water (Fig. 1.9), suggesting that in the late development stage with high water-cut, the core resistivity logging interpretation derives multiple solutions, and the logging interpretation model should be constructed with full consideration to various possible factors. Peijun (2013) analyzed the log responses of the swept layer in Shuanghe Oilfield in the extra-high water-cut stage, and established a swept layer logging interpretation model for Shuanghe Oilfield in the late development stage with extra-high water-cut under multiple displacement modes. Counted the biggest problem is the accuracy of swept layer logging interpretation now which is too low to satisfy the reperforating of adjustment wells, selection of horizons for acid fracturing, and characterization of remaining oil. It is necessary to strengthen the fundamental experiment study of petrophysical properties of swept layer in the reservoir conditions, and deeply analyze the responses derived from various logging operations in the process of reservoir water flooding. Furthermore, depending on the geological and development characteristics of different reservoirs, the swept layer logging techniques should be improved and should come with innovations built-in, and more thorough and accurate swept layer logging interpretation models should be constructed, so as to achieve higher-level interpretation results.

#### *1.2.2.4 Research on dominant flow paths*

Dominant flow path refers to the low-resistance flow channel formed locally in the reservoir due to geological and development factors, through which the injected water flows preferentially in the late stage of water flooding development, resulting in an ineffective circulation of massive injected water (Ming and Zhiping, 2007). Cheng et al. (2012) studied the distribution pattern of the dominant flow path within point bars and its control over remaining oil distribution, taking Block S17–19 in Fuyu Oilfield of Jilin Oilfield as an example (Fig. 1.10). Yubao et al. (2014) used the plane sand packing model, real sandstone model, and other experimental techniques and methods to illustrate the macro- and microflow characteristics, and distribution pattern and formation mechanism of remaining oil after water flooding in the late development stage of the reservoir with high water-cut. Jiang et al. (2014) took the lower VIII in Shuanghe Oilfield as an example to analyze the changes of the dominant flow path in the reservoir before and after water flooding and inferred such dominant flow paths as high-permeability channels formed under the control of sedimentary facies belt or as a result of the migration of micro-particles during the long-term water flooding. Yikun et al. (2015) investigated the variation of reservoir properties and the seepage characteristics of oil and water after a complex fault block oilfield enters the middle-high water-cut stage, where he had used natural cores showing three water-cut stages in the high-permeability channel of two blocks in Jidong Oilfield. At present, the researches on dominant flow path are mainly the identification and description of dominant flow path, and the analysis of its influence on the

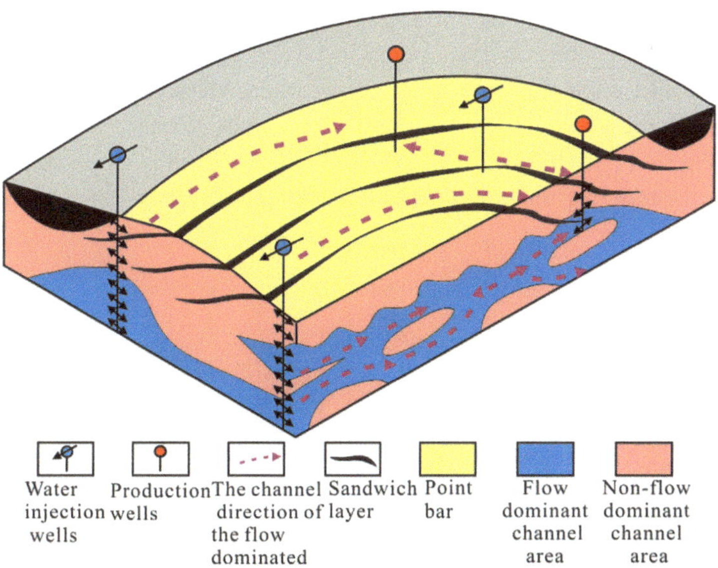


**Figure 1.10** Distribution of dominant flow paths within point bars (Cheng et al., 2012).

distribution of remaining oil, with geological, numerical simulation, and mathematical calculation methods applied to explain reservoir geological characteristics found. The biggest consequence of the existence of a dominant flow path is the ineffective circulation of injected water in the oilfield developed by water flooding. It is ubiquitous and especially severe in high water-cut oilfields. Such ineffective circulation causes the economic loss to several hundred million Yuans every year, and it has become the most prominent problem in a water flooding oilfield to realize beneficial development in the late stage. Therefore, the accurate identification of the dominant flow path has also become a very important objective in fine reservoir description of high water-cut oilfields. The implementation of water plugging, profile control, and other measures based on the identification of the dominant flow path can prevent the injected water from inefficiently circulating along the dominant flow path, thereby improving the development performance and enhancing the oil recovery made.

### 1.2.2.5 Geological modeling based on multiple-point geostatistics

The final result of reservoir description is a quantitative reservoir geological model as the basis of reservoir simulation, reservoir engineering, and oil recovery process (Qiu and Chen, 1996). This geological model includes a structural model, a reservoir model, and a fluid model. In the fine reservoir description of a high water-cut oilfield, a reservoir prediction model will be constructed. Currently, the most popular method is to predict cross-well attributes by the interpolation operation with a variable function under the constraint and control of the sedimentary microfacies model. This method was used in the geological modeling for an area of Western Sag, the Liaohe Basin (Fig. 1.11). Although this method can establish a three-dimensional geological model of the reservoir easily and quickly, its prediction accuracy

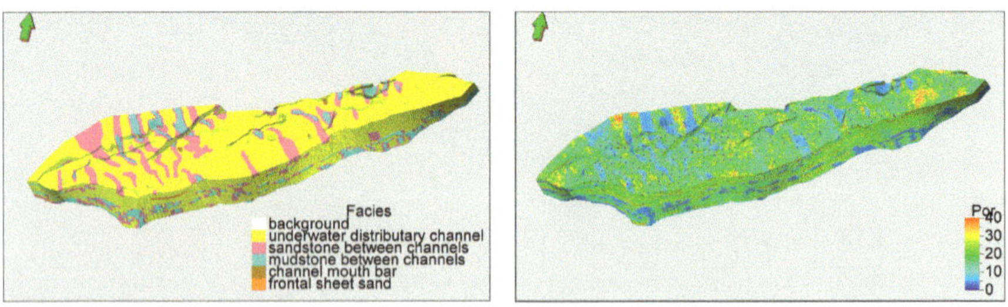


(A) (B)

**Figure 1.11** Geological model for an area of Western Sag, the Liaohe Basin. (A) is the sedimentary facies model. (B) is the porosity model.

of reservoirs between wells is not close to the expectations for the continental sedimentary reservoir with strong heterogeneity. Therefore, geological modeling methods with obvious advantages in the prediction of three-dimensional development characteristics of the reservoir are required, and the multipoint geostatistical modeling technique is just such a typical one (Jiahua and Tao, 2013). The application of multipoint geostatistics to stochastic modeling began in 1992. In the multipoint geostatistical modeling, training images, instead of variation functions, are applied to express the spatial structure of geological variables; thus, it overcomes the shortcomings of traditional geostatistics, which cannot reproduce the geometric shape of the target. Moreover, as it still takes pixels as simulation units and adopts a sequential algorithm (noniterative algorithm), it sticks close to hard data and has the specialty of high-speed; therefore, it overcomes the shortcomings of object-based stochastic simulation algorithm (Shenghe and Kewen, 2005). Many researchers have conducted instructive and effective research on the multipoint geostatistical modeling method (Wei et al., 2008a,b; Shaohua et al., 2009; Yanshu et al., 2014; Lihui et al., 2015). Nevertheless, there are still many unaddressed issues, such as stability of training image, continuity of object, and comprehensive seismic information.

The software implementation of geological modeling is also a critical consideration in multipoint geostatistical modeling. Petrel, RMS, and other foreign software are commonly used. Some enterprises, universities, and institutions in China are also engaged in the development of software. So far, certain achievements have made their presence felt, but there are still many problems hindering the popularization and application of domestic software. Such problems are mainly manifested in three aspects. First, the research in China started relatively late. Restricted by personnel, funds, and other factors, the domestic software are far behind the foreign ones in operational prowess and results display. Second, without an effective commercialization support, the domestic software are deficient in popularity, promotion, and update. Third, there is no perfect training and after-sales service system, thus impeding the application and promotion of the domestic software. Reservoir geological modeling is one of the cores in fine reservoir description of high water-cut oilfields. Under the current context of low oil prices, the localization of geological modeling software will be practically significant for oilfields to reduce costs and improve efficiency.

The abovementioned problems are just some key concerns for fine reservoir description of high water-cut oilfields. In fact, there are many bottlenecks in oilfield operations that need to be further explored. One example is the characterization of fractures in fault block and in fractured reservoirs using seismic data (Junhua, 2012), which mainly includes an accurate prediction of interwell fractures and 3D geological modeling of fractures. Another example is the fine description of a single sand body and the description of interwell sand body connectivity. It is urgent to clarify how the interwell sand body connectivity is analyzed by combining the data from the geological static monitoring and dynamic monitoring, and of production performance. The reservoir heterogeneity is also a typical bottleneck, given the research focuses on the flow field heterogeneity, but it rarely involves the fluid heterogeneity.

## 1.3 Fine reservoir description of low-permeability oilfield

Fine reservoir description has become a normal task in oilfields. In 2003 PetroChina started to carry out fine reservoir description extensively, and regulated the reservoir description as a mandatory strategy for all oilfields to be developed (Huanqing, 2006; Huanqing et al., 2006; Huanqing and Xiaomin, 2008). As the fastest-growing segment of PetroChina in reserves and production increase recently, low-permeability oilfields have become the focus of efforts to fine reservoir description. So far, low-permeability reservoirs have been discovered widely in almost all oilfields in China, including Daqing and Shengli in East China, to Changqing in Central China, and to Xinjiang in West China. These oilfields are very different from high water-cut oilfields and complex lithologic oilfields in geological and development characteristics. Through literature review and practical researches, here, we will present a comprehensive summary on fine reservoir description of low-permeability oilfields to provide a reference for the efficient development of these oilfields.

### 1.3.1 Current status

In China, a number of different types of low-permeability and ultralow-permeability oilfields have been successfully developed, such as Chaoyanggou Oilfield in Daqing (a low-permeability oilfield with high viscosity of crude oil), Xinli Oilfield and Xinmin Oilfield in Jilin (two ultralow-permeability oilfields with fractures in reservoir), Maxi Oilfield in Dagang (a deep low-permeability oilfield), Bonan Oilfield in Shengli (a large low-permeability oilfield with geological reserves of over 100 million tons), Ansai Oilfield and Jing'an Oilfield in Changqing (two ultralow-permeability oilfields with extremely complex ground conditions), Wendong intersalt reservoir in

Zhongyuan and Gaskule Oilfield in Qinghai (two abnormally high pressure, low-saturation, and low-permeability oilfields), Shanshan Oilfield in Tuhu (a low-permeability oilfield developed by integral fracturing), and Wuerhe reservoir in Block 8 of Karamay (a massive conglomerate low-permeability oilfield) (Daopin, 1999). The low-permeability oilfield is one of the most important targets in oilfield development, both in China and abroad. The so-called low-permeability oilfield is a relative concept, without a unified standard and boundary agreed domestically and internationally. Since different countries exhibit different resource conditions, and economic and technical conditions in different periods, the classification standards are different. Daopin (1997) classified the reservoirs with a permeability ranging from 0.1 to 50 mD as low-permeability reservoirs. From the universal and adaptable perspective, we also support Li Daoping's concept, namely, the low-permeability reservoir has a permeability between 0.1 and 50 mD.

The research on the low-permeability oilfield almost kept the same pace with the development of oilfield historically, and many researchers have worked on it. Shouyu (2005) summarized the research content and process of the low-permeability reservoir description (Fig. 1.12), which is still greatly instructive and referential to the fine description of the low-permeability reservoir. Based on a number of literatures (He et al., 1990; Zongxiong, 1994; Daopin, 1997, 1999, 2003; Hoffman and Kovscek, 2004; Huanqing, 2006; Huanqing et al., 2006; Mallon and Swarbrick, 2008; Lianbo, 2008; Huanqing and Xiaomin, 2008; Wenrui, 2009; Qigu and Gongyang, 2010; Lemiski et al., 2011; Wenxiang et al., 2011; Zhipeng et al., 2012; Mørk, 2013; Odunowo et al., 2013; Clarkson et al., 2014; Shanley and Cluff, 2015; Clarkson and Qanbari, 2016; Zhang et al., 2016) and based on practical efforts, we find great differences in researches on fine reservoir description of the low-permeability oilfield between China and other countries (Table 1.6). Generally, compared to research taking place outside China, the researches in China are superior in degree of emphasis and research level, but inferior in some aspects like numerical simulation and laboratory test.

### 1.3.2 Core contents

Generally, fine reservoir description covers all aspects of reservoir geologic study, including fine stratigraphic classification and correlation, characterization of sedimentary microfacies and reservoir architecture, reservoir evaluation, geological modeling, and flow unit characterization. In the

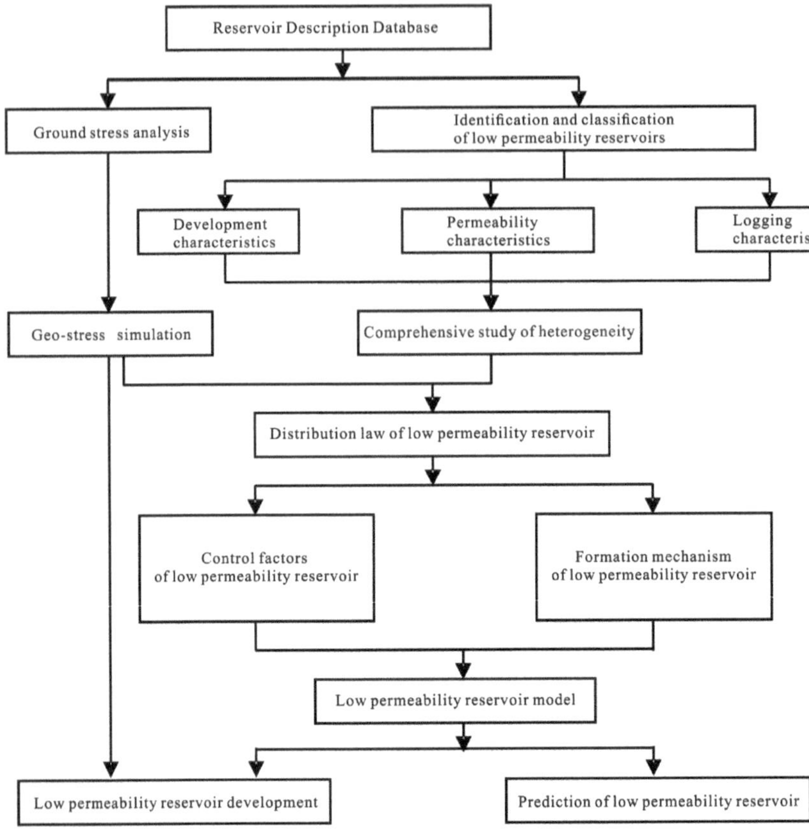


**Figure 1.12** Research content and process of low-permeability reservoir description (Shouyu, 2005).

fine reservoir description of the low-permeability oilfield, pore space and flow capacity are the focal points, although the aspects mentioned earlier are also included. Specifically, the fine reservoir description of the low-permeability oilfield involves the fine prediction and characterization of sand body with multiple methods, comprehensive fracture characterization based on multiple information, research of micropore structure of the reservoir, classification of flow units in the low-permeability reservoir, and low-permeability reservoir protection.

#### *1.3.2.1 Fine prediction and characterization of sand body with multiple methods*

In view of geological genesis, the rocks in the low-permeability oilfields are various, including the clastic rock, the carbonate rock, the volcanic

**Table 1.6** Researches on fine reservoir description of low-permeability oilfields in China and other countries.

| | Merit | Weakness |
|---|---|---|
| Other countries | 1. Identification of low-permeability sandstone continuity using outcrop or adjacent well data<br>2. Analysis of the porosity and permeability anisotropy through experiment<br>3. Design and evaluation of hydraulic fracturing<br>4. Study on the physical properties of low-permeability reservoir after steam flooding or polymer flooding<br>5. Study on diagenesis of low-permeability reservoir<br>6. Evaluation of fluid saturation in low-permeability sandstone reservoir | 1. Seismic data are not considered for the characterization of low-permeability reservoirs<br>2. Quantitative study of diagenesis is inadequate, and deeper analysis should be made on diagenetic facies<br>3. Dynamic fracture characteristics in the development process are not adequately considered |
| China | 1. Study on diagenetic facies of low-permeability sandstone reservoir<br>2. Logging evaluation technology of low-permeability reservoir<br>3. Study on microstructure and pore structure of low-permeability reservoir<br>4. Static and dynamic characterization of reservoir fractures<br>5. Division of reservoir flow units<br>6. Analysis of sensitivity and fracturing damage for low-permeability reservoirs | 1. Fine prediction and characterization of sand body based on outcrop, modern sedimentation, and seismic data need to be strengthened<br>2. The coincidence rate of interpreted swept layer for low-permeability sandstone reservoir is low<br>3. Inadequate research is made on the change law of reservoir in the development process |

*Source:* Based on Huanqing, C., Shunming, L., Xiaojuan, D., 2018. Advances in the study of fine reservoir description in low permeability oilfield. Science Technology and Engineering 18 (32), 129–142.

rock, and the metamorphic rock. From the necessities of adhering to space limitations, this book will only take the clastic rock as an example to discuss the fine reservoir description of the low-permeability oilfield. Fine prediction and characterization of sand bodies are always one of the cores in fine reservoir description, especially for the low-permeability oilfields. In several PetroChina blocks in the Ordos Basin, sand body identification is considered as the most effective technique, from the perspective of reservoir geology, stress on reserve and production increase by progressive development, and low oil price. The sand bodies identified imply the initiation of drilling and, thereby, the ensuing economic benefits.

#### 1.3.2.1.1 Prediction and characterization of sand body

There are many sand body description techniques, including field outcrop observation (Fig. 1.13), modern sediments investigation (Fig. 1.14), statistical analysis of dense well patterns, flume experiment, and seismic inversion prediction, which are applied as a whole to identify the depositional genetic models and to quantify the development scale of various sand bodies. In the needs of fine reservoir description of the low-permeability oilfields, the most commonly used technique for description of sand body is the dense well pattern data, since the log data are relatively accurate and accessible. Field outcrop or modern sediment data are also used; many research efforts (Yangquan et al., 1995; Xinghe, 2004) in this aspect have provided an abundant basis for understanding the low-permeability sand bodies with different sedimentary origins. Moreover, with the improvement of seismic acquisition and processing technology, high precision

(A) (B)

**Figure 1.13** Triassic Yanhe Formation in the Ordos Basin (Yanhe Section). (A) River diversion causes sand body superposition in different periods and (B) river stacking in different sedimentary periods.

(A) (B)

**Figure 1.14** Modern sediments in Erguna River, Inner Mongolia. (A) Quliu River Beach in Head Section of Montenegro and (B) abandoned channel of meandering river in Linjiang section.

sand body prediction using 3D seismic data has gradually appeared in the fine reservoir description of the low-permeability oilfields. Typical application examples include Xinjiang, Daqing, and Shengli oilfields. However, it should be noted that problems still exist in the large-scale application of seismic data to the inversion and prediction of meter-scale sand body in the low-permeability reservoirs due to the limitations of data quality and techniques. The fine characterization of sand body mainly depends on well data, especially the data of dense development well pattern.

#### 1.3.2.1.2 Analysis of geological genesis

The book argues that, for the purpose of sand body description, it is necessary to first analyze the geological genesis of the sand body, and then to identify the genetic model of sand body, and, thereafter, ascertain the horizontal and vertical distribution laws of sand bodies with different origins, so that the sand body can be described and predicted accurately from the macroscopic viewpoint. In the geological genesis analysis, the question of the sedimentary type of reservoir versus one that is not a sedimentary reservoir should be determined. The reservoirs in the low-permeability oilfields are diverse in sedimentary types, including alluvial fan, submarine fan, sublacustrine fan gravity flow, river, and delta (Daopin, 1997). First of all, the sedimentary facies type should be determined according to the comprehensive analysis of rock type, mudstone color, grain size, bedding structure, paleontological assemblage, and other features. Field outcrop and modern sediments can provide vast information on the geological genesis of the sand body. With this information, together with core data and seismic data, the sedimentary genetic model of reservoirs can be clarified. Under the guidance of this model, the fine description and prediction of sand bodies are completed based on the dense well pattern data and seismic data and in combination with the production performance data. The geological genesis analysis of the low-permeability oilfields is consistent with that of other types of oilfields.

#### 1.3.2.1.3 Sand body characterization based on the sedimentary microfacies analysis

The sedimentary microfacies analysis is the most commonly used method for sand body characterization and has been widely applied in oilfield development. The sedimentary microfacies are classified by genesis and then characterized in space so as to predict the sand bodies. Such a study can describe the distribution pattern of sand bodies and help researchers to understand the properties of sand bodies more profoundly. For example, the shapes of two genetic types of sand bodies, that is, underwater distributary channel and river mouth bar, in the delta front can be characterized by sedimentary microfacies, with their vertical sedimentary rhythms identified: positive rhythm for the former, and negative rhythm for the latter. Similarly, the sand bodies of underwater distributary channel have much better physical properties than have the sand bodies of underwater distributary interchannel, which can also be recognized by the sedimentary microfacies analysis.

#### 1.3.2.1.4 Analysis of sand body connectivity

Sand body connectivity is a hot but difficult topic in fine reservoir description. The difficulty is manifested in five aspects. First, the formation of underground reservoir sand bodies is controlled and influenced by structure, sedimentation, diagenesis, and other factors, which together determine their connectivity. It is very difficult to accurately identify the spatial development scale of sand bodies with different genesis and by accounting for the weight of different factors. For example, the sand bodies encountered by two adjacent wells are disconnected, which possibly means that they are two isolated sand bodies, or one sand body that is separated by the closed fault, or by the lithologic flow barrier caused by diagenesis, or by other possible causes. Second, the analysis of connectivity within sand bodies requires a high precision of data. However, a considerable amount of data used in oilfields can hardly meet this requirement. For the well-to-well tracer test data, generally, the tracer injection is conducted and observed at the level of pay zone or layer, while the sand body corresponds to the level of the bed, which is a fifth-order cycle (Huanqing et al., 2014a,b). As a result, one set of tracer monitoring data for injection wells and production wells may correspond to two or more sand bodies. However, it is difficult to determine whether these sand bodies are connected between injection wells and production wells. Third, the analysis of sand body connectivity involves both dynamic and static data of the reservoir. It is very difficult to effectively combine and verify these data. In most of the oilfields in China, abundant dynamic and static data have been accumulated through a long history of development efforts of oilfields and their screening and sorting are extremely important. The dynamic data of oilfield development, for example, are sensitive to human factors, traced to the operator, in view of precision and, also, greatly sensitive to engineering factors; thus, the researchers should make the data quality control. Fourth, under the current trends of low oil price, the quantity of research projects and test data decrease dramatically in response to the requirement for cost reduction and efficiency improvement in oilfields, which directly leads to much less dynamic and static data for analysis of sand body connectivity to be conducted. Some scholars have discussed the sand body connectivity. Qigu and Gongyang (2010) investigated the sand body connectivity in the Wuliwan area of the Ordos Basin, depending on the lithology, sedimentary sequence, and electrical characteristics of logs, as well as the logging interpretation results (Fig. 1.15). Qihong et al. (2013) combed the research methods of dynamic connectivity between wells for the low-permeability reservoir and established a new numerical approach—the multivariable linear regression

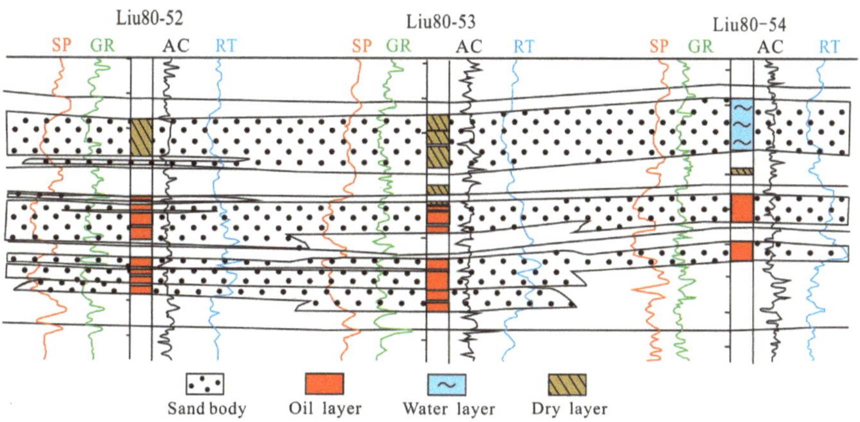


**Figure 1.15** Profile of sand body connectivity in the Wuliwan area of the Ordos Basin (Qigu and Gongyang, 2010).

model—which was proved as a simple and practical approach worthy of broad application on sites. Fifth, the sand body connectivity is related to the optimization and performance improvement of injection-production well pattern and to the results of operations like progressive extension of mature oilfield. Thus, sufficient attention needs to be paid to it. In the analysis of sand body connectivity, the most important consideration is to fully combine a variety of data for verification and comparison so as to achieve the results close to the actual geological conditions underground.

#### 1.3.2.2 Comprehensive fracture characterization based on multiple information

All oil and gas reservoirs generally have fractures, which differ in extent and validity and, consequently, may have distinct capacities of hydrocarbon storage and fluid flowing (Nan et al., 1996). Effective pore space and migration channels are one of the prerequisites for oil and gas production from the low-permeability reservoir, which is typically tight. Therefore, both natural and artificial fractures are the research focus. The so-called fracture refers to the discontinuous surface generated by rock breaking, and it is a result of the process when the rock is broken under force (Huanqing et al., 2016a,b,c,d). Lianbo et al. (2010) systematically summarized the methods for analyzing fractures in the low-permeability oil and gas reservoirs, including geological analysis, identification, and evaluation of fractures using conventional log data, identification and evaluation using imaging log data, seismic detection, structural fracture prediction, and reservoir engineering analysis. Huanqing et al. (2011a,b,c) divided the fractures in volcanic reservoirs of the first member of Yingcheng Formation ("Ying 1 Member") in Xudong area, the Songliao Basin, into structural fractures, condensing shrinkage fractures, explosive fractures, dissolved fractures, stylolitic fractures, weathering fractures, and other types, by genesis. With the combination of dynamic and static data under consideration, and multiple information such as core, thin section, conventional and FMI log data, and seismic data combined, that together with the support of seismic coherence analysis and ant tracking techniques, they were successful in thoroughly characterizing various types of fractures. The results show that this method combining both logging and seismic data can realize the characterization of fractures in volcanic reservoir. Bo and Jianchun (2013) took the ultralow-permeability reservoir in the northern area of Z23, Wuhaozhuang sub-sag, Zhanhua Sag, the Bohai Bay Basin as an example, and discussed the fracture parameter optimization of large hydraulic fracturing using the fine geological model.

Shibin et al. (2012) analyzed the characteristics and genesis of the low-permeability microfracture reservoirs in the Neogene Youshashan Formation in Nanyishan Oilfield, the Qaidam Basin, through the methods of core observation, thin section identification, and X-ray diffraction whole rock-mineral analysis, and found that the area is generally poor in reservoir physical properties, with microfractures as the main flow channels and micropores as the main pore space. Bhattacharya et al. (2012) reported that the unified fracture design can help create the picture of the optimal hydraulic fractures to maximize the oil production in the low-permeability reservoir, and exemplified how to improve the fracture extension length in the unified fracture design. Yang et al. (2012) studied the application of real-time monitoring technology on artificial fractures by microseismic wave in the low-permeability reservoir stimulation, with the use of the Carboniferous low-permeability volcanic reservoirs in the south of Block 9 in Karamay Oilfield as an example, and he finally ascertained the orientation, length, height, and occurrence of artificial fractures, and arrived at the direction of the maximum horizontal principal stress in the fractured interval, which could provide the basis for optimizing the parameters of hydraulic fracturing and making the development adjustment plan for similar reservoirs. Youjing et al. (2015) took the Wangyao area in Ansai Oilfield as an example to analyze the dynamic fracture characteristics of the low-permeability reservoir, and he came to the conclusion that the dynamic fractures are new fractures produced by the rock breaking when bottom-hole pressure exceeds the breakdown and extension pressure of the rock, or they are effective fractures produced by activation of originally closed and filled natural fractures, as the area near the injection well holds pressure during the long-term water flooding process. Dynamic fractures change the percolation characteristics of displacement of oil by water in the ultralow-permeability reservoir, and significantly aggravate the heterogeneity and, also, seriously affect water flooded volume.

With reference to previous studies and combining these with our own practice, we believe that, currently, the reservoir fracture research efforts mainly involve the fracture's geological genesis analysis, laboratory simulation and numerical simulation of fracture genetic mechanism, characterization of fracture development, geological modeling, and analysis of oil storage capacity and flow capacity of fracture. The research methods include field outcrop observation (Fig. 1.16), core observation and description, microscopic thin section observation (Fig. 1.17), geological statistics of well logging interpretation, ant tracking of seismic data,

(A) (B)

**Figure 1.16** High-angle fractures in the Triassic Yanhe Formation, the Ordos Basin (Yanhe Section). (A) and (B) are both features of outcrops of faults. The difference is that the Angle is different. (A) is the profile directly opposite the photographer, and (B) is the profile diagonally opposite the photographer.

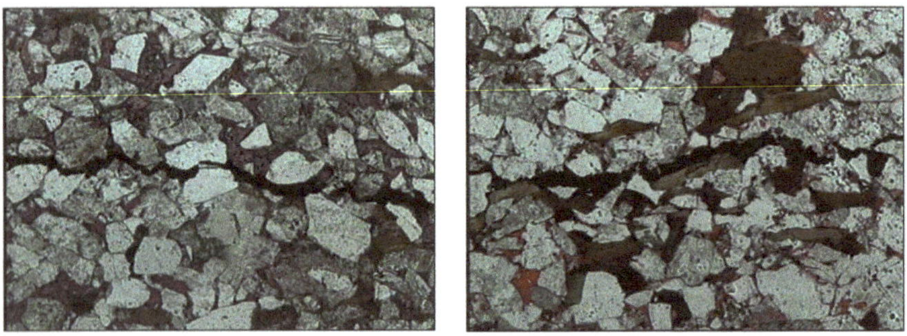

(A) (B)

**Figure 1.17** Thin section photomicrograph of fractures in the Chang 6 pay zone in an area of the Ordos Basin. (A) Well A9, ×5, depth:1415.16 m and (B) well A2, ×5, depth:1419.74 m.

coherence analysis, and laboratory simulation and numerical simulation. In general, though substantial progress has been made in the areas of genesis analysis, occurrence identification, and scale characterization of the fracture, still bottlenecks exist in interwell prediction and 3D modeling of fractures. The ultimate purpose of fine reservoir description is to establish a three-dimensional geological model of the reservoir, which includes the fracture model as an important part; yet, so far, there is no quantitative method to accurately describe the development characteristics of the fractures in three dimensions. Moreover, the development of fractures is directly related to the effective development of the low-permeability oilfield, and so the interwell prediction and 3D geological modeling of fractures will certainly be the major decisive points of departure in recording fine reservoir description of the low-permeability oilfields.

#### 1.3.2.3 Study of reservoir pore structure

The pore structure of reservoir refers to the geometry, size, distribution, and connection of pores and throats in rocks (Shenghe and Qihua, 1998). As much as it's influenced by the physical properties of the reservoir, the study of microscopic pore structure is particularly important in the fine reservoir description of the low-permeability oilfield. Haiyan et al. (2012) explored the microscopic pore structure characteristics, and classification and evaluation methods of the low-permeability reservoir in Sulige Gas Field in a study based on the data of the mercury-injection test, scanning electron microscope, and casting thin section of coring wells. Zhenyu et al. (2013) took Blocks Su 48 and Su 120 in Sulige Gas Field as examples to identify the microscopic pore structure and flow characteristics of the low-permeability tight gas reservoir, indicating that both blocks are dominated by secondary pores and that they contain the reservoirs with high irreducible water saturation, low equal-permeability point, and narrow co-permeation zone. Hui et al. (2013) quantitatively evaluated the micropore-throat heterogeneity of the ultralow permeability sandstone by the constant-rate mercury-injection experiment, with the use of the Chang 8 reservoir in Xifeng Oilfield in the Ordos Basin as an example, and he found that the heterogeneity of micropore throat is mainly reflected in the throat characteristic parameters, which restrict the reservoir quality and affect the development effect. Given large pore-throat ratio and wide distribution range are the outstanding features of the ultralow permeability sandstone reservoir and, also, the main causes for poor development effect, they concluded that the reservoirs with different permeability levels should be treated specifically according to the throat size and distribution range during the development. Huanqing et al. (2013a,b,c) divided the pores in the low-permeability volcanic reservoir of Ying 1 Member in Xudong area, the Songliao Basin, into the primary pore and the secondary pore from the perspective of geological genesis, and then further subdivided them into the seven subtypes, including the vesicle, intergranular pore, and intragranular pore. From the available data, Chen Huanqing et al. analyzed the correlation between pore structure and reservoir permeability depending on the pore structure parameters (e.g., average pore radius, sorting coefficient, relative sorting coefficient, homogeneous coefficient, structural coefficient, and average pore-throat size) characterizing the pore structure, and then arrived at the conclusion that the median pore-throat radius is the parameter that can best reflect the permeability. Then, coupling this with permeability and porosity, and based on SPSS-cluster analysis platform, they divided the pore structure of the reservoir into four types, quantitatively, and, finally, described the horizontal and vertical distribution laws of different pore structure types and their impact on development. Yao et al. (2016) took the Chang 9 pay zone of Yanchang Formation in the Zhijing-Ansai area of the Ordos Basin as an example to analyze the microscopic pore structure of low-permeability sandstone reservoir and discovered that the diversity and heterogeneity of pore structure are the main causes of reservoir heterogeneity.

From a review of the previous studies, we find that the reservoir microscopic pore structure is mainly investigated with a view to understanding the pore structure development characteristics and its relationship with oil and gas distribution, using the techniques like microscopic thin section observation (Fig. 1.18), analysis based on mercury-injection data, and statistical analysis through experiments. Currently, the constant-rate mercury-injection experiment, CT scanning technology, and core modeling technique are relatively advanced. Given there are many parameters to characterize the microscopic pore structure of the reservoir, it follows That the parameters should be selected, depending on the specific block, to quantitatively evaluate the characteristics of the microscopic pore structure of the reservoir. Furthermore, the geological genesis of microscopic pore structure is analyzed by combining the geological and dynamic data; based on this, the genetic mechanism of microscopic pore structure is

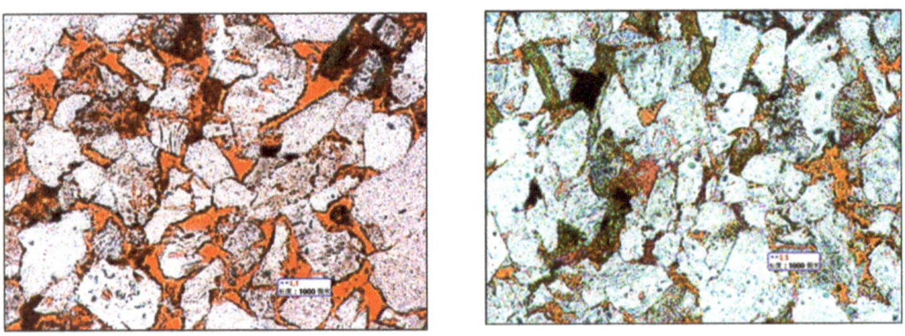

(A) (B)

**Figure 1.18** Thin section photomicrograph of pore structure of reservoir in the Triassic Yanchang Formation in an area of the Ordos Basin. (A) Good pore structure, Well A19, ×10, depth:1460.09 m and (B) poor pore structure, Well A2, ×10, depth:1428.87 m.

clarified through laboratory experiments or numerical simulation. Ultimately, of course, the influence of microscopic pore structure on oil and gas storage capacity and flow capacity of the reservoir is determined.

#### *1.3.2.4 Classification of flow units in the low-permeability reservoir*

It is well known that the fluids in the low-permeability reservoir follow the non-Darcy flow pattern (Fig. 1.19) (Wenrui, 2009). Therefore, a thorough analysis of such flow patterns is very important as a part of the fine reservoir description for the purpose of economic and effective development of low-permeability oilfields. Flow unit is an expressive form of fluid flow. Flow unit, a concept proposed by Huanqing, C. (2019), is defined as a portion of the reservoir which is continuous, vertically and horizontally, with similar internal permeability, porosity, and bedding characteristics (Zhilin, 2000; Yang and Jianmin, 2005; Huanqing et al., 2010, 2011a,b,c). Qigu and Gongyang (2010) divided the flow units, by genesis, in the low-permeability reservoir in the Wuliwan area of the Ordos Basin, into three categories. Fengjuan et al. (2012) divided the flow units in low-permeability sandstone reservoirs based on the entropy TOPSIS method, which provided a new idea for the classification and evaluation of flow units. Xinlong et al. (2013) reviewed the mechanism of microflow in low-permeability reservoir and its application, and they put forth the suggestion that the pore structure, clay mineral occurrence, wettability, and charge performance of rock minerals are the critical factors influencing the microflow in low-permeability reservoir, which can

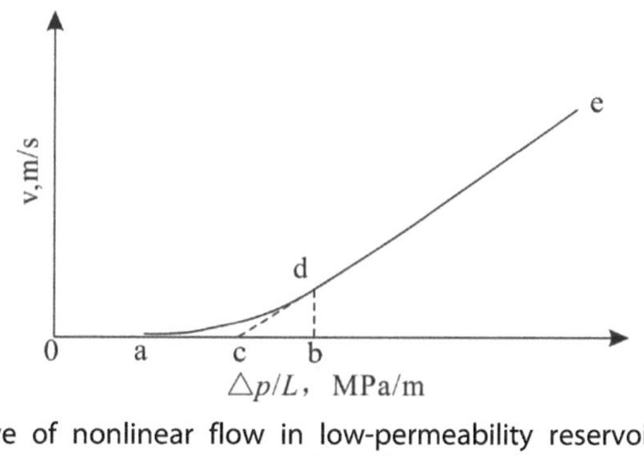


**Figure 1.19** Curve of nonlinear flow in low-permeability reservoir (Wenrui, 2009). Note: ΔP/L represents the pressure gradient; V represents the flow velocity; Point a represents the start-up pressure gradient; when the pressure gradient reaches Point b, fluids begin to flow linearly; Point c is the intersection of the straight line segment de and x-axis, indicating the pseudo start-up pressure gradient.

be controlled to change the flowing characteristics of fluids in low-permeability reservoir, thereby maintaining the oil, gas, and water production. Maolei et al. (2013) established a large physical model for the ultralow permeability sandstone and proposed a new approach to identify the flowing law of fluids in the ultralow-permeability reservoir. Panja et al. (2013) analyzed the grid characteristics of numerical simulation of hydraulic fracturing of the low-permeability reservoir, and he found that it is difficult to make an accurate prediction of reservoir production by using the conventional grid system when the permeability is largely different between adjacent horizons, while the fine grid near wells and faults can facilitate the fluid prediction between different fault blocks.

Essentially, the study of reservoir flow units is conducted to figure out whether the reservoir has enough flow capacity to support the production of an oilfield with low permeability generally. As from what has been introduced earlier, we find that the researchers have different viewpoints on the concept of flow unit. This book supports Hearn's definition (1984). The accuracy of flow unit characterization is directly decided as a 2-step method: (1) the selection of flow unit classification parameters, and (2) the selection of flow unit classification method. Given sufficient production performance data, it is recommended to verify the results of flow unit characterization. In case of any deviation, the flow unit classification parameters and methods can be adjusted until the study results are highly consistent with the production performance data. Therein, the selection of parameters should be able to reflect the reservoir flow capacity. Note that more parameters selected doesn't mean better; in this regard, it is different from reservoir evaluation. For the selection of data, the data of different natures should be normalized first to prevent the error in order of magnitude, and the weight coefficients of different parameters should be set based on the geological knowledge of researchers additionally to make the flow unit classification as reasonable as possible (Fig. 1.20). Huanqing et al. (2016a,b,c,d) studied the flow units in the low-permeability volcanic reservoir in Ying 1 Member in Xudong area, the Songliao Basin. First, two parameters, porosity and permeability, which could fully reflect the flow capacity of reservoir, were selected from the coring wells. Then, the cluster analysis software was selected for cluster analysis and discriminant analysis, and thereafter to finally obtain the discriminant formula. The discriminant formula was used to discriminate and analyze noncoring wells, and accordingly the flow units were divided. Finally, the classification results of flow units in the whole study area were determined.

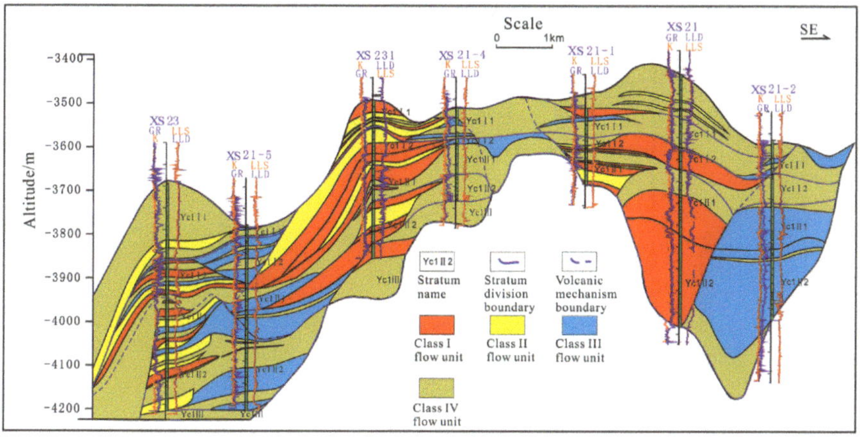


**Figure 1.20** Profile of flow units in low-permeability volcanic reservoir in Ying 1 Member in Xudong area, the Songliao Basin (Huanqing et al., 2016a,b,c,d).

### *1.3.2.5 Protection of low-permeability reservoir*

In the development process of oil and gas reservoirs, the injection of foreign fluids results in the change of reservoir properties, including the decrease of reservoir flow capacity and the decrease of production capacity, causing the formation of damages to take place on the reservoir. For the ultralow-permeability oilfields, where the porosity and permeability are very low, stimulation measures such as acidizing and fracturing are generally required to facilitate the production. Meanwhile, water or gas is injected to keep the formation pressure stable. All these operations may cause huge damages to the reservoir. As the abovementioned engineering measures have been normally applied in the development of the low-permeability oilfield, effective protection of reservoir for minimizing the reservoir damage becomes an extremely important theme in fine reservoir description.

At present, the protection of the low-permeability reservoir has attracted the attention of many researchers. Xingxin and Shuxin (1992) studied the clay minerals and protection of sandstone reservoirs. Fanhui and Jianchun (2011) analyzed the damage and stimulation of low-permeability reservoir in CIII pay zone in Donghe Oilfield of the Tarim Basin and were the first to reveal the main causes for reservoir damage as complex mineral composition, poor pore-throat structure, strong sensitivity, oil-wet rock surface, and improper drilling/completion fluids and stimulation fluids. Chunning et al. (2011) took the Ansai extra-low-permeability oilfield as an example to investigate the characteristics and influences of scaling minerals in water injection formation (Table 1.7) and

**Table 1.7** Scaling minerals in water injection and their influence on core porosity and permeability in Ansai Oilfield of the Ordos Basin (Chunning et al., 2011).

| Well | Depth/m | Neogenic mineral/% | | Porosity/% | | | Permeability/$10^{-3}$ μm$^2$ | | |
|---|---|---|---|---|---|---|---|---|---|
| | | Calcite | Barite | Before water flooding | At present | Decline/% | Before water flooding | At present | Decline/% |
| Jian 16–151 | 1157.68 | 1.2 | | 14.38 | 13.18 | 8.34 | 4.117 | 2.761 | 32.94 |
| | 1162.98 | 2.1 | 0.2 | 15.04 | 12.74 | 15.29 | 0.445 | 0.338 | 24.06 |
| | 1163.62 | 1.3 | | 14.34 | 13.04 | 9.07 | 2.858 | 1.842 | 35.55 |
| | 1167.07 | 1.7 | | 14.12 | 12.42 | 12.04 | 1.224 | 0.739 | 39.65 |
| Jian 16–152 | 1155.32 | 1.1 | | 12.63 | 11.53 | 8.71 | 0.297 | 0.231 | 22.17 |
| | 1160.36 | 1.0 | | 11.81 | 10.81 | 8.47 | 1.689 | 0.960 | 43.19 |
| | 1165.83 | 1.3 | | 11.06 | 10.76 | 2.71 | 0.779 | 0.577 | 25.98 |
| Jian 16–153 | 1181.01 | 1.4 | 1.0 | 15.51 | 12.11 | 16.54 | 0.414 | 0.291 | 29.74 |
| | 1183.12 | 2.1 | 0.6 | 16.55 | 13.85 | 16.31 | 3.457 | 1.986 | 42.55 |
| | 1196.49 | 1.0 | 0.3 | 13.73 | 12.43 | 9.47 | 1.225 | 0.788 | 35.71 |
| | 1199.32 | 0.6 | 0.3 | 10.78 | 9.88 | 8.35 | 0.490 | 0.404 | 17.47 |

they concluded that the injected water, being $Na_2SO_4$ type, is incompatible with the formation water, being $CaCl_2$ type with $Ba^{2+}$, so that abundant scaling minerals (calcite and barite) were formed in reservoirs after long-term water injection. They also discovered a good correlation between the amount of scaling minerals and the decrease of porosity and permeability. After 20 years of water flooding, the average decreases of porosity and permeability were 15% and 31.73%, respectively. In view of reservoir damage, calcium carbonate scale is larger than barium sulfate scale. Renyuan et al. (2013) studied the effect of steam flooding on the core wettability in the low-permeability carbonate heavy oil reservoir, and they found that steam flooding could improve the wettability of oil-wet reservoir, thus improving the recovery efficiency by dialysis. Jingang et al. (2013) took the Chang 8 sandstone reservoir in Zhenjing Oilfield as an example to investigate the methods for evaluating how fracturing fluids damage the reservoir, and then carried out the dynamic damage experiment to determine the rate of damage of fracturing fluid filtrate to the permeability of matrix core sample, and, the influence of dynamic filtration of gel breaker on the flowback recovery of fracturing rock sample, and they also explored the compatibility of fracturing fluid with formation fluid and working fluid, and the wettability of fracturing fluid and crude oil, and measured the emulsification rate and residue of fracturing fluid. Chunjie et al. (2013) took the middle-deep high pressure and the low-permeability reservoir in the third member of Shahejie Formation in Wendong Oilfield of the Dongpu Sag as an example to analyze the reservoir sensitivity in detail, and reported that the most direct factors affecting the reservoir sensitivity included reservoir physical properties, reservoir pore structure, and clay minerals. Lijuan (2014) analyzed the influence of pressure sensitivity on elastic production capacity of low-permeability reservoir, and he suggested that in the early stage of actual oilfield development, by relying on elastic production simply, the decrease of formation pressure would reduce the permeability and then the reservoir deliverability, due to the pressure sensitivity.

Reservoir damages are believed to have resulted from fracturing and acidizing, and water flooding development. In a few of the oilfields, injection of $CO_2$ and other gases possibly leads to the change of reservoir properties. Before any large-scale fracturing and acidizing operation, typical test blocks should be selected for damage evaluation of fracturing fluids, and corresponding measures could be taken if necessary to minimize the damage. For water flooding development, the study of reservoir damage mainly focuses on the sensitivity. The influence of injected water on clay minerals in reservoirs should be fully taken into account, and appropriate injection pressure and rate should also be maintained. For the purpose of effective reservoir protection, various experimental and numerical simulations are absolutely necessary. Of course, the study of diagenesis based on the microscopic observation of thin sections and the statistics of various test data is also required.

### 1.3.3 Trend

Fine reservoir description is a systematic discipline, covering almost all aspects of development geology. Along with the progress of development, relevant data increase greatly. Moreover, various new experimental techniques, physical simulations and mathematical calculation methods are increasingly applied in fine reservoir description. In addition to the aspects mentioned earlier, the following should be emphasized since these may represent the orientation and growth points for fine reservoir description of the low-permeability oilfield.

#### *1.3.3.1 Description of microstructure*

In the Ordos Basin, for example, there is no large fault system with its having a simple structure. However, in the fine reservoir description, special attention is required to be paid to the microstructure, as it is an important controlling factor of the remaining oil. Microstructure, also called the sedimentary microstructure or reservoir microstructure, refers to a general name for local structural features and uncertain micro-faults shown by small changes taking place in the structural undulation shape of the top of reservoir on the oilfield structure (Jinliang and Jun, 2011). In low-permeability oilfields with different genetic types, the research focus of microstructure is not the same. For low-permeability oilfields dominated by lithologic reservoirs such as Changqing Oilfield, emphasis should be given on small high points, small nose structures, small low points, small grooves, and small monoclines. For low-permeability oilfields dominated by structural or structural-lithologic reservoirs such as the Dagang Oilfield and the Jidong Oilfield, more efforts should be devoted to the characterization and recognition of microstructures such as small faults and small fault terraces.

#### 1.3.3.2 Analysis of reservoir geological genesis including diagenesis

Low-permeability reservoirs have various genetic types, of which diagenesis is a typical one and should be dwelt on thoroughly (Fig. 1.21). Shengli et al. (2013) analyzed the genesis of low-permeability reservoirs in the Chang 4 + Chang 5 Members of Yanchang Formation in the Jiyuan area, the Ordos Basin, and they are convinced that diagenesis is one of the main genetic factors. Jing et al. (2013) quantitatively studied the reservoir diagenetic facies of Yanchang Formation in the Longdong area, the Ordos Basin, and this study effectively guides the distribution prediction of high-quality reservoirs. In addition to the conventional analysis of diagenetic types and division of diagenesis stages, the study of diagenetic facies has been widely used in the fine description of the low-permeability

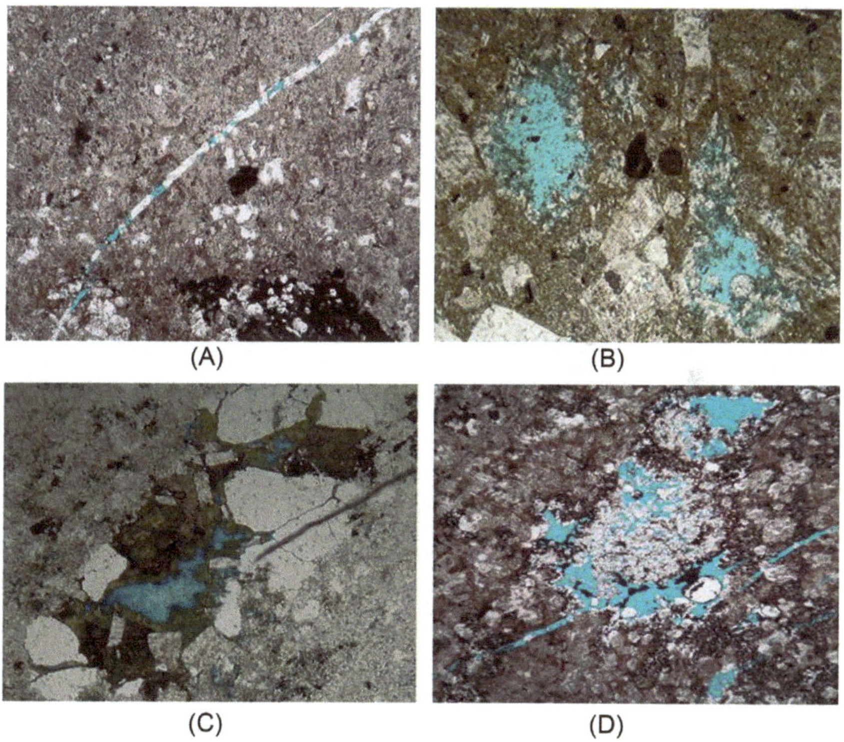


**Figure 1.21** Thin section photomicrograph of diagenesis in the low-permeability volcanic reservoirs in Ying 1 Member in Xudong area, the Songliao Basin. (A) Quartz semifilled fractures, Well A3, 3899.47 m, × 40, (–); (B) moldic pores generated from dissolution of lithoclasts in volcanic breccia, Well A2, 3701.06 m, × 40, (–); (C) vesicle-filled chlorite and spherulitic micropores, Well A12, 3730.58 m, × 40, (–); (D) matrix dissolved pores in spherulitic rhyolite, Well A3, 3899.47 m, × 40, (–).

reservoirs. Its most outstanding characteristic is to make the research of diagenesis develop toward quantification. Moreover, the close combination of diagenetic facies maps and sedimentary microfacies maps for different layers (beds) contributes greatly to the prediction of favorable development zones and provides a solid geological basis for progressive extension and designing of the development adjustment plan for mature oilfields.

#### 1.3.3.3 Experimental study on the mechanism of $CO_2$, $N_2$, or polymer flooding for oil recovery in low-permeability oilfields

Zekri et al. (2013) investigated the petrophysical changes of heterogeneous oil-bearing, low-permeability carbonate reservoirs caused by $CO_2$ flooding in the Abu Dhabi Oilfield, United Arab Emirates, and his results showed that the $CO_2$ injection under supercritical state reduced the porosity and permeability, leading to the reduction in oil-water interfacial tension. Park et al. (2015) carried out an experiment to analyze how the polymer concentration would affect the permeability of polymer flooded low-permeability reservoirs, and concluded that the hydrodynamic action made polymer molecules adsorbed in a bed and such absorption led to the thickening of walls of pore throats, thereby hindering the oil flowing in the development process. For low-permeability oilfields, the tertiary recovery by steam or polymer flooding can enlarge the swept volume, improve the displacement effect, and ultimately enhance the oil recovery. This technology has been applied in Daqing, Changqing, Shengli, and other oilfields, with good results, and it is also put into pilot test in Qinghai Oilfield and the likes. Compared to water flooding, steam flooding or polymer flooding exhibits very different displacement mechanisms and interactions with the reservoir. Therefore, relevant research efforts should be strengthened in fine reservoir description to provide support for the improvement of development effect and oil recovery of low-permeability oilfields. Overall, the injection of $N_2$, $CO_2$, or other gases to develop low-permeability oilfields can increase the swept volume, enhance the oil recovery, and avoid or reduce the damage of injected agent to the reservoir as far as possible. However, some new problems may appear subsequently; for example, the increase in development cost caused by gas sources or the corrosion of $CO_2$ on pipelines and casings are noticeable disadvantages. Thus, it needs to consider using glass steel casing in surface engineering projects.

#### 1.3.3.4 Economic effectiveness evaluation of fine reservoir description under low oil price

Economic effectiveness evaluation can ensure the successful application of the results of fine reservoir description to the development adjustment and progressive extension of mature oilfields. Low-permeability oilfield is uniquely characterized by low porosity and low permeability, resulting in poor economic benefits of development. This, however, does not mean that a low-permeability oilfield cannot be developed economically and effectively. Gaofeng et al. (2015) proposed a method to screen low-permeability reservoirs for $CO_2$ flooding in terms of the economic benefits of $CO_2$ flooding. For example, in the Ordos Basin, of the many low-permeability reservoirs in the Triassic Yanchang Formation, a majority of the reservoirs are shallow which means the drilling cost is generally low. Therefore, dense well patterns and horizontal wells become relatively economical and effective choices to improve the development effect of oilfields. In summary, the core issue in low-permeability oilfield development is the economics of it. Many techniques applicable to high water-cut or complex lithologic oilfields do not work in low-permeability oilfields due to, fundamentally, economic effectiveness.

#### 1.3.3.5 Basic research on horizontal well design and implementation of other engineering measures in fine reservoir description

Among the determinants of economic and effective development of low-permeability oilfields, engineering factors take the lion's share having a weight more than 50%. The engineering factors include water injection in early stage, artificial fracturing, horizontal well drilling, and other prerequisites of economic benefits. As a very effective measure for economic and effective development of low-permeability oilfields, horizontal well drilling has been successfully applied in Changqing, Daqing, Dagang, and other oilfields. Daopin (2003) compared the horizontal well and vertical well deployment schemes (Fig. 1.22). Scheme 2 was the best simply according to development index. However, the comprehensive economic evaluation results showed that Scheme 3 based on inverted five-spot pattern, that is, one horizontal well (for oil recovery) surrounded by four vertical wells (for water injection), was the best; while Scheme 4 based on mixed five-spot pattern was the worst. As the basic support for horizontal well design and implementation, the fine reservoir description of low-permeability oilfield is indispensable. Only through the fine reservoir description and the accurate prediction of sand horizon and scale can key

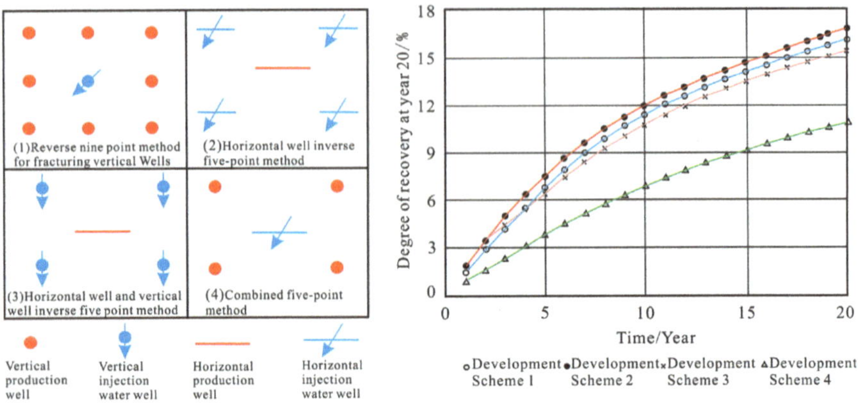


**Figure 1.22** Placement pattern and production performance of vertical wells and horizontal wells in Changqing extra-low-permeability oilfield, the Ordos Basin (Daopin, 2003). Note: If a horizontal well is equivalent to three vertical wells, the well placement is 16 wells/km$^2$ in square reverted nine-spot well pattern, 12 wells/km$^2$ in horizontal well pattern, and 8 wells/km$^2$ in vertical–horizontal mixed well pattern.

parameters, including the location and length of the horizontal section of a horizontal well, be reasonably determined. As the horizontal well drilling technology improves, the horizontal well can realize a production performance more than three times of the vertical well from the perspective of economics, albeit the former's higher drilling cost than the latter's. Therefore, more and more horizontal wells are deployed for developing low-permeability oilfields. Bhattacharya and Nikolaou (2016) presented the design of multilateral horizontal wells for low-permeability unconventional gas-bearing reservoirs by using the comprehensive optimization method. This method avoids the mistakes that creep in the design drawn on the basis of experience, and is a very effective optimization method for multilateral horizontal well design. To design a horizontal well, the spatial distribution of sand bodies should be understood and, then, the development law of underground faults and fractures should be clarified. Different types of microstructures also play a key role in deciding the well location and the length of horizontal section. All of these questions need to be solved by fine reservoir description. In addition, many development measures for low-permeability oilfields, such as leading water injection and fracturing, need fine reservoir description as the research base on which founds the design of a horizontal well.

At present, low-permeability reservoirs have almost become the foremost prospect to watch out for in the face of the reserves and the production increasing in all oilfields in China, and will play an increasingly important role in the future. Accordingly, the fine reservoir description of low-permeability oilfields will always be the focus in research efforts of oilfield development.

## 1.4 Fine reservoir description of the complex lithologic reservoir

The complex lithologic reservoir is widely developed in the Songliao Basin, Bohai Bay Basin, Ordos Basin, Junggar Basin, Tarim Basin, Sichuan Basin, and other major petroliferous basins in China. Fine description of this critical reservoir type is of great significance for understanding the reservoir geological characteristics, enhancing the oil recovery, and tapping the potential of the remaining oil in oilfields in the middle-late stage of development.

### 1.4.1 Current status and key problems

#### *1.4.1.1 Current status*

As the name implies, a complex lithologic reservoir is a reservoir composed of complex types of rocks. The term arches over to include a wide range of rocks, including volcanic rock, carbonate, glutenite, and metamorphic rock. According to the current understanding, almost all reservoir rocks that are different from conventional clastic rocks are classified as complex lithologic reservoirs. The study of the complex lithologic reservoir began early, but its width of scope is not sufficient, neither in present definition nor in future extent, due to its geological complexity, limited distribution, and difficult development. In this aspect, many scholars have devoted certain efforts (Chenglin et al., 1997; Shiqing et al., 2007; Haicheng et al., 2008; Caineng et al., 2011; Xiongyan et al., 2012; Xiaobo et al., 2013; Gang et al., 2014; Xiaoqing et al., 2015; Jin et al., 2016). Chenglin, et al. (1997) published the Special Hydrocarbon Reservoirs, which expounds the hydrocarbon reservoirs of magmatic rock, metamorphic rock, weathering crust, coal measure clastic rock, and conglomerate, providing an important reference to the study of the complex lithologic reservoir. Shiqing et al. (2007) discussed the rational development mode of complex lithologic fault block oil reservoir rich in bottom water and indicated that sidetracking of the old well could tap the potential of remaining oil between wells, on the one hand, and in high parts, improve well pattern, and effectively control the bottom water coning, on the other, so that it was a measure for economic and efficient development of the complex small fault block reservoir with bottom water. Haicheng et al. (2008) took Well C3 in Yumen Oilfield as an example to exhibit the application of NMR in the stimulation of complex lithologic reservoirs. Caineng et al. (2011) published the Unconventional Petroleum Geology, which introduces the origin and the exploration and development characteristics of unconventional oil and gas in detail, including tight sandstone oil and gas, coalbed methane, shale gas, carbonate fracture-cavity oil and gas, volcanic reservoir oil and gas, and metamorphic reservoir oil and gas. Based on the fundamental principles of optimization algorithm and classification algorithm, Xiongyan et al. (2012) established the lithology prediction model by using the self-organizing map neural network cluster, constructed the fluid prediction model by combining the decision tree and support vector machine (SVM), and optimized the key SVM parameters using the genetic algorithm, grid algorithm, and quadratic algorithm in order to accurately identify the lithology and multiphase fluids in the complex reservoir. Xiaobo et al. (2013) selected natural gamma, neutron, density, acoustic, and other logs sensitive to lithology to identify the complex lithology of the Lucaogou Formation tight reservoir in the Malang Sag, revealing 12 lithologies like mudstone, tuffaceous mudstone, limy mudstone, and argillaceous dolomite. Gang et al. (2014) predicted the productivity of complex lithologic reservoirs in the Hailaer Basin, achieving consistent results with the production performance. Xiaoqing et al. (2015) presented the fine characterization of the thick complex lithologic reservoir and its influence on development, with the use of the Sha 1 and Sha 2 reservoirs in A Oilfield in the Shijiutuo area, the Bohai Bay Basin, as examples. Jin et al. (2016) presented the methodology of reservoir prediction in areas with complex lithology in salt lake basins through a case study of the Qianjiang Sag, the Jianghan Basin, and developed the "four-definition" reservoir prediction method. To sum up, the existing studies in China on fine reservoir description of complex lithologic reservoirs involve the objects, including volcanic rock, carbonate rock, glutenite, and metamorphic rock, in the aspects of lithology identification, analysis of sedimentary genetic models, fine logging interpretation of reservoir physical properties, inversion prediction of reservoir development characteristics, fracture characterization, geological modeling, and reservoir fluid prediction. The research methods include field outcrop observation (Fig. 1.23), laboratory analysis and test experiment, core observation and description, well

(A) (B)

**Figure 1.23** Field geological profile of complex lithologic reservoir in Huanlegu, Dunhuang, Gansu Province: (A) desert alluvial fan deposition and (B) windy dune deposits.

logging interpretation, and seismic inversion based on a combination of log data and seismic data. At present, the research hotspots are tight oil/gas and shale oil/gas, with the support of CT scanning analysis of core, reservoir fracture characterization based on static and dynamic data, and 3D porosity modeling, etc.

#### *1.4.1.2 Key problems*

The study of the complex lithologic reservoir is very difficult due to its complexity of geological genesis. Here, we summarize the key problems in fine reservoir description of the complex lithologic reservoir in six aspects.

##### 1.4.1.2.1 Lithology identification and classification

The complex lithologic reservoir involves various types of rocks, and each type can be subdivided into subtypes. Due to the complexity of geological genesis and the different development characteristics in different blocks and horizons, the rock types in the complex lithologic reservoir are difficult to identify and classify correctly. The commonly used methods include core observation and description, logging interpretation, various analyses and tests, and seismic interpretation and prediction. For logging interpretation, the rock types will be interpreted using log data by blocks and horizons due to the complex variation of lithology.

##### 1.4.1.2.2 Analysis of reservoir geological genesis

Before understanding the characteristics of reservoir development, the reservoir genetic mechanism is first analyzed deeply. For volcanic rocks, the volcanic eruption mode (fissure eruption, central eruption, or fissure-central eruption) should be determined first to allow the researchers to identify the distribution law of reservoirs accurately at the macro level.

#### 1.4.1.2.3 Fine stratigraphic classification and correlation

The complex lithologic reservoir is greatly different in fine classification and correlation from the conventional strata. Corresponding solutions are needed in this regard. For the fine classification and correlation of volcanic reservoirs, it is necessary to identify and trace the volcanic rocks based on the large-scale fine isochronous stratigraphic framework established before the stratigraphic breakdown happens. The simple application of the isochronous stratigraphic classification method for clastic rocks may easily lead to the diachroneity of volcanic rocks in fine stratigraphic classification.

#### 1.4.1.2.4 Quantitative description of fracture development

The description of fractures using quantitative methods is very difficult to realize. Fractures are well developed in carbonate rocks, volcanic rocks, metamorphic rocks, and other complex types of rocks, apart from the fact that they are important channels for oil and gas accumulation and migration and have a very important impact on oilfield development. However, a method that can realize the fine quantitative characterization of fractures is a major lacuna. The difficulties therein are mainly reflected in fracture modeling and prediction of interwell fractures.

#### 1.4.1.2.5 Fine logging reinterpretation

The accuracy of fine logging reinterpretation for complex lithologic reservoirs needs to be improved. For the glutenite reservoir in Xinjiang Oilfield, the accuracy of logging interpretation is less than 70%, which greatly undermines the fine reservoir description. Thus, more efforts should be made to improve the fine logging reinterpretation techniques.

#### 1.4.1.2.6 Geological modeling

The geological modeling of the complex lithologic reservoir is insufficient, and the accuracy of interwell reservoir prediction needs to be improved. For example, the geological modeling of fractures and cavities in carbonate reservoirs is still in a nascent stage. Albeit the progress in this aspect, a great difference exists between the geological modeling results and the verification results using production drilling data in terms of the spatial development of fractures and cavities.

## 1.4.2 Core contents

### *1.4.2.1 Identification and classification of complex lithology and lithofacies*

Identification of lithology and lithofacies is critical and fundamental in fine reservoir description of the complex lithologic reservoir. Yanjie et al. (2007) studied the application of SVM-based logging identification of complex lithology in some intervals in Bayantala Oilfield, revealing the average coincidence of up to 96%. Gang et al. (2007) discussed the logging identification of complex lithology based on the constrained least-squares algorithm, which was applied in the carbonate strata in the Sichuan Basin, with high operation speed and high identification accuracy. Yu et al. (2010) studied the litho-electric properties of the complex lithologic reservoir in the Carboniferous volcanic rocks in Kelamei Gas Field, where 11 types of volcanic rocks and five types of rock structures were identified, and revealed through verification with drilling cores that the logging identification results were in good agreement with the drilling core analysis results. Lichun et al. (2013) studied the logging identification of lithology for complex lithologic tight oil reservoir in the Lucaogou Formation in Jimusaer Sag and established an adaptive lithology identification chart, which facilitates the logging identification of lithology for a complex lithologic reservoir. Qiuqiang et al. (2013) identified the complex lithology in a block of the Congo Basin through well logging facies–lithofacies analysis, with good geological effect achieved. Zhang et al. (2014) discussed the identification of complex lithology based on well logging facies analysis, with the use of the Tongbomiao Formation in Tanan Oilfield of the Hailaer-Tamtsag Basin as an example. Xianguo et al. (2015) took the Cretaceous strata in South Tarim Sag as an example to explore the logging identification of lithology for the complex lithologic reservoir. The logging identification of lithology for noncoring wells was realized, with an accuracy of 98.1%. Lithology identification is an extremely challenging part in the fine reservoir description of the complex lithologic reservoir as compared to that of the conventional clastic reservoir, and it can hardly be realized satisfactorily by conventional techniques. Complex lithologic reservoir, naturally a concept opposite to conventional clastic rocks, presents an added difficulty and particularity in research than do conventional clastic rocks. Now, complex lithology and lithofacies are commonly identified by geological method, logging interpretation method, and seismic inversion method with their advantages and disadvantages. The geological method can determine the reservoir lithology based on core observation and description, microscopic thin section identification, and other test data. The logging interpretation method realizes the lithology interpretation depending on the electrical properties reflected by different types of rocks. The seismic inversion method inverts the lithology based on the seismic reflection characteristics and seismic reflection wave velocity corresponding to each lithology. The logging interpretation method is the most popular with researchers, but the results varied greatly in accuracy when the method was applied in different areas. More efforts should be spent to carry out the core observation and description, perform the analysis and test experiments and, therefore, improve the accuracy of lithology/lithofacies prediction based on seismic data. The seismic inversion method is often used for carbonate reservoirs, while the geological method and logging interpretation method is widely applied to lithologies other than the carbonate rocks. The use of the logging interpretation method, especially, has been successfully promoted to volcanic and metamorphic rock reservoirs (PetroChina Exploration & Production Company, 2009). It's our observation that using the logging interpretation method to identify and analyze lithology will be a major means of fine reservoir description of the complex lithologic reservoir. We successfully identified glutenite in an area at the northwestern margin of the Junggar Basin through precise core observation and description, together with fine reinterpretation of dense well patterns. The target reservoirs in the study area mainly contain sandy conglomerate, cobblestone, medium conglomerate, microconglomerate, and sandstone (Fig. 1.24). According to the correspondence between lithology and reservoir architecture, we completed the classified identification and interwell prediction of reservoir architecture on the basis of lithology identification and by considering the electrical properties. This could provide evidence for the implementation of development and adjustment measures for glutenite reservoirs in the middle-late stage based on reservoir architecture characterization. According to the results of reservoir architecture characterization, which are mainly the quantitative data of spatial scale of different architectures, we analyzed the interwell connectivity under different reservoir architectures in the current well pattern. Furthermore, we analyzed the correspondence between injection wells and production wells based on the dynamic monitoring and production performance data and, then, proposed the program for adjusting the deployment of injection and production wells, thereby facilitating the development and EOR of the

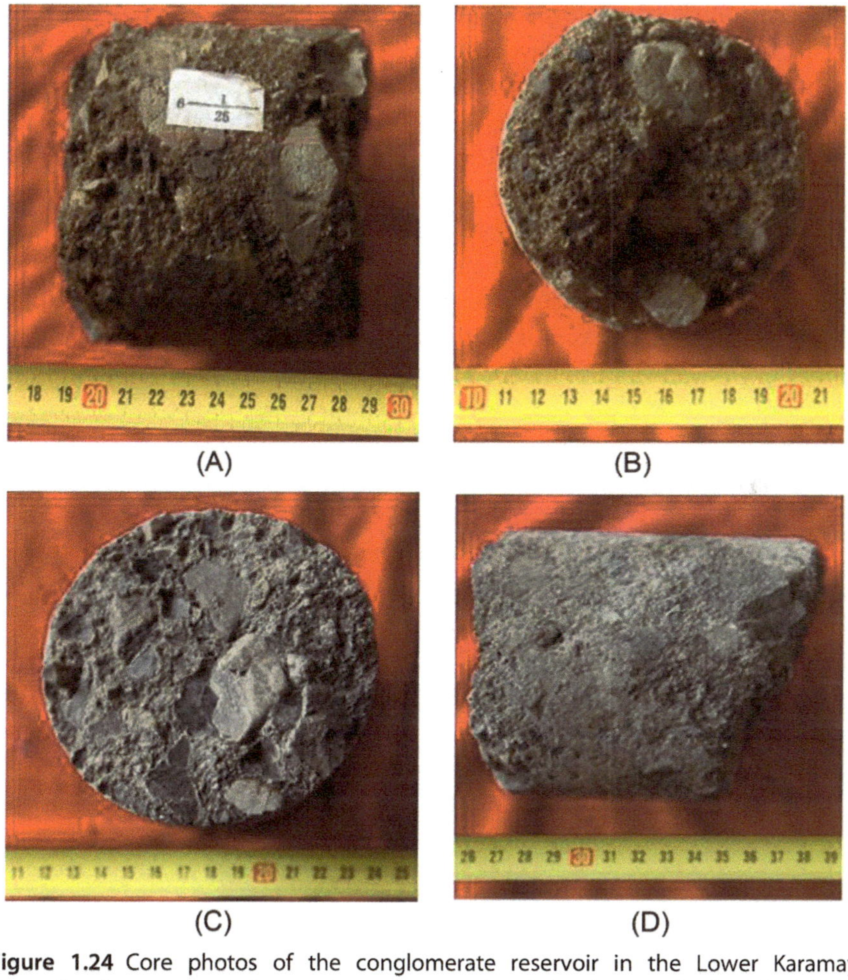


**Figure 1.24** Core photos of the conglomerate reservoir in the Lower Karamay Formation in an area at the northwestern margin of the Junggar Basin. (A) Well J1, light brown sandy conglomerate, 402.40–402.51 m; (B) Well J1, light gray sandy medium conglomerate, 407.80–407.90 m; (C) Well J1, light grayish brown medium conglomerate, 411.80–411.90 m; (D) Well J7, light grayish brown sandy conglomerate, 426.47–426.57 m.

high water-cut glutenite reservoir. For the purpose of complex lithology identification, two options are open to consideration: (1) development of new geological, geophysical, and geochemical methods, and (2) organic combination of existing identification methods to fully exert their respective merits.

### *1.4.2.2 Analysis of geological genetic mechanism of reservoir*

The particularity of a complex lithologic reservoir is derived from the particularity and complexity of its genesis. For the fine reservoir description of complex lithologic reservoir, the geological genetic mechanism of the reservoir should be analyzed first to accurately understand the genetic mechanism and inherent causes thereof. Zhong et al. (2012) studied the application of seismic sedimentology in the complex lithologic area, taking the second member of Shahejie Formation in Qibei Sag as an example, and concluded that the inversion of logging parameters could help effectively distinguish complex lithology. In the study of the geological genesis of carbonate reservoir in the Qiongdongnan Basin, we completed the stratigraphic sequence division and sedimentary microfacies classification based on the lithology and typical data, paleontological data, and seismic data from 12 coring wells (Fig. 1.25), and also summarized the sedimentary model in the study area according to the logging and the core data to provide the basis for reservoir prediction (Fig. 1.26). As shown by logging and core observation of Well A2, the carbonate platform shoal near a bulge in the Qiongdongnan Basin mainly reflects the characteristics of coral reef islands and bioclastic beach, with the former in dominance (Fig. 1.26). Although the analysis of reservoir geological genesis is reported in some studies of fine reservoir description, its importance has not gained enough attention. This book believes that for the analysis of reservoir geological genetic mechanism, the top priority is given to the basic geologic studies, including field outcrop observation, modern sediment investigation, core observation and description, microscopic thin section identification, and core analysis and test. Moreover, CT scanning of cores, flume experiments, and other experimental studies should be added and optimized to clarify the reservoir geological genetic mechanism from the micro- and macro perspectives. If possible, a geological knowledge base should be established using the dense well pattern data; on the basis of statistical analysis of the quantitative data of spatial development of different sedimentary types, the sedimentary genetic model should be ascertained to get a deeper understanding of the reservoir genetic mechanism. In the fine reservoir description of different types of complex lithologic reservoirs, the geological genesis analysis is helpful to figure out the research focus, for which relatively applicable techniques can be adopted. For example, in the fine description of the volcanic reservoir, the geological genesis analysis is conducted to determine if the volcanic eruption

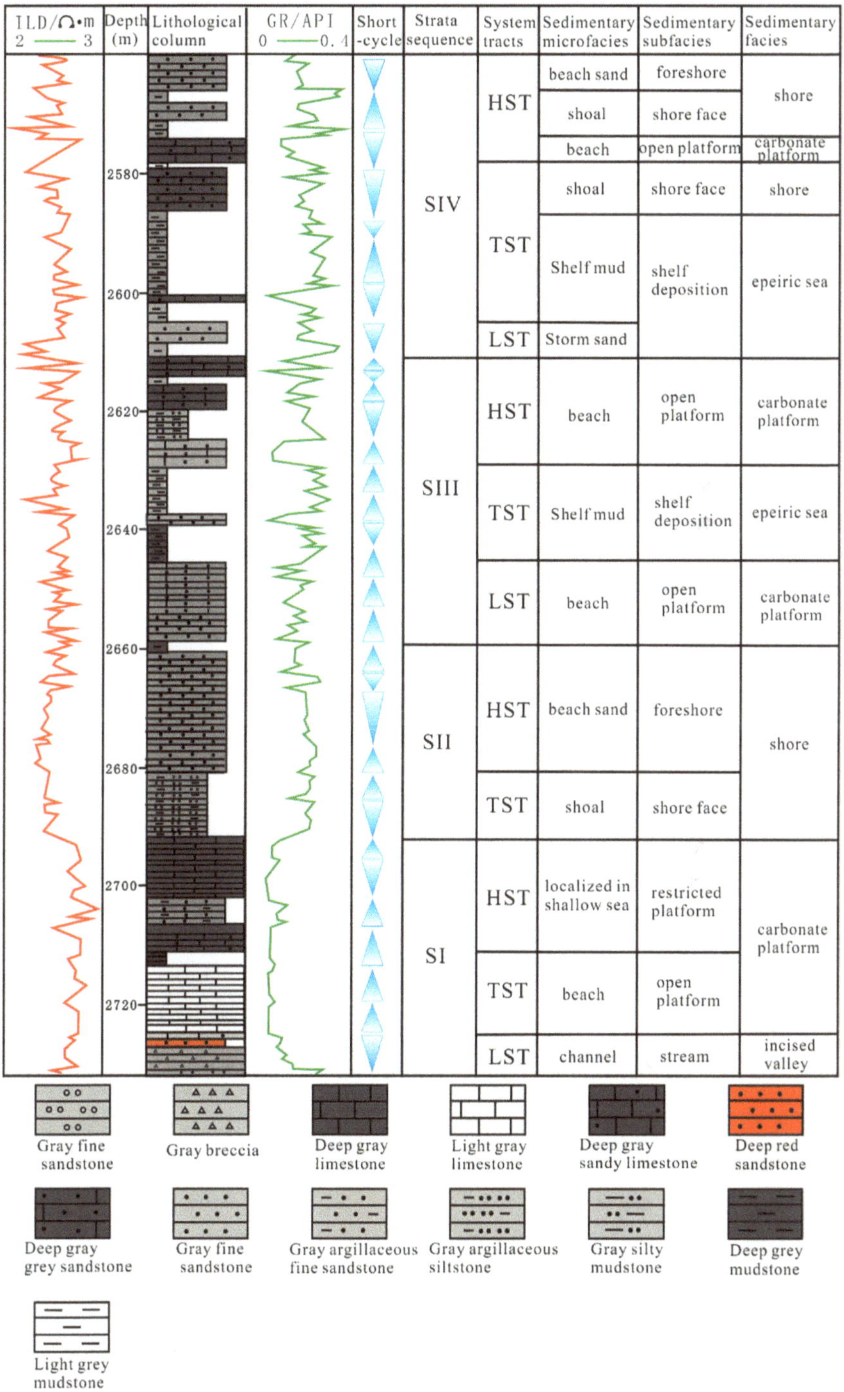


**Figure 1.25** Section showing the characteristics of sedimentary facies in the Paleogene Lingshui Formation in Well A2, the Qiongdongnan Basin (Huanqing et al., 2009).

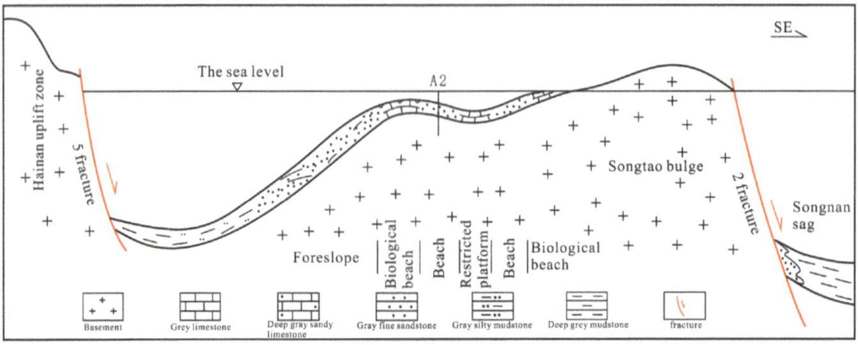


**Figure 1.26** Sedimentary model of carbonate platform in a bulge, the Qiongdongnan Basin. *Modified from BGP of CNPC, 1998.*

mode is fissure eruption, fissure-central eruption or central eruption; then, the location of the volcanic crater is clarified, the relationship between volcanic facies and reservoir properties is analyzed, and the location of favorable reservoirs is predicted. For the carbonate reservoir, the geological genesis analysis can be made to judge whether the main pore space is pore, cavity, or fracture; accordingly, proper techniques are taken to complete the characterization of favorable reservoirs. Similarly, for the metamorphic reservoir, the geological genesis analysis is conducted to determine whether the pore space is dominated by structural fracture or secondary pore; then, the reservoir can be described in a manner corresponding to its history. Of course, the techniques and databases used may vary greatly depending on the reservoir types. For fracture-cavity reservoirs, the description is mainly based on seismic data and takes place through conventional seismic data interpretation and various forms of seismic inversion and prediction. For the reservoirs where the pore space is controlled by the pore structure and lithology change, the description is realized by various logging interpretation methods, in close combination with the data of geological cores and various analyses and tests. Anyway, there is no distinct boundary between the techniques for different reservoir types. The selection of such techniques mainly depends on the research objective or pore space type. In the fine reservoir description of the volcanic reservoir, for example, the early geological genesis analysis to define the volcanic eruption mode is helpful for the researchers to determine the specific location of the volcanic crater. The volcanic facies is generally volcanic conduit facies at the crater and becomes explosive facies, effusive facies, and volcanogenic sedimentary facies successively with the increase of distance from the crater. Different types of volcanic facies correspond to reservoirs with distinct storage properties. Based on the geological genesis analysis, the researchers can gain a relatively accurate macroscopic understanding of the spatial location of favorable reservoirs and, thereby, provide a solid foundation for reservoir characterization.

#### *1.4.2.3 Identification and description of pore space*

The development characteristics of pore space are one of the important contents of reservoir geologic study. For the glutenite reservoir, which features a "double moldic" structure that is obviously different from conventional clastic rocks due to the formation mechanism, the accurate characterization of its pore space is of great significance to the effective development. In the description of the pore structure of the Lower Karamay Formation glutenite reservoir in an area at the northwestern margin of the Junggar Basin, the genetic mechanism and distribution law of the glutenite reservoir were analyzed based on various data like thin section and SEM (Fig. 1.27), laying a foundation for the closer-spacing and effective development of the reservoir. The pore space of the target reservoir in the study area comprises residual intergranular pores, intragranular dissolved pores, intergranular dissolved pores, intercrystalline pores, microfractures, and fissures. The residual intergranular pores are mainly the original pores between grains that are not filled and cemented completely. The intragranular dissolved pores are generated from the dissolution of feldspar and rock debris. The intergranular dissolved pores are derived from the dissolution of cement, such as matrix and calcite. The intercrystalline pores are pores between authigenic quartz or calcite crystals. The microfractures are caused by compaction, and the fissures are caused by tectonism and diagenesis. By pore types, the primary pores are represented by residual intergranular pores, and the secondary pores by intragranular dissolved pores and intergranular dissolved pores, which account for over 70% of the pore space. In addition, there are a small number of intercrystalline pores, microfractures formed by fracturing, and fissures. The glutenite reservoir generally has a double moldic structure. The so-called double moldic structure is a configuration that the pores of rock skeleton formed by conglomerate are often partially or completely filled with sand particles, while the pores formed by gravel and sand particles are partially filled with clay particles. The grain size, content, and association of gravel, sand, and clay particles vary in different sedimentary environments, forming the complex rock structure and pore structure of

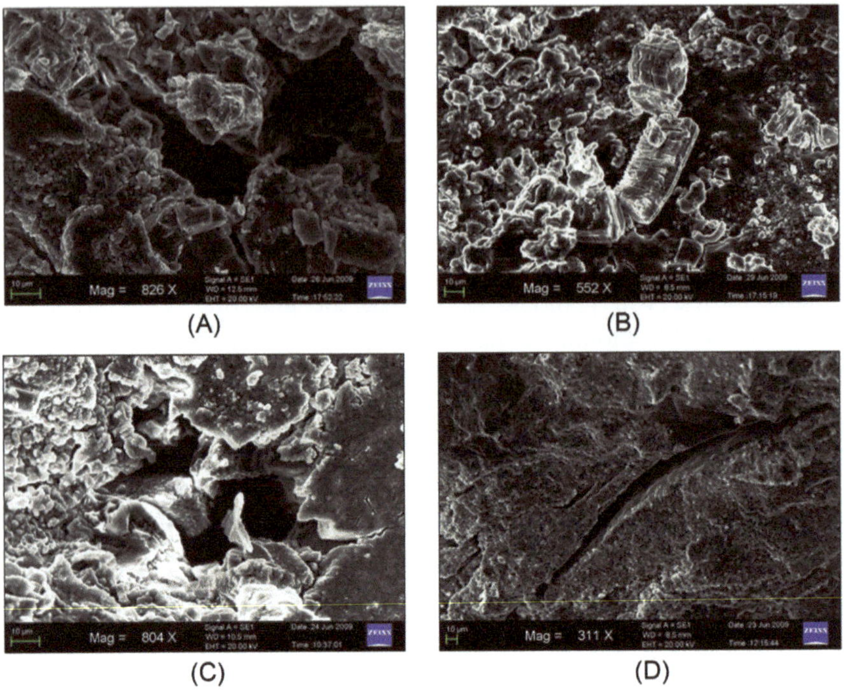


**Figure 1.27** SEM characteristics of pore space of the conglomerate reservoir in the Lower Karamay Formation in an area at the northwestern margin of the Junggar Basin. (A) Well J7, intergranular pore filled with vermicular kaolinite, glutenite, 427.85 m; (B) Well J8, intergranular pore, oil-immersed vermicular kaolinite, pebbly sandstone, 415.6 m; (C) Well J3, intragranular dissolved pore, brownish gray glutenite, 103.79 m; (D) Well J3, microfracture, brownish gray oil-bearing sandy microconglomerate, 396.09 m.

the conglomerate reservoir (Qingchang et al., 1997). Large-diameter cores are often used to analyze the physical properties of the conglomerate reservoir. The double moldic structure can also be quantitatively characterized by mathematical geological modeling with the help of a thin casting section, SEM, and mercury-injection experiment. There are many types of pore spaces in complex lithologic reservoirs, such as vesicles and fractures in the volcanic reservoir, fractures and cavities in carbonate reservoir, and primary pores, secondary pores, and microfractures in glutenite reservoir. The pore space should be classified and described according to the specific rock type. The spatial distribution of different types of pore space should be described to procure their quantitative characterization. Huanqing et al. (2012a,b,c) divided the pore space of the volcanic reservoir in Ying 1 Member in Xudong area, the Songliao Basin into primary and secondary types, which were subdivided into 12 subtypes, thereby providing the basis for quantitative evaluation of pore space (Table 1.8).

#### *1.4.2.4 Fine logging interpretation of reservoir physical properties*

Due to the limited data of coring wells, and given the need to analyze reservoir physical properties, the most commonly used method is still the logging interpretation (Fig. 1.28) (Huanqing et al., 2016a,b,c,d). Junhua et al. (2008) considered the application of the optimization method in the logging interpretation of complex lithology and revealed a consistency between the porosity calculated and the porosity determined by core analysis. Feng et al. (2014) took the western area of the Qaidam Basin as an example to study the T2 cutoff value of complex lithology; in the study, the reasonable centrifugal pressure was determined depending on the actual rock samples and according to the industry standard, and then the T2 cutoff values for different lithology were measured by the centrifugal method. Yuming and Can (2014) conducted the NMR analysis of core samples, where the NMR logging interpretation model was established through the test and analysis of porosity, irreducible water saturation and porosity, the cutoff value of lateral relaxation time, and permeability, which guided the reservoir evaluation and improved the calculation accuracy of reservoir parameters. Xingli et al. (2015) evaluated the permeability of complex lithologic reservoirs using the ECS log data and revealed the basic consistency between the calculated permeability and the core permeability according to the processing of some well data in the Bohai area. Shaoyang et al. (2016) used the Bayes discriminant method to evaluate the porosity based on lithology identification, and they demonstrated by the application in M Oilfield, Iraq, that the error between the porosity calculated after lithology identification and that by core analysis was very small, which provided the theoretical basis for reservoir interpretation and evaluation. In the stage of fine reservoir description, due to the addition and supplement of sealed coring wells and smaller-spacing wells, more and more data become available for logging interpretation of reservoir physical properties. Moreover, with the progress of oilfield development, data from diversifying dynamic monitoring and production data can be acquired, which also provide favorable conditions for fine logging interpretation of reservoir physical properties. At present, the study of logging interpretation of reservoir physical properties mainly focuses on improving the interpretation accuracy of the model. The relatively effective way to do this is to complete the interpretation by blocks and horizons, and the fine logging interpretation is the most basic

**Table 1.8** Pore space types of the volcanic reservoir in Ying 1 Member in Xudong area, the Songliao Basin (Huanqing et al., 2012a,b,c).

| Pore space type | | | Main rock types | Genesis | Features and identification marks |
|---|---|---|---|---|---|
| Primary type | Pore | Vesicle | Rhyolite, breccia lava | When the volcano erupts, the gas capped by molten magma fails to escape in time. After the gas escapes in the cooling of magma, vesicle forms | Circular, elliptical, and flattened and elongated pore walls may be irregular but smooth. Most of them are isolated, and few are in bead-shaped arrangement |
| | | Intergranular pore | Volcanic breccia, fusion breccia | The pores between relatively coarse-grained volcanic debris are shrunk by welding and compacting as not filled | They have irregular shape, and are generally developed between pyroclastic components |
| | | Intragranular pore | Volcanic breccia, welded breccia, breccia lava | They are brought by the rigid and plastic debris of pyroclastic rock | They are mainly developed in coarse volcanic debris depending on their own types (vesicle, intraamygdule pores, dissolved pores, etc.) |
| | | Micropore | All kinds of volcanic rocks | Crystallization of the lava matrix occupies the space, and pores between the fine-grained volcanic debris are not filled | They are developed between microcrystalline minerals of lava matrix or between fine-grained volcanic debris (volcanic ash and dust) |
| | Fracture | Shrinkage fracture | All kinds of volcanic rocks | Cooling contraction of magma and Geologic Origin contraction of pyroclastics | Concentric circular, parallel arc shapes |

| | | | | | |
|---|---|---|---|---|---|
| | | Explosive fracture | All kinds of volcanic rocks | They are formed by the gas-liquid explosion caused by the upwelling force and explosive force of magma during volcanic eruption | Fractures are not oriented; some curved ones extend outwards from some center in irregular radiated shape, which is consistent with adjacent boundary. Network fractures in volcanic breccia, fractures between volcanic breccia, intercrystalline fractures and vertical tensional fractures are commonly observed |
| Secondary type | Pore | Intragranular dissolved pore | Rhyolite, breccia lava, welded breccia, volcanic breccia | Feldspar crystals (phenocryst, and crystal chip) and feldspar, volcanic dust and glassy, etc., in the debris generated by surface water leaching or dissolution of groundwater | They are distributed within the feldspar and rock debris or on their edge. They are distributed along cleavage fracture with irregular shapes |
| | | Matrix dissolved pore | Rhyolite, welded breccia, volcanic breccia, tuff | They are formed by lava matrix and fine-grained volcanic ash and dust by leaching of surface water or dissolution of groundwater | They have an irregular shape and are small and developed in lava matrix or fine-grained volcanic debris |
| | | Intergranular dissolved pore | Welded breccia, breccia lava, volcanic breccia | They are formed by surface water leaching or groundwater dissolution along intergranular pores | They have an irregular shape and are often connected with other pores and fractures. The single pore is large |

(*Continued*)

**Table 1.8** (Continued)

| Pore space type | | | Main rock types | Genesis | Features and identification marks |
|---|---|---|---|---|---|
| | Fracture | Structural fracture | All kinds of volcanic rocks | They are generated by tectonic stress | They often appear parallel in groups. When appearing directional, several groups intersect and cut through crystal or pyroclastic particles. They are often connected with other pores |
| | | Dissolved fracture | All kinds of volcanic rocks | They are formed by surface water leaching or groundwater dissolution of particles and matrices | Directionally different fractures have irregular shapes |
| | | Stylolitic fracture | Welded breccia, breccia lava, volcanic breccia | They are formed by compaction and pressure dissolution | They extend in a zigzag. The suture fracture often cuts the phenocryst and matrix of lava or cuts the volcanic debris |

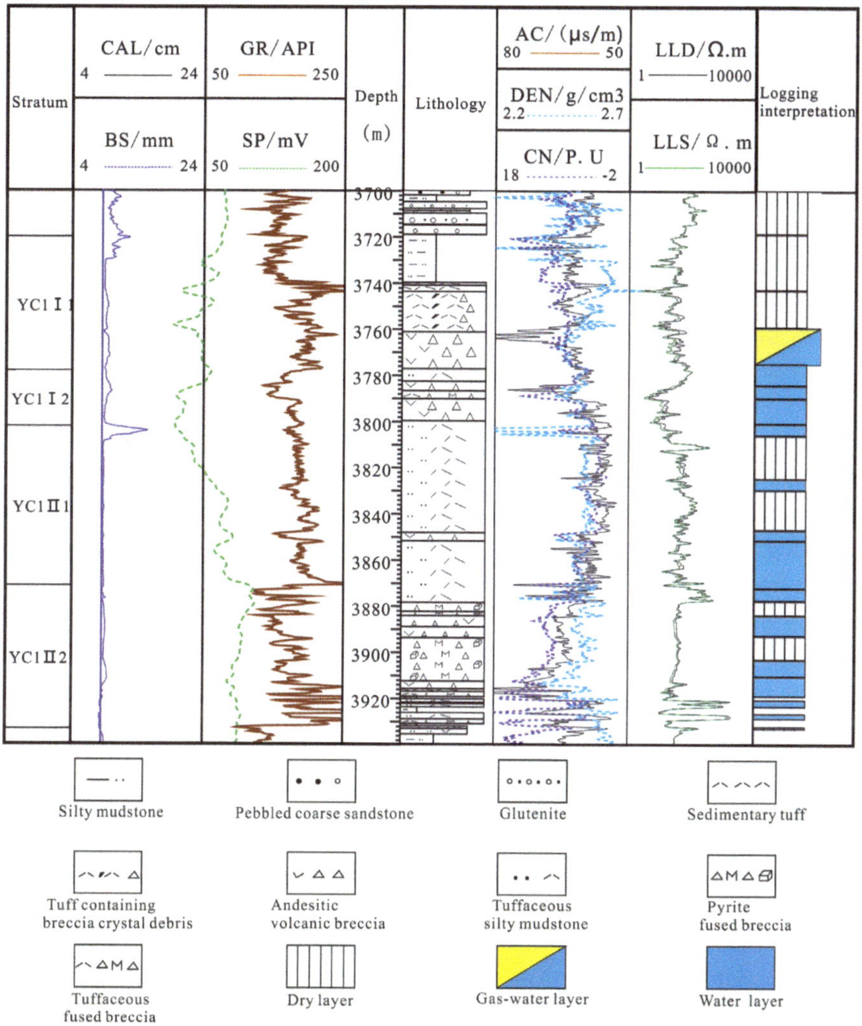


**Figure 1.28** Stratigraphic classification and logging interpretation results of the volcanic reservoir in Ying 1 Member in Xudong area, the Songliao Basin (Huanqing et al., 2016a,b,c,d).

means for interpreting the physical properties of the complex lithologic reservoir. Generally, volcanic reservoirs are dual-porosity media (fracture + pore) reservoirs. Therefore, the quantitative calculation of reservoir physical properties should also include two parts: calculation of matrix porosity and quantitative calculation of fracture parameters. For acid volcanic rocks, NMR logging is the most direct and effective method to calculate the matrix porosity, and it can measure the porosity and permeability of volcanic rocks relatively directly. For intermediate-basic volcanic rocks, the NMR logging is inefficient due to a large error captured; instead, it can couple with conventional logging methods to comprehensively determine the porosity (PetroChina Exploration & Production Company, 2009). For carbonate rocks, the application of conventional logging interpretation is much more challenging for the cavity reservoir with strong spatial heterogeneity, which is more than what the fractured reservoir displays; in practice, imaging logging and NMR logging are often used (Liqiang, 2002). For fluid identification, the electric imaging-apparent formation water resistivity spectrum method can be used. For the water layer, the difference between the imaging characteristics of the pore and the background value of the reservoir is small, its color is darker than that of the gas layer, the apparent formation water resistivity spectrum has a small distribution range and narrow band, and the main peak deviates to the small. For the oil and gas layers, the conclusion is converse. For a metamorphic reservoir, where fractures are the main pore space, imaging logging and conventional logging methods are often used jointly to analyze the physical properties of the reservoir.

#### *1.4.2.5 Geological modeling of complex lithologic reservoir*

Reservoir geological modeling is the core of fine reservoir description, so is it for the complex lithologic reservoir. Under the control of the genetic model, the geological modeling of the complex lithologic reservoir is very different from that of conventional clastic rock reservoirs in content, methods, and model characteristics. For the geological modeling of the volcanic reservoir, which distributes around the volcanic crater, leading to greatly variable reservoir properties corresponding to different rock types, a geological model should be established under the control of volcanic eruption mode and in close combination with the drilling data and seismic data to accurately predict the distribution of reservoirs between wells. When characterizing the volcanic gas reservoir, Qiquan et al. (2011) first identified and characterized different volcanic edifices (volcanic rocks) and, then, established the geological model (Fig. 1.29). For conventional clastic reservoirs, the interwell interpolation method based on the dense well pattern is commonly used, which is realized by the fitting of the variogram. Some scholars have made important progress in the research on the multipoint geostatistical method based on training images. For the geological modeling of complex lithologic reservoirs such as volcanic

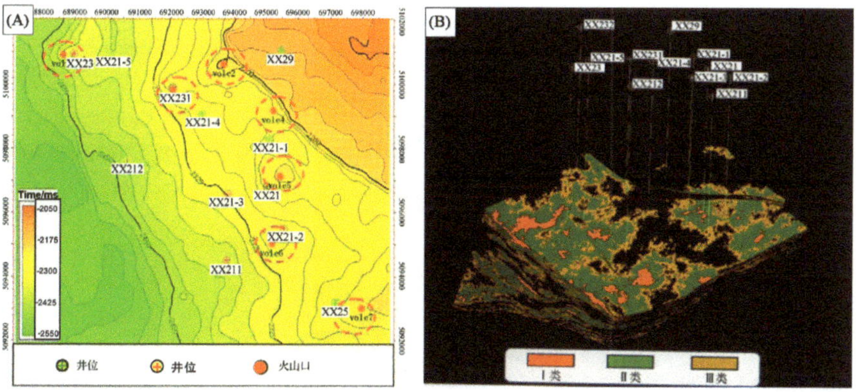


**Figure 1.29** Geological model of the volcanic reservoir in the Yingcheng Formation in an area of Xushen Gas Field, the Songliao Basin (Qiquan et al., 2011). (A) structural model. (B) reservoir model. Note: Two well locations in (A) indicate wells drilled in different periods.

reservoirs and carbonate reservoirs, we should pay more attention to the role of seismic data and depict the location of pore space (volcanic conduit, fractures, cavities, etc.) by using attribute slicing and pre-stack inversion to realize the interwell reservoir prediction and establish the 3D geological model of the complex lithologic reservoir.

## 1.4.3 Trend

### *1.4.3.1 Fine stratigraphic classification and correlation*

Fine stratigraphic classification and correlation is always a focus of fine reservoir description (Xiurong et al., 2001; Huanqing and Xiaomin, 2008). Different from the conventional clastic reservoirs, the complex lithologic reservoirs, typically volcanic, metamorphic, and carbonate reservoirs, exhibit an extremely special distribution of strata. To realize the fine stratigraphic classification and correlation of such reservoirs, the genetic mechanism of strata should be defined first and, then, the spatial distribution law recognized. Wenling et al. (2016) discussed the isochronous stratigraphic correlation technology based on the logging-seismic data for the complex lithologic reservoir in the Wangxuzhuang Oilfield in Dagang and, in this study, the logging-seismic stratigraphic classification scheme and the correlation criteria of finer-classified layers were reviewed before the isochronous stratigraphic correlation technology controlled by acoustic guided seismic constraint mode was established; thus, the AC log is integrated into the geological stratigraphic classification, thereby greatly improving

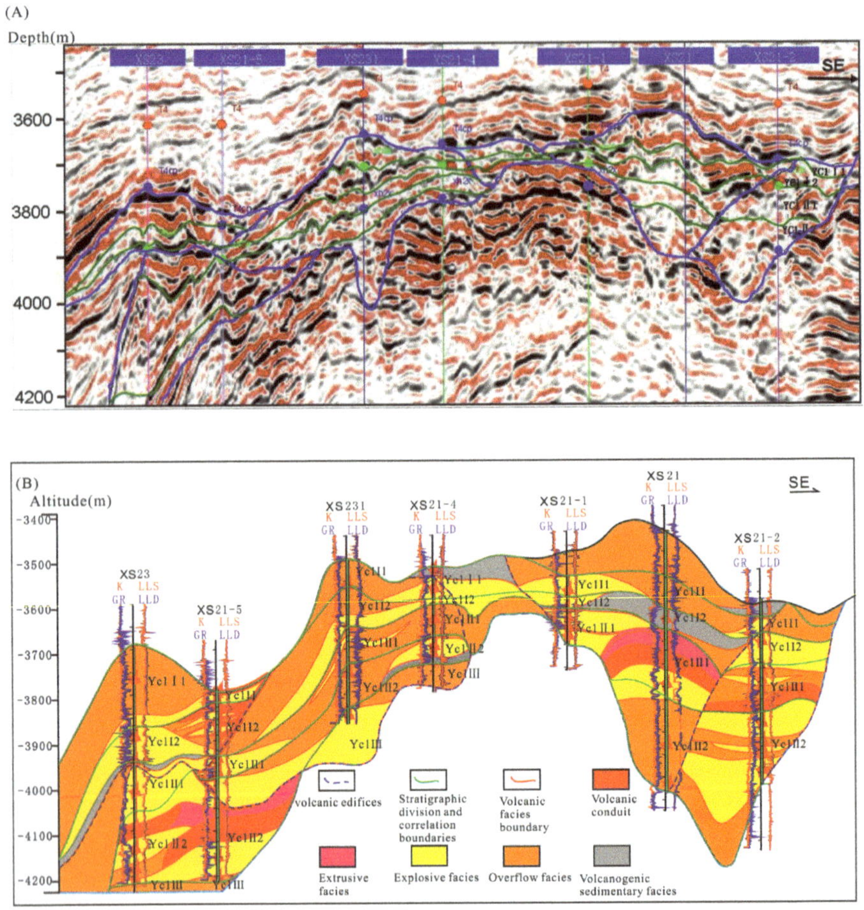


**Figure 1.30** Stratigraphic and lithofacies classification of the volcanic reservoir in Ying 1 Member in Xudong area, the Songliao Basin (Huanqing et al., 2016a,b,c,d).

the stratigraphic classification of reservoirs with high calcareous content, tightness, or large mudstone interval. Huanqing et al. (2016a,b,c,d) put forward a fine stratigraphic classification and correlation method featured with "two steps and two combinations" for the volcanic reservoir in the Yingcheng Formation in Xudong area of the Songliao Basin, and they established the high precision isochronous stratigraphic framework for the target reservoir in the study area (Fig. 1.30). In this method, characteristically, the step of tracking and identification of volcanic edifice and volcanic rock is added to the process of stratigraphic correlation to avoid the diachroneity showing up as the correlation boundary-crossing volcanic rock, and where the well logging and seismic data are closed combined to guarantee the unity and accuracy of the large-scale isochronous stratigraphic framework and the fine small-scale stratigraphic classification and correlation boundary in the process of stratigraphic correlation. In conclusion, as there is no standard method to follow for stratigraphic classification and correlation of complex lithologic reservoirs, we should work out specific solutions depending on the genetic characteristics of different lithologies.

### *1.4.3.2 Characterization of microscopic pore structure*

The pore structure of reservoir refers to the geometry, size, distribution, and connection of pores and throats in rocks (Shenghe and Qihua, 1998; Huanqing et al., 2013a,b,c, 2016a,b,c,d). Pore structure belongs to the microscopic study of the reservoir; thus, it is also called the microscopic pore structure of the reservoir. The microscopic observation and analysis of thin sections is the most direct and effective method to study the pore structure of a reservoir. Based on extensive microscopic thin section identification, we analyzed the origin, type, and distribution of pore structure of glutenite reservoir in the Lower Karamay Formation in an area at the northwestern margin of the Junggar Basin (Fig. 1.31). Since the pore structure of the complex lithologic reservoir is unique and also because it is the direct factor controlling the storage properties of the reservoir, studying the pore structure has always been attractive for researchers. At present, the study of reservoir pore structure tends to gradually change from qualitative to quantitative. In practice, one or more parameters that can fully reflect the characteristics of pore structure can be selected from available parameters to classify and evaluate the pore structure through cluster analysis or using other mathematical calculation methods; then, the spatial development characteristics of different types of reservoir pore structure are described, and favorable development zones are finally predicted. Huanqing et al. (2013a,b,c) successfully applied this technique to the volcanic reservoir in Ying 1 Member in Xudong area, the Songliao Basin and, in this work, based on the statistical analysis of 206 samples from 23 coring wells, three parameters, that is, effective porosity, total permeability, and median pore-throat radius, were selected from numerous parameters reflecting the pore-throat size; then, with SPSS software, the cluster analysis was conducted to divide the pore structure of the target reservoir into Types I, II, III, and IV (Table 1.9), while their respective distribution location, and as per law on the plane of different layers, were analyzed (Fig. 1.32), which provided a basis for well placement design in the reservoir development plan.

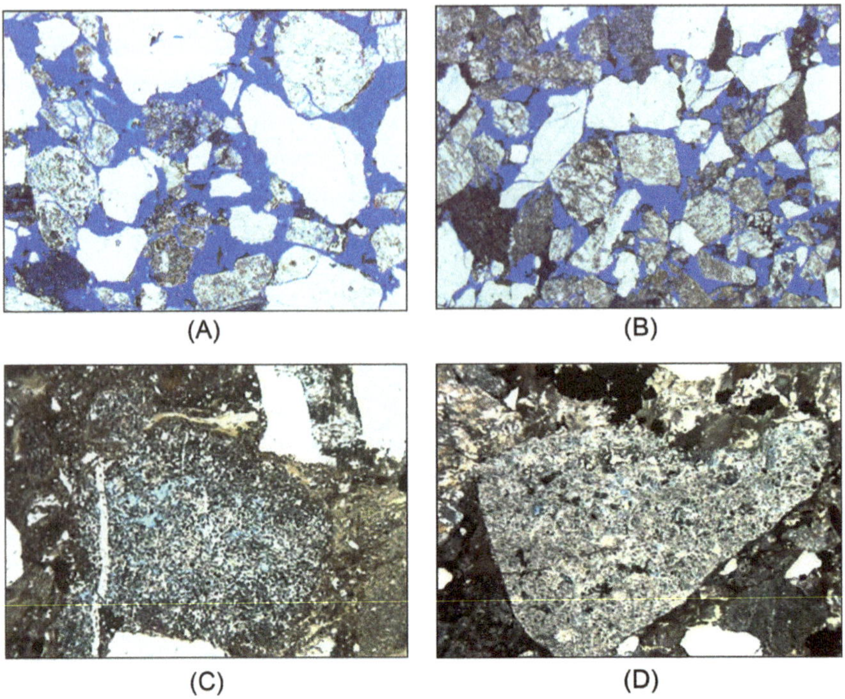

**Figure 1.31** Pore structure of conglomerate reservoir in the Lower Karamay Formation in an area at the northwestern margin of the Junggar Basin. (A) Well J3, intergranular pore, brownish gray glutenite, 416.20 m, × 25; (B) Well J3, intergranular pore, brownish gray microconglomerate, 405.59 m, × 25; (C) Well J55, intragranular dissolved pore, gray argillaceous small conglomerate, 382.49 m, × 40; (D) Well J56, intragranular dissolved pore, small conglomerate, 404.44 m, × 25.

**Table 1.9** Evaluation results of pore structure based on physical properties of volcanic reservoir in Ying 1 Member in Xudong area, the Songliao Basin.

| Type | Porosity $\Phi$/% | | | Permeability $K/10^{-3}\ \mu m^2$ | | | Median pore-throat radius $R_{50}/\mu m$ | | |
|---|---|---|---|---|---|---|---|---|---|
| | Max. | Min. | Avg. | Max. | Min. | Avg. | Max. | Min. | Avg. |
| I | 14.4 | 10.5 | 13.0 | 1.19 | 0.04 | 0.20 | 0.299 | 0.062 | 0.161 |
| II | 9.9 | 7.7 | 8.8 | 1.12 | 0.02 | 0.10 | 0.251 | 0.021 | 0.113 |
| III | 7.5 | 4.9 | 6.3 | 1.35 | 0.01 | 0.05 | 0.128 | 0.018 | 0.06 |
| IV | 4.8 | 2 | 4.0 | 1.21 | 0.01 | 0.04 | 0.086 | 0.016 | 0.03 |

*Source:* Data from Huanqing, C., Yongle, H., Chengfang, S., 2016. The single element characterization and quantitative evaluation of volcanic rocks in songliao basin. Petroleum Industry Press. Beijing.

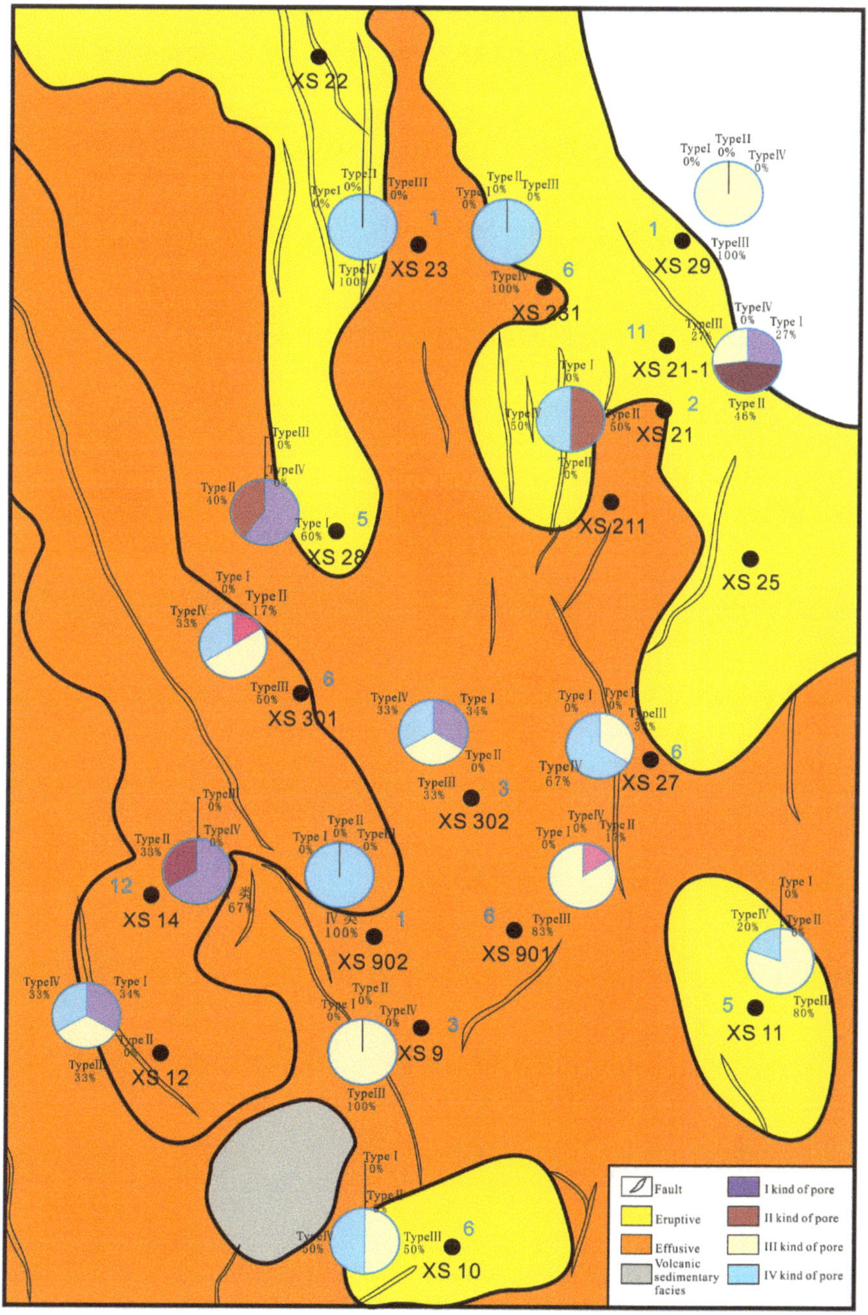


**Figure 1.32** Distribution of samples for pore structure classification of YC1I1 in Ying 1 Member in Xudong area, the Songliao Basin (The number labeled next to the well represents the total number of samples participating in the cluster analysis).

### *1.4.3.3 Reservoir fracture characterization technology*

A fracture refers to the reservoir walls having a discontinuity surface formed by rock breaking. It is the result of wall surface breaking after a force is exerted on the rock. Fracture is an important pore space for the oil and gas reservoir, especially the fractured reservoir, and is also the favorable flow pathway (Fig. 1.33). Xueru et al. (1999), in his monograph "Research Technology of Structural Fractures in Metamorphic Reservoir," took the Liaohe buried-hill reservoir as an example to state systematically the description of structural fractures in metamorphic reservoirs, the identification of structural fractures in metamorphic reservoirs, and the prediction and evaluation of structural fractures in metamorphic reservoirs. Shiyi et al. (2004) made a comprehensive summary on the development technology of fractured reservoirs, which involved glutenite, volcanic rock, carbonate rock, mudstone, and other complex lithologic

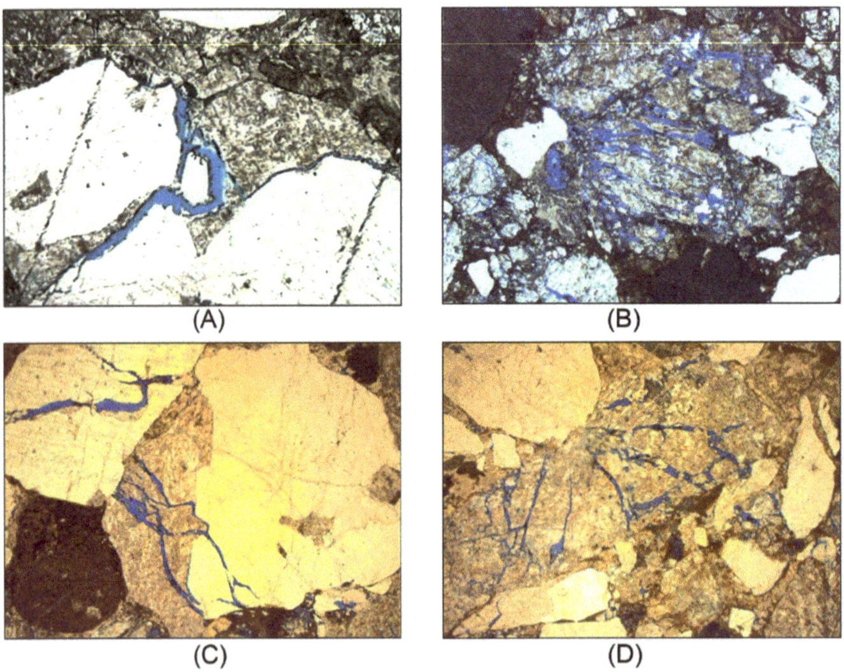


**Figure 1.33** Pore structure of conglomerate reservoir in the Lower Karamay Formation in an area at the northwestern margin of the Junggar Basin. (A) Well J57, microfracture, glutenite, 443.19 m, ×25; (B) Well J7, microfracture, glutenite, 429.07 m, ×25; (C) Well J8, microfracture, glutenite, 430.73 m, ×25; (D) Well J8, microfracture, glutenite, 421.42 m, ×25.

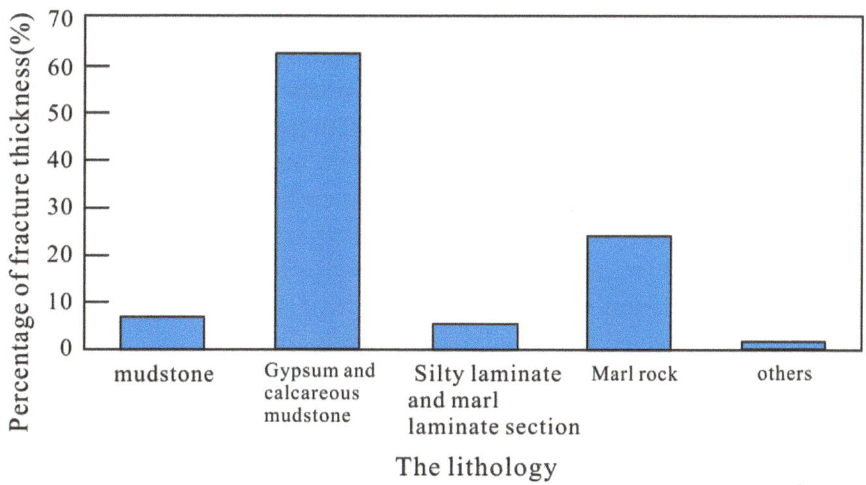


**Figure 1.34** Relationship between fracture development intensity and lithology of $E_3{}^2$ in Shaxi Oilfield (Shiyi et al., 2004).

reservoirs (Fig. 1.34). The particularity of the complex lithologic reservoir makes fractures as its critical pore space, so do volcanic, carbonate, and metamorphic reservoirs. Although reservoir fracture characterization has become one of the hotspots in fine reservoir description, many unsolved problems still exist, such as the prediction of interwell fractures and 3D modeling of reservoir fractures. The current research methods of reservoir fracture mainly include field outcrop investigation, core observation and description, microscopic observation, conventional logging interpretation, imaging logging, seismic attribute slice, and coherence analysis and comparative analysis of production performance data. All methods have their respective merits, but also their various deficiencies. This book argues that the fracture characterization should use a variety of information for cross verification in order to determine the fracture development location and law and to avoid the multiple solutions and uncertainty of research results.

#### *1.4.3.4 Comprehensive quantitative evaluation of reservoir*

This book thinks that the evaluation of complex lithologic reservoirs should begin with the genesis analysis to find out the main factors controlling the formation of the reservoir and then will depart into an analysis of suitable characteristic parameters. For this, reasonable calculation methods should be selected according to the available data to evaluate the reservoir quantitatively. Without the geological genesis analysis, the parameters for reservoir classification and evaluation cannot be selected accurately, and the reservoir evaluation will become a simple digital game, which cannot guide and help oilfield development. If only the geological genesis analysis is emphasized, the requirement for quantitative reservoir evaluation for oilfield development will not be met. In practice, geological genesis analysis and quantitative evaluation of reservoirs should be combined organically. The selection of the reservoir evaluation method is very important. Huanqing et al. (2015a,b) reviewed and compared different reservoir evaluation methods and highlighted their advantages and disadvantages. Reservoir evaluation is a systematic job. Due to the diversity of data and differences in their purpose, the reservoir evaluation methods are also diverse. In actual application, the methods should be organically and complimentarily combined, together with the putting to work useful data of various types as much as possible, to accurately evaluate the reservoir properties.

#### 1.4.3.5 Identification of fluid in complex lithologic reservoir

The characterization of fluid distribution in complex lithologic reservoirs plays an important role in the calculation of reserves and the design of oilfield development and adjustment plan. Jianfeng et al. (2008) explored the well logging interpretation and evaluation technologies for the complex lithologic reservoir in the Erlian Basin, and they developed the logging interpretation and evaluation technology based on the petrophysical analysis and the establishment of well logging interpretation models for different sedimentary microfacies. Yongmei et al. (2010) identified the complex lithologic features of oil and gas reservoirs in the Wuba area of the Junggar Basin. Weilin et al. (2012) conducted fluid identification for the reservoirs dominated by volcaniclastic rocks in Tanan Oilfield and Nanbei Oilfield in the central fault depression zone of the Tamtsag Basin, and they established the method of fluid identification based on block, horizon, lithology, and reservoir category from a consideration of reservoir type and pattern, which provided more accurate fluid identification of the complex lithologic reservoir. Yongguang (2015) studied the fluid identification method of the complex lithologic reservoir in the Nantun Formation in the Hailaer Basin, and he concluded that lithology and porosity are the main controlling factors for reservoir fluid identification. The gas-bearing property of the volcanic reservoir in Yingcheng Formation in Xushen Gas Field of the Songliao Basin was analyzed previously, and the results showed that the gas-bearing property varied greatly in the horizontal direction, and all the large effective

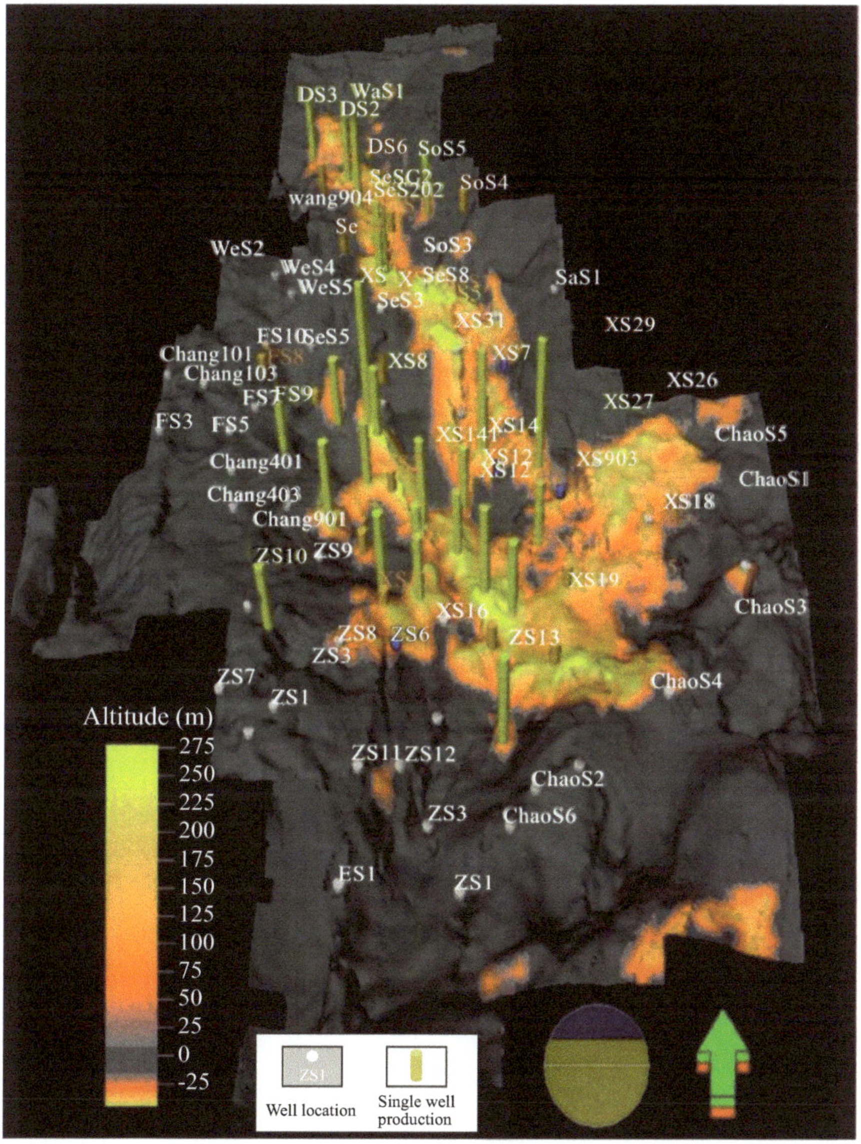


**Figure 1.35** Superposed map of the top structure of volcanic rock, a volcanic cone, and gas production of a single well in the Cretaceous Yingcheng Formation, the Songliao Basin (PetroChina Exploration & Production Company, 2009).

thickness areas were located in the structurally high volcanic edifice, which was more conducive to the accumulation of natural gas (Fig. 1.35) (PetroChina Exploration & Production Company, 2009). According to the current research efforts on the identification of fluid in complex lithologic reservoirs, well logging and seismic data were found good to be relied on, and well logging interpretation is the main research method. Although the seismic method is also used sometimes, the accuracy of seismic-based fluid prediction is far beyond the needs of fine reservoir description of the complex lithologic reservoir. Further efforts are needed in data acquisition, data processing, and interpretation methods and techniques. In particular, for complex lithologic reservoirs, the distribution of remaining oil is highly dispersed as a whole and relatively enriched locally, especially in the middle-late development stage, due to the geological complexity (Dakuang, 2010). As to the fluid identification for complex lithologic reservoirs, many issues involving the characterization of remaining oil are to be addressed. In practice, on the basis of logging-seismic data as well as all kinds of test data, dynamic monitoring data, and production performance data, great efforts should be devoted to strengthening the activities like fine microstructure interpretation, reservoir architecture characterization for the sand body, and logging interpretation of swept layer in the complex lithologic reservoir, so as to improve the accuracy in characterization and precision of remaining oil in the complex lithologic reservoir.

# CHAPTER 2

# Characteristics of fine reservoir description contents

According to the focus on different development stages of oil and gas fields, fine reservoir description can be divided into two parts: fine reservoir description for the early development stage and fine reservoir description for the middle–late development stage. If we consider research growing out from a focus for one of these stages, the contents of fine reservoir description are different subject to the specific object and aims of the study. Simply speaking, the main tasks of research in the early development stage is to re-understand the reservoir geology after the implementation of the oilfield development plan, further ascertain the main geological characteristics such as structure, strata, deposition, reservoir, and distribution of oil, gas, and water, and establish or improve the reservoir geological model so as to provide a basis for oilfield development, production, and adjustment, as well as reserves recalculation. Whereas, the main tasks of the middle–late development stage focused research is to make full use of static and dynamic data to delineate microstructural characteristics; subdivide sand beds; carry out research on sedimentary microfacies, reservoir architecture, and reservoir physical properties; further improve the geological model; and carry out numerical simulation, quantify the spatial distribution of remaining oil, and predict reservoir dynamic change so as to provide a basis for further enhancing oil recovery. Generally, the contents of fine reservoir description research focused in the middle–late development stage are much more detailed than that focused on the early development stage. This chapter introduces the contents of fine reservoir description corresponding to different development stages.

## 2.1 Fine reservoir description in the early development stage

The fine reservoir description for the early development stage mainly includes the three aspects of due diligence to be carried out: structure, reservoir, and oil-gas-water relationship (Fig. 2.1). In view of the structure to be studied, the logging–seismic data are calibrated in a time-depth domain on the basis of the synthetic seismogram, and the faults are

*Fine Reservoir Description*
DOI: https://doi.org/10.1016/B978-0-323-95401-3.00002-0

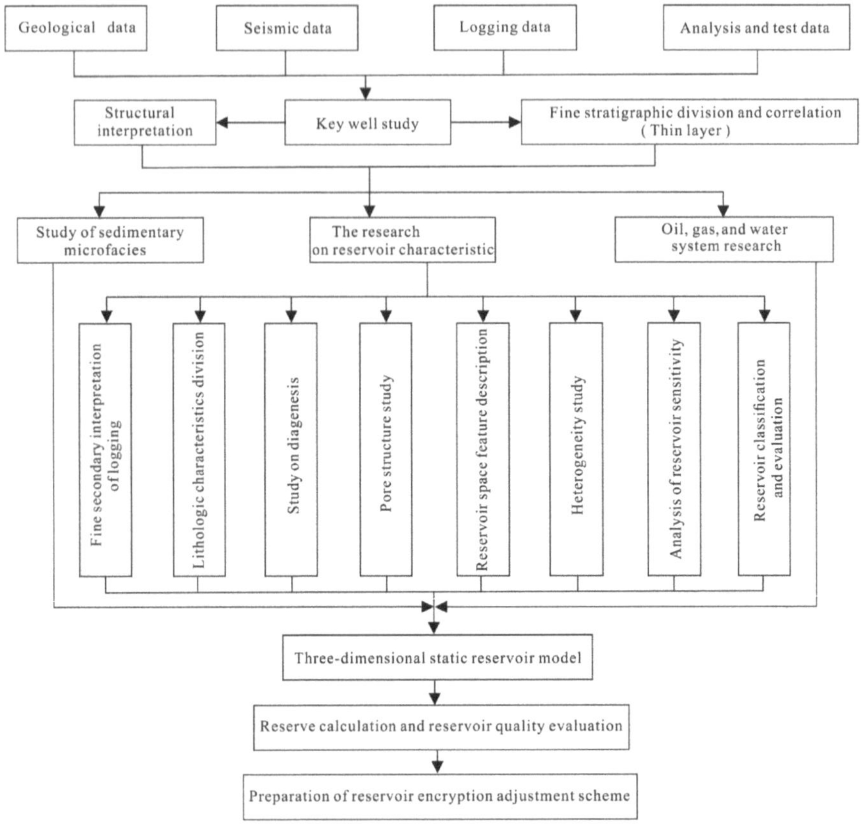


**Figure 2.1** Contents and process of fine reservoir description in the early development stage.

interpreted on seismic profiles or the breakpoints are traced in wells. According to the results of fault interpretation on seismic profiles and well-based breakpoint identification, together with the geological setting and the evolutionary history of the fault system, the faults are combined in a plane and the development law of fault system in space is depicted. In view of the reservoir to be developed, the description mainly includes the recognition of the stratum distribution and the fine stratigraphic classification and correlation to be undertaken, which should be based on at least layers, to meet the requirement for more specific formation in oil and gas field development and production. It also includes sedimentary microfacies analysis, fine logging reinterpretation, characterization and quantitative classification and evaluation of the reservoir, flow unit characterization, and reservoir geological modeling. In the early development stage, when the development plan of the oil and gas fields has just been implemented, the remaining oil research is not a requirement. Sedimentary microfacies analysis requires one to verify and determine the sedimentary facies/subfacies by virtue of geology, discern the core mudstone color, the log characteristics, the sedimentary structure, the grain size, the paleontological fossils, the heavy minerals, and other data, and to establish logging facies identification criteria and complete sedimentary microfacies classification through analysis of log lithoelectric characteristics. The sedimentary genetic model is constructed through the depiction of the inter-well connectivity and spatial superimposition of sand bodies of different sedimentary types based on the analysis of sedimentary microfacies in the single well, profile, and plane. Then, the spatial development characteristics of sedimentary sand bodies are predicted. The fine logging reinterpretation mainly interprets reservoir porosity and permeability data through the four-property relationship analysis, which provides data basis for subsequent heterogeneity analysis, reservoir evaluation, geological modeling, and flow unit characterization, so that the reservoir heterogeneity (including the heterogeneity between reservoirs and within a reservoir) can be deeply understood. A reservoir evaluation mainly includes analysis of reservoir lithology, genesis analysis and classification of reservoir pore structure, and analysis of pore space types. The reservoir sensitivity analysis is carried out to determine the influence of velocity, salt, acid, alkali, water, and pressure sensitivities on reservoir development; then, the corresponding countermeasures are put forward to improve the reservoir development effect. Quantitative reservoir classification and evaluation are realized through optimizing the relevant parameters and by choosing appropriate mathematical calculation methods. The main purpose of the flow unit characterization is to analyze reservoir flow capacity and to quantitatively identify the spatial distribution of reservoir units with different flow capacities. The efforts at geological modeling are to establish the corresponding models of reservoir structure, sedimentary facies, and physical properties and flow units; to predict the above-mentioned structures and reservoir distribution rules between wells; and, to understand the spatial development rules of structures and reservoirs. A reservoir physical property modeling generally adopts the method of "facies-controlled modeling" to establish the reservoir physical property model under the control of sedimentary microfacies. The oil-gas-water relationship analysis is to clarify the distribution of oil, gas, and water underground. Overall, the fine reservoir description for the early development stage emphasizes comprehensive coverage of all aspects of due diligence to be carried out and involves a wide range of areas while requiring the most comprehensive understanding of the structure of the target formation and the basic characteristics of reservoir development, though the research depth is far less demanding than the fine reservoir description for the middle–late development stage.

## 2.2 Fine reservoir description in the middle–late development stage

Taking PetroChina as an example (Huanqing, 2006; Huanqing et al., 2006; Huanqing and Xiaomin, 2008), with efforts spanning over a decade, the fine reservoir description of all oilfields has constantly been improving. At present, some oilfields have carried out the secondary, and even the third, fine reservoir description. Fine reservoir description needs to run through the whole process of oilfield development as the most important basic work. Its results have played a significant role in the design of oilfield development plans, a progressive extension of mature oilfields, adjustment of oilfield development, design and scale application of horizontal wells, and major development tests. Therefore, many researchers have been paying attention to fine reservoir description. PetroChina stipulates that reservoir description is the first step for all oilfields to be developed. Fine reservoir description has become the most basic research work for oilfield development. With the continuing oilfield practices, many large oilfields in eastern and western China have entered the middle–late development stage, when various production contradictions are prominent such as water-cut rise. The oilfield development has encountered unprecedented challenges in the current context of low oil prices (Huanqing, 2016). The book believes that in order to overcome these challenges, we must first clarify the changes and characteristics of fine reservoir description research for the middle–late development stage and grasp the focus of work. Previous studies have also summarized and analyzed the contents of fine reservoir description for different development stages (Longxin and Yinan, 1999; Yongmin et al., 2004; Ailin, 2010). With the progress of technologies and the change of contradictions in production practice, the contents of fine reservoir description in the middle–late development stage have become very different, so they need to be updated and summarized comprehensively. We currently participate in the revision of PetroChina specifications for fine reservoir description. Based on the follow-up study of fine reservoir description in

PetroChina oilfields in recent years and our own research practice, we believe that the fine reservoir description for the middle–late development stage is significantly different from that for the early development stage, especially in the contents. The fine reservoir description for the early development stage mainly serves the interests of undertaking the design of preliminary development plan and infilling, with the purpose that it helps to clarify the basic characteristics of the reservoir. Its most prominent feature is to fully understand the reservoir characteristics. However, it doesn't call for in-depth research but only seeks the most basic characteristics of structure and reservoir. Its contents are comprehensive but not thorough, for these mainly include structural interpretation, fine stratigraphic classification and correlation, sedimentary microfacies analysis, reservoir quantitative evaluation, flow unit characterization, and geological modeling. In contrast, the fine reservoir description for the middle–late development stage is mainly designed to address the key bottlenecks in current oilfield development, typically represented by remaining oil characterization (Fig. 2.2). For the purpose of remaining oil characterization, a series of specific, deep, and thorough research contents are set up in the aspects of structure and reservoir. Moreover, the research has become finer thanks to the availability of more data. Therefore, it is necessary to sort out the contents of fine reservoir description in the middle–late development stage comprehensively and systematically so as to provide assistance for related activities. Nevertheless, there are still many challenges in the way to accomplishing the fine reservoir description for the middle–late development stage. Besides, the fine reservoir description in China is greatly different from that abroad (Huanqing et al., 2016a,b,c,d), and there is no ready-made experience for reference. With this background, the researchers should summarize relevant research results to guide domestic production.

A fine reservoir description is a systematic discipline that involves all aspects of structural interpretation, reservoir prediction, and remaining oil characterization. However, the fine reservoir description in the middle–late development stage is quite different from that for the early development stage. We indicate that the fine reservoir description for the middle–late development stage includes the core contents as follows.

## 2.2.1 Fine description of small faults and microstructures

If we consider the needs of production, such as remaining oil characterization, the study of second- and third-order faults in reservoir structure

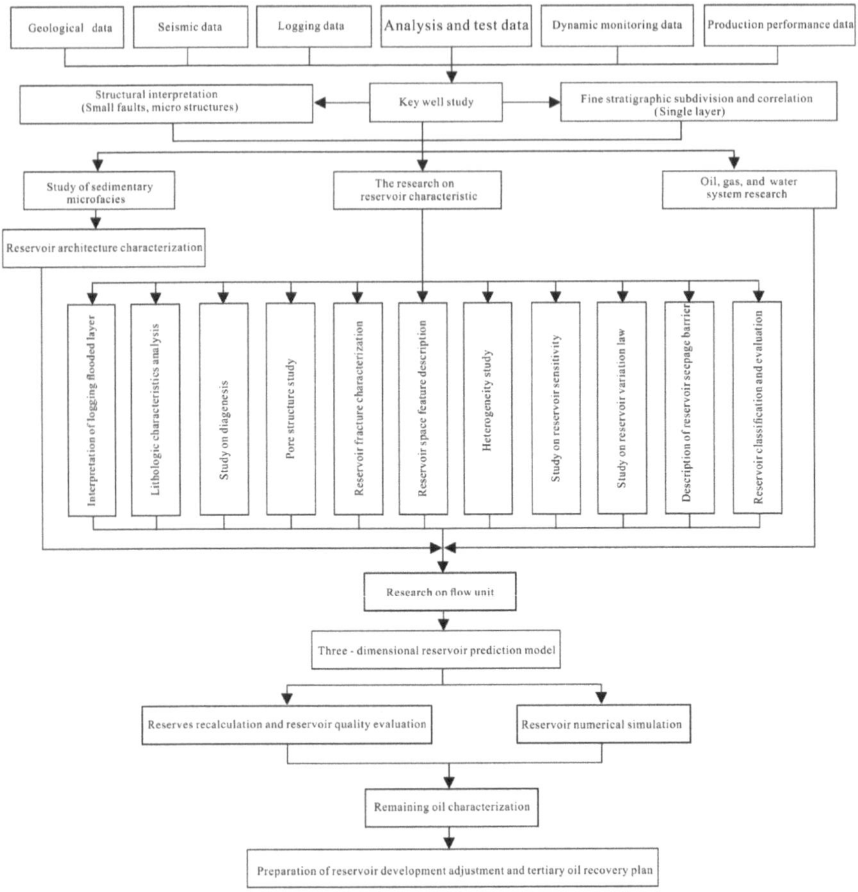


**Figure 2.2** Contents and process of fine reservoir description in the middle—late development stage.

research for the middle—late development stage is no longer adequate. The description of the fault system needs to be more elaborate, especially, since research must delve into the fourth- or even fifth-order, i.e., the low-order faults, going by the needs of production. The so-called "low-order faults" generally refer to the faults of fourth-order and below, which control the oil-water relationship and the distribution of remaining oil in the middle—late development stage. Yang and Jianmin (2007) defined the low-order faults in detail. The fourth-order faults are small, with a fault throw of only 20—50 m, or more than 100 m occasionally. They generally extend not more than 1—2 km, or only hundreds of meters sometimes. They are variable in strike, with the direction roughly parallel to the primary and secondary tectonic lines in the region in slight dominance. Most of the fourth-order faults were formed by superimposing the local stress field and have developed over a short period of time. They usually do not cut through the second- and third-order faults. With a steep fault plane, they mainly distribute in local structures and serve as the basis for dividing natural fault blocks. The fourth-order faults can form the boundary of oil-bearing fault blocks to make the latter have its own oil-water system. A fourth-order fault can separate the fault blocks on both sides to prevent them from withstanding mutual interference during development. The fifth-order faults are located in the interior of the oil-bearing fault block. They are small faults derived from the fourth-order faults, with a short extension of only a few hundred meters and a small fault throw of a few meters to a dozen meters. The fifth-order faults do not control fault blocks and deposits but only complicate the fault blocks and oil-water relationship. Li Yang (2011) presented the research process of low-order faults. The low-order faults and high-order faults are closely related because they formed under the same regional stress field. The development and distribution characteristics of low-order faults are controlled by higher-order faults. Therefore, according to the theories of tectonic kinematics, and geometry and kinetics, the structural model of a fault block is established through tectonic physical simulation experiment, by mathematical simulation of tectonic stress field, and using analysis of mechanical genetic mechanism to predict the development direction and fault system of low-order faults. The high-resolution processing of 3D seismic data, post-stack processing of seismic data volume, and mergence of 3D seismic data are completed to provide a reliable database for identifying small faults. With breakpoints determined through well-to-well correlation, the spatial combination and strike of low-order faults are preliminarily determined by using the coherence cube and multiscale edge detection techniques, and the faults of different orders are determined and combined by full three-dimensional interpretation, unified interpretation of fault horizons, and cross-well seismic techniques. Then, the results are verified using the comprehensive production performance data (Fig. 2.3). Moreover, Li Yang elaborated on the advances of the current low-order faults from the four aspects of fineness: (1) the fault block structural style is the theoretical basis for the genesis mechanism analysis of the low-order faults, (2) breakpoints are determined through fine well-to-well correlation under the dense well pattern of mature oilfield, (3) coherence cube analysis and multiscale edge detection are used to identify the

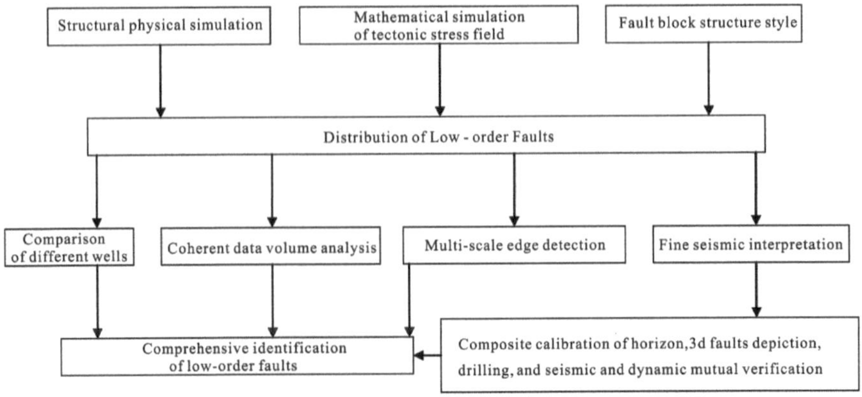


**Figure 2.3** Block diagram for describing low-order faults (Yang, 2011).

strike of low-order faults in the plane, and (4) the spatial distribution of low-order faults is determined finally through comprehensive analysis. At present, there are still many problems in the identification of small faults. Such problems are mainly attributed to the limited research level and data accuracy. Another critical reason behind this is the unclear understanding of the genetic mechanism of the fault system, which leads to widening gaps of different fault system interpretations made by different interpreters according to different data. The problem exists in various oilfields of PetroChina, but it cannot be solved in a short time, although this situation has attracted the attention of researchers. Chen Huanqing et al. (2011a,b,c) successfully used the ant tracing method to characterize the fault system of volcanic rocks in the Xudong area of the Songliao Basin. Although this method can characterize the fault system at a fifth-order accuracy, its results contain some false parts, which need to be verified by other methods. With consideration to the remaining oil characterization, microstructure has gradually come to the attention of the researchers. The concept of microstructure was defined by Li Xingguo (1994) as "The top and bottom of the reservoir are uneven, but present local fluctuations with a small area ($<0.3\ km^2$) and amplitude ($<20$ m). These local fluctuations are called microstructures." The microstructures are believed to have formed from the processes caused: (1) by factors irrelevant to tectonism and under the control of the sedimentary environment, differential compaction, and paleotopography, etc.; and (2) by factors relevant to tectonism, such as small fault nose and small fault trench along the downthrown side of the fault. Microstructures can be divided into three categories: (1) the relatively convex part of the sand bed in the positive microstructural system, including small high point, small nose structure, and small fault nose structure; (2) the relatively concave part of the sand bed in the negative microstructural system, such as small low point, small trench, small fault trench, and small syncline; and (3) the normal inclined part of the sand bed in the slope microstructural system, usually located between the positive and negative microstructures, or existing independently, including also a small terrace (Xingguo, 1994). Daqing and Liaohe oilfields have achieved good results in tapping the remaining oil potential by microstructure research, and other oilfields are also actively exploring it. The biggest problem in the microstructure research is the accuracy, which is largely limited and affected by the data accuracy.

## 2.2.2 Fine stratigraphic classification and correlation based on high-resolution sequence stratigraphy

Fine stratigraphic classification and correlation has always been one of the most important contents of oil and gas field development research, and it is the basis of all development geology. For the reservoirs in the middle–late development stage, the minimum unit of work has changed from layer to bed to depict sand body, corresponding to the fifth-order depositional cycle, i.e., the fifth-order sequence. There is controversy on the definition of bed. A number of scholars believe that bed corresponds to the fourth-order sequence. This book holds that this view is wrong. In the study of oil and gas field development, a layer corresponds to the fourth-order sequence (Huanqing et al., 2014a,b), which is consistent with Yinan and Ziqi (1996). Mostly different from the traditional "cyclic correlation, hierarchical control" method, the high-resolution sequence stratigraphy endows the stratigraphic genetic mechanism in stratigraphic classification and correlation and achieves higher correlation accuracy, making it satisfy the fine reservoir description in the middle–late development stage. When the theory of high-resolution sequence stratigraphy is used to guide the fine stratigraphic classification and correlation, the most important thing is to determine the principle of fine stratigraphic classification in addition to defining the scheme of stratigraphic classification and correlation. It is worth mentioning that the scheme of bed classification is not arbitrarily determined. The preliminary scheme is determined by observing the shape of the logging curve, depositional cycle, and other information concerns. Then, the scheme is extended to the wells on the skeleton section network through critical wells. If there are problems, the scheme is modified and then extended to the skeleton section network.

The scheme is finally determined through repeated practice and modification. Next, the results of fine stratigraphic classification and correlation on the skeleton section network are extended to the whole study area, and thus the high-precision isochronal stratigraphic framework of the whole area is established. The direction of the skeleton section network is kept parallel or vertical to the provenance to ensure that the researchers have a correct understanding of the distribution law of strata when correlating strata. In the fine stratigraphic classification and correlation based on high-resolution sequence stratigraphy, the three aspects of development geology are essential. The first is the logging–seismic combination for establishing large-scale isochronal stratigraphic framework (Fig. 2.4) (Huanqing et al., 2014a,b). Time-depth conversion of seismic data is realized through seismic calibration by coring wells, and then a large-scale isochronal stratigraphic framework is established (in this case, "large-scale" refers to the third-order sequence, which corresponds to the long-term base-level cycle in high-resolution sequence stratigraphy). The second is the establishment of an accurate model of short-term base-level cyclic response of typical wells (Fig. 2.5). Within the large-scale isochronal stratigraphic framework, coring wells with complete strata and data are selected. Combining this with lithology, logging, analysis, and testing data, the short-term base-level cyclic response model is established, and the fine stratigraphic classification and correlation scheme is determined.

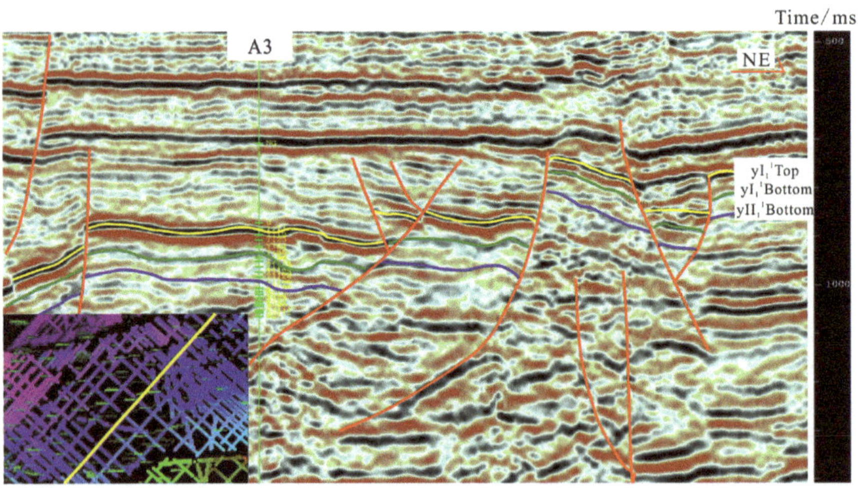


**Figure 2.4** Establishment of large-scale sequence stratigraphic framework by logging–seismic combination in an area of Western Sag, the Liaohe Basin.

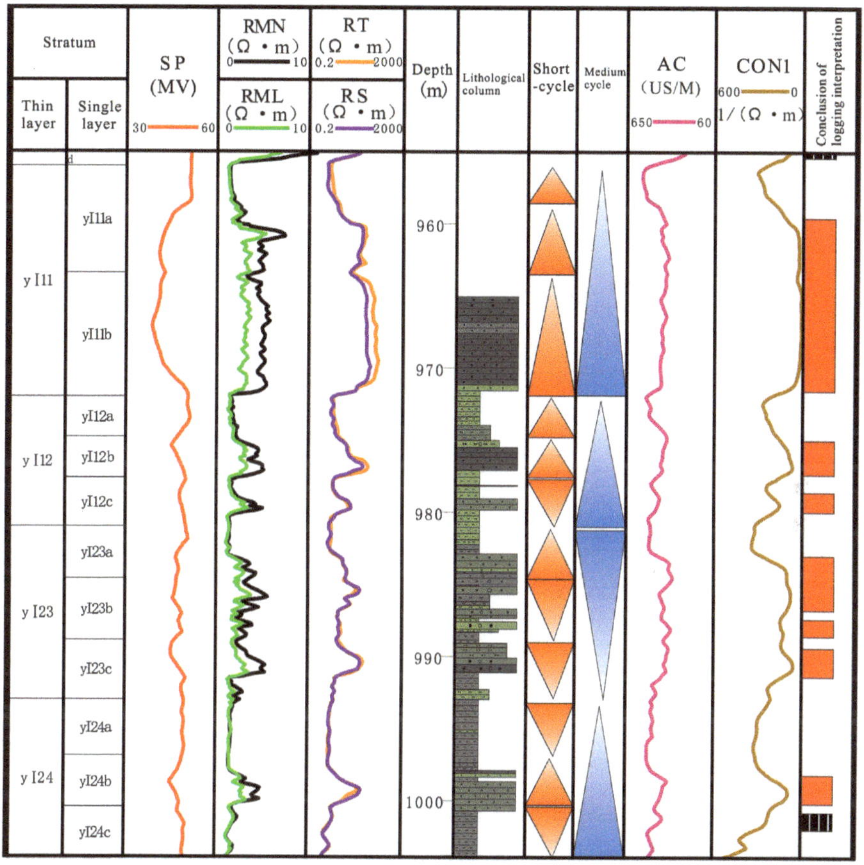


**Figure 2.5** Short-term base-level cyclic response model of Well A9 in an area of Western Sag, the Liaohe Basin. Note: RMN-micropotential; RML-microgradient; CON1-conductivity.

The third is the close combination of stratigraphic classification and correlation based on high-resolution sequence stratigraphy and by the traditional "cyclic correlation, hierarchical control" method. Currently, the traditional method is still popular, while the high-resolution sequence stratigraphy is seldom applied due to the limitations of data availability and research level. This situation seriously hinders the progress of fine reservoir description. The improvement of research level is inseparable from the advancement of methodology and technology. Sequence stratigraphy is known as the third revolution of sedimentology (Yangquan et al., 2015). The application of high-resolution sequence stratigraphy theory and method should be vigorously advocated and promoted to improve

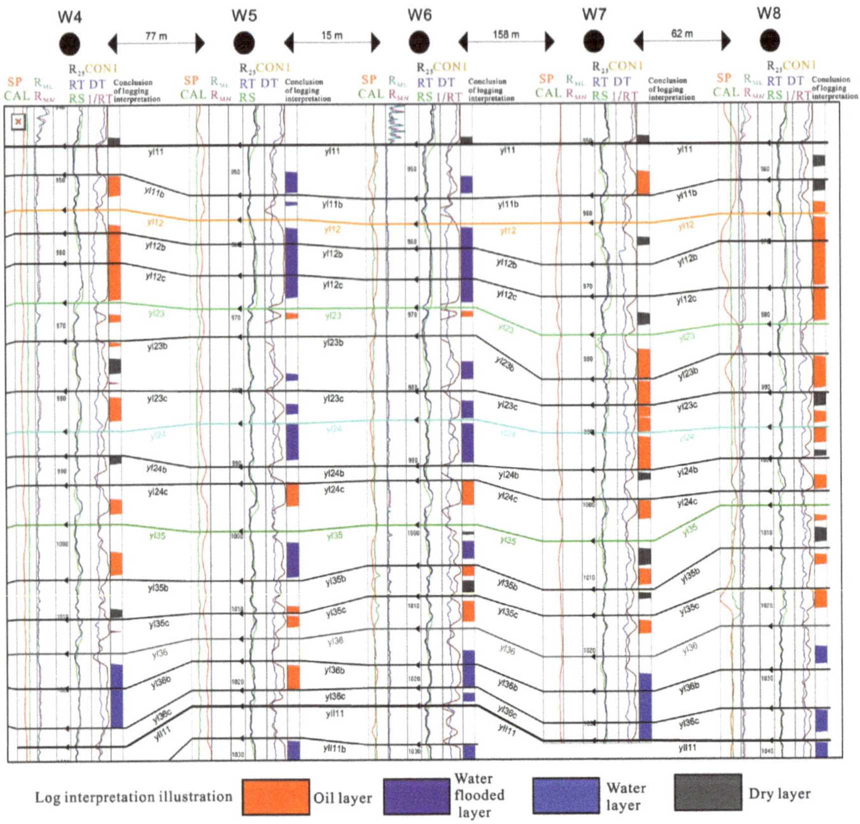


**Figure 2.6** High-resolution sequence stratigraphic analysis of a section in an area of Western Sag, the Liaohe Basin.

the research level and meet the needs of fine reservoir description. Finally, a high-precision isochronal stratigraphic framework is established to meet the requirements of oilfield development and production practice (Fig. 2.6).

## 2.2.3 Reservoir architecture characterization based on sedimentary microfacies analysis

With the progress of oil and gas field development, the traditional sedimentary microfacies analysis becomes no longer valid for sand body prediction and intercalation characterization in fine reservoir description; instead, reservoir architecture characterization is growing as a research focus for future studies. Reservoir architecture refers to the pattern, scale, direction, and overlapping relationship of different orders of reservoir architectural elements (Shenghe, 2010). Reservoir architecture characterization mainly includes two parts: (1) description of the spatial development of different orders of reservoir architectural elements (sand body prediction) and (2) fine description of the three-dimensional distribution of different orders of reservoir architecture boundaries (intercalation research). To characterize the reservoir architecture, the reservoir architecture classification should be determined first. Currently, the commonest classification is the eight-order classification proposed by Miall (1996). Log data are mostly used, and the four-order and five-order reservoir architecture classifications are generally adopted. It should be emphasized that a finer classification is not necessarily a better one. This book holds that two elements should be fully considered in determining the reservoir architecture classification hierarchy: (1) the results of reservoir architecture characterization should meet the requirements of oilfield development and production practice; and (2) the available data should be taken into full account. Excessive pursuit of classification hierarchy without the data for its basis will inevitably lead to wrong results. Reservoir architecture characterization is essentially the extension of sedimentary microfacies analysis. Unlike the sedimentary microfacies, however, the reservoir architecture characterization provides higher accuracy and, in view of contents, focuses on sand body and also on different orders of architecture boundaries, i.e., intercalation. We have successfully studied the architecture of an alluvial fan reservoir in the Lower Karamay Formation in an area at the northwestern margin of the Junggar Basin on the basis of sedimentary microfacies analysis (Fig. 2.7). Firstly, based on the sedimentary analysis, the target conglomerate reservoir interval was divided into four subfacies: upper fan inner zone, upper fan outer zone, middle fan, and lower fan. Then, according to lithology, log pattern, and sedimentary model, the target interval was divided into 13 types of architectural elements, i.e., trough flow gravel, trough beach sand-gravel, sheetflood inner sand, sheetflood inner fine grain, laminar flow gravel, sheetflood outer sand, sheetflood outer fine grain, braided flow channel, braided flow sand-gravel bar, cross flow sand, cross flow fine grain, runoff channel, and inter-channel fine grain (Table 2.1). Trough flow gravel, laminar flow gravel, braided flow channel, and braided flow sand-gravel bar are dominant (Huanqing et al., 2015a,b,c,d) concerns for fine reservoir description. Taking full account of the accuracy of log data and the needs of development and production practice, the architecture of target reservoir in the study area was divided into four orders. The results played a vital guiding

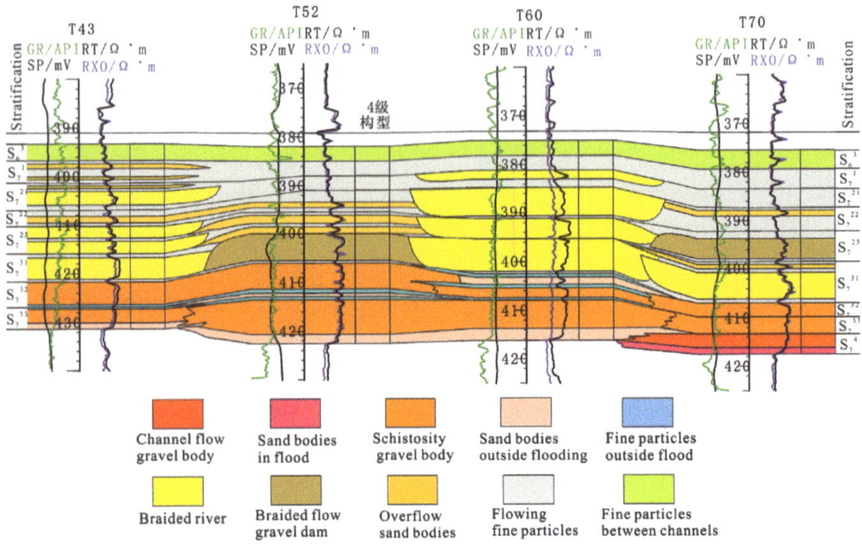


**Figure 2.7** Architecture profile (parallel to the provenance) of alluvial fan conglomerate reservoir in the Lower Karamay Formation in an area at the northwest margin of the Junggar Basin.

**Table 2.1** Evaluation of pore structure of volcanic reservoir in Ying 1 Member in Xudong area, the Songliao Basin (Huanqing et al., 2016a,b,c,d).

| Type | Porosity $\Phi$/% | | | Permeability $K/10^{-3}\ \mu m^2$ | | | Median pore throat radius $R_{50}/\mu m$ | | |
|---|---|---|---|---|---|---|---|---|---|
| | Max. | Min. | Avg. | Max. | Min. | Avg. | Max. | Min. | Avg. |
| I | 14.4 | 10.5 | 13.0 | 1.19 | 0.04 | 0.20 | 0.299 | 0.062 | 0.161 |
| II | 9.9 | 7.7 | 8.8 | 1.12 | 0.02 | 0.10 | 0.251 | 0.021 | 0.113 |
| III | 7.5 | 4.9 | 6.3 | 1.35 | 0.01 | 0.05 | 0.128 | 0.018 | 0.06 |
| IV | 4.8 | 2 | 4.0 | 1.21 | 0.01 | 0.04 | 0.086 | 0.016 | 0.03 |

role in the deployment of development and adjustment of wells in practice. Reservoir architecture characterization is more advanced than traditional sedimentary microfacies analysis. It realizes a greatly improved accuracy of sand body characterization based on the genetic mechanism analysis, which makes the prediction of spatial development characteristics of sand bodies more accurate and fine, and also makes the prediction of inter-well sand body connectivity more accurate. Meanwhile, the fine characterization of different orders of reservoir architecture boundary has greatly improved the accuracy of describing the spatial development law of intercalations with different thicknesses. Generally, the reservoir architecture characterization can be summarized into four steps. The first step is to make fine stratigraphic classification and correlation using the above data and establish a fine layer classification and correlation database. The second step is to establish the reservoir architecture classification system according to the lithofacies analysis and sedimentary analysis within the fine isochronal stratigraphic framework. Determination of the architecture classification hierarchy (four orders or five orders?) is a key step in the study of reservoir architecture. Only when a reasonable hierarchy is determined can the follow-up work be carried out smoothly. The third step is to summarize the architecture configuration pattern (mode). This mainly includes two aspects of reservoir architecture. First, the reservoir architecture boundary characterization is achieved through the analysis of layer boundary and the characterization of reservoir intercalation. Second, the development of architectural elements in the sand body of different geneses, including the scale, origin, and internal structure of different architectural elements, is identified by integrating the characteristics of reservoir architecture boundary, diagenesis, and microscopic reservoir characteristics like pore structure. The fourth step is to, on the basis of the third step, collect the architecture configuration pattern of the reservoir and analyze the distribution law of the remaining oil. Ultimately, a set of reservoir architecture characterization technology is formed, which can provide guidance for the effective development of reservoir and for the tapping potential of remaining oil.

## 2.2.4 Quantitative evaluation of reservoir pore structure

The reservoir pore structure has always been the focus of reservoir geology study (Heqing et al., 2017). The research of reservoir pore structure for the middle—late development stage is more thorough and detailed than that for the early development stage. The research methods are gradually changing from core slice observation and identification and statistics of core physical property test data to pore throat constant velocity mercury injection test, CT scanning pore structure analysis, digital core, and so forth. Chen Huanqing et al. (2013a,b,c,d) introduced in detail several research methods for pore structure, including the statistical analysis of physical data such as casting thin section observation and mercury injection, diagenesis, logging analysis, fractal dimension model, mathematical methods, new techniques like numerical simulation and three-dimensional imaging, geological (or geophysical) model, and cluster analysis. From the overall trend, it can be seen that the research of reservoir pore structure advances gradually from qualitative or semi-quantitative to quantitative, accompanied by the need for improving accuracy. Chen Huanqing et al. (2016a,b,c,d) completed the classification of different pore structures of volcanic reservoirs in Ying 1 Member in Xudong area, the Songliao Basin, by cluster analysis. The pore structures were divided into Types I to IV (Fig. 2.8, Table 2.1). The quantitative research results of reservoir pore structure provide a solid basis for the design and adjustment of the development plan. There are many parameters for reservoir quantitative evaluation, including average pore size (Rs), maximum connected pore throat radius (Rd), displacement pressure (pd), median pore throat radius (R50), median capillary pressure (p50), minimum unsaturated pore volume percentage (Smin), average pore throat radius (Rm), average radius of main flow pore throat (Rz), difficult-to-flow pore throat radius (Rn), pore throat sorting coefficient (Sp), relative sorting coefficient (Dr), homogeneity coefficient ($\alpha$), pore throat skewness (Skp), pore throat peak state (Kp), and mercury ejection efficiency (We) (Shenghe and Qihua, 1998). Quantitative evaluation of pore structure is mainly achieved by means of geostatistics. It is not the situation that the more evaluation parameters, the better is the statistical analysis. Suitable evaluation parameters should be selected according to the data as the basis of evaluation and the research purpose. It should be noted that quantitative evaluation of reservoir pore structure must be based on a detailed analysis of the geological genesis of pore structure. The main factors controlling the formation of reservoir pore structure are identified through systematic

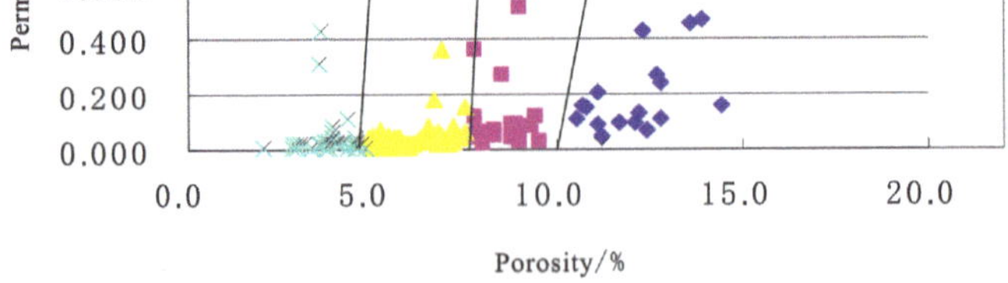


**Figure 2.8** Pore-permeability relationship of different pore structures in a volcanic reservoir in Ying 1 Member in Xudong area, the Songliao Basin (Huanqing et al., 2016a,b,c,d).

geological genesis analysis; then, the parameters that can fully reflect these factors are selected, coupled with appropriate calculation method, to realize the quantitative evaluation of reservoir pore structure. If there is no systematic geological genesis analysis, the research undertaken will be blindfolded as far as the selection of pore structure evaluation parameters is concerned. The evaluation results can only have the characteristic of digital games in essence founded on chance and whose authenticity and credibility a bet on their veracity at the best and not a verifiable outcome of the study.

### 2.2.5 Reservoir fluid heterogeneity analysis

Fluid heterogeneity plays a very important role in the development and production of oilfields in the middle−late development stage. Especially with the implementation of development measures such as water flooding or steam flooding, the distribution and migration of underground oil, gas, and water become more complex, which should be described accurately and understood thoroughly. Chen Yongsheng (1993) divided the heterogeneity of oilfields, from macro to micro, into two categories: flow field heterogeneity and fluid heterogeneity. At present, the study of flow field heterogeneity mainly includes reservoir intercalation analysis, interlayer heterogeneity analysis, intralayer heterogeneity analysis, plane heterogeneity analysis, and micro-heterogeneity analysis. There are obvious differences in physical, chemical and physical and chemical properties between oil, natural gas and water as an oil displacement agent, which, in the process of water (gas) and other oil displacement agents (gas), affect the change and development of the uneven characteristics of water displacement process caused by the heterogeneity of flow field from beginning to end, and in most cases, these differences aggravate the contradiction. In the process of water flooding, the difference in various properties of fluids differentiating between crude oil, natural gas, or water is called fluid heterogeneity (Yongsheng, 1993). At present, most researchers focus on the flow field heterogeneity but pay little attention to the fluid heterogeneity. Yin Wei et al. (2001) used the thin-layer chromatography−flame ion detection (TLC-FID) technology to study the reservoir fluid heterogeneity characteristics such as the group composition of reservoir extracts and the distribution of asphalt pad in Block Qian 12 of Liaohe Oilfield, which provided a reliable basis for the formulation and adjustment of the oilfield development plan. Yang Chiyin (2004) summarized four controlling factors of fluid heterogeneity of the Qianmiqiao buried hill condensate gas reservoir in Huanghua Depression:

(1) bidirectional hydrocarbon supply, (2) multi-phase reservoir formation, (3) overall late reservoir formation, and (4) buried hill fault development and strong reservoir heterogeneity. Huang Haiping et al. (2010) studied the application of reservoir fluid heterogeneity in heavy oil evaluation and development and production in the Peace River area, Western Canadian Basin (Fig. 2.9). The results showed that drastic changes in chemical composition and physical properties of crude oil are very common in heavy oil reservoirs. Understanding the causes and controlling factors of these changes is very important for deepening reservoir geological evaluation, optimizing heavy oil development plan design, and enhancing heavy oil recovery. It is concluded that reservoir fluid heterogeneity mainly uses geochemical and hydrodynamic techniques to depict the distribution and genetic mechanism of underground oil, gas, and water. There are some related studies in oil and gas exploration. However, the reservoir fluid heterogeneity seems to be much less attractive than other contents of fine reservoir description for the middle−late development stage. Currently, it is mainly investigated using the geochemical methods and analytical methods, and less of the quantitative ones. For the mature oilfields in the middle−late development stage, the properties and distribution of underground oil, gas, and water have undergone tremendous changes along with decades of development. Compared with the original state of the reservoir, the mature oilfields are more complex. In order to achieve the goal of enhancing oil recovery and of enhancing the tapping potential of remaining oil, it is necessary to work on fluid heterogeneity and figure out the relevant methodologies and, also, to quantitatively depict the development law of fluids with different properties.

### 2.2.6 Reservoir change analysis

Oilfield development is a dynamic process. Along with it, some changes occur in the reservoir, including pore structure and clay minerals. These changes have a tremendous impact on subsequent reservoir development, thus directly related to the enhancement of oil recovery. Therefore, it is of great theoretical and practical significance for the effective development of oil and gas reservoirs to understand the changes of reservoirs in the process of development. Some scholars have worked on this aspect (Zhizhang and Jinhua, 1999; Kato et al., 2008; Wenqi et al., 2010; Lanxiong, 2011; Higgs et al., 2013). Overall, however, the attention given to it is not high enough, and there are some problems. According to the literature review, the studies abroad mainly focus on reservoirs with water flooding and gas (steam, $CO_2$, $N_2$, etc.) flooding, typically clastic rock reservoir and carbonate reservoir. In China, the studies mainly focus on reservoirs with

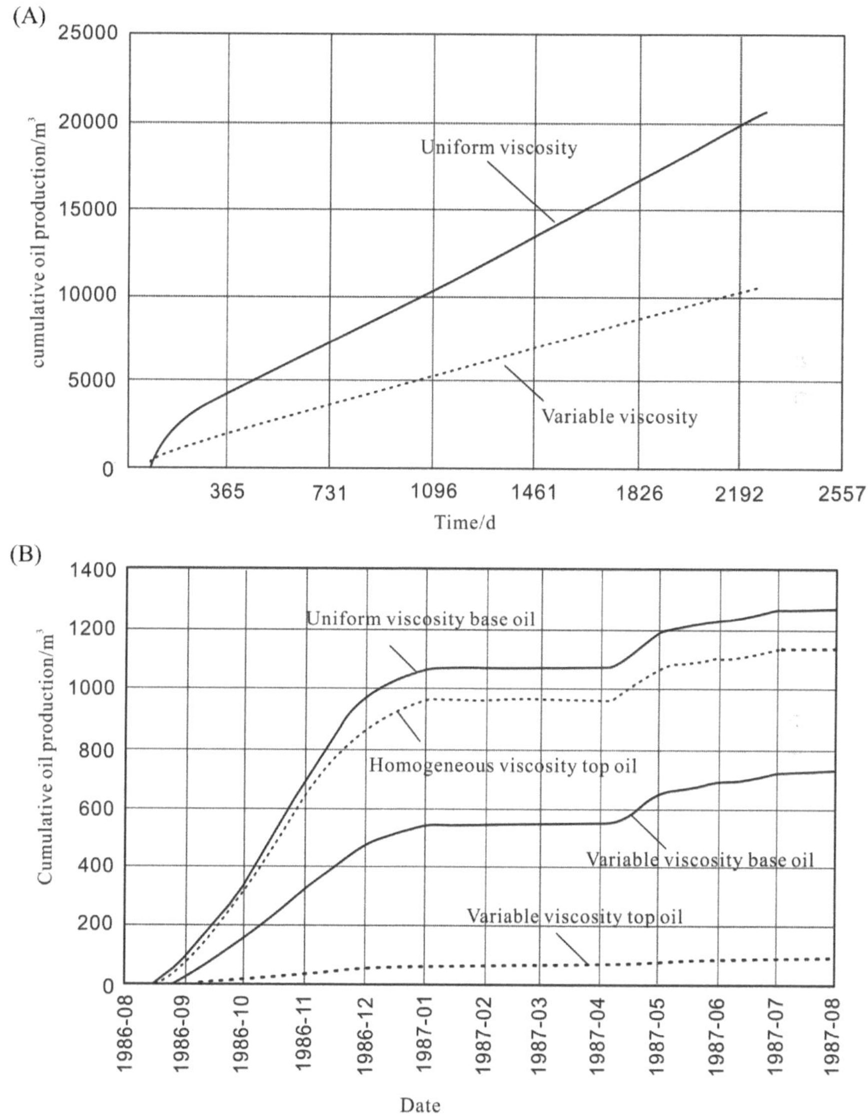


**Figure 2.9** Effects of heterogeneity of crude oil composition on cumulative production (Haiping et al., 2010). (A) Cumulative oil productions calculated by SAGD-based uniform viscosity and variable viscosity simulation; (B) Cumulative oil productions calculated by CSS-based uniform viscosity and variable viscosity simulation. *CCS*, carbon capture and storage; *SAGD*, steam-assisted gravity drainage.

water flooding and steam flooding for thermal recovery, as well as a reservoir with in-situ combustion, mainly on the clastic reservoir, but less on carbonate reservoir. In China, reservoir changes have been successfully analyzed in some oilfields, such as Daqing, Shengli, Liaohe, Dagang, and Xinjiang. In view of the research method used, the popular methods include the observation and comparison of thin sections under a microscope before and after development and the experimental analysis of reservoir velocity sensitivity, water sensitivity, salt sensitivity, acid sensitivity, and pressure sensitivity. Quantitative methods such as physical simulation and numerical simulation can be considered in a future application. Wang Zhizhang and Jinhua (1999) indicated that the analysis of reservoir variation rules and mechanism in the development process is to establish a quantitative model of layer-scale or pore-scale flow field that can reflect the current reservoir characteristics through experimental simulation and numerical simulation, which is then used to reveal the change rules and mechanism of reservoir parameters in three-dimensional space and to predict the remaining oil enrichment area, laying a foundation for reservoir evolution analysis and development mode optimization. Taking the Qigu Formation heavy oil reservoir in Block 9 of Karamay Oilfield as an example, Wang Zhizhang made a statistical analysis on the physical properties of reservoirs before and after steam flooding (Table 2.2). Except for porosity, both permeability and saturation change greatly after steam flooding. We analyzed the changes of Yulou reservoir in an area of Western Sag, the Liaohe Basin during steam stimulation and steam flooding for thermal recovery (Fig. 2.10). The results show that in heavy oil thermal recovery reservoirs, the porosity and permeability decrease even more after steam flooding. Understanding the law of reservoir changes in the development process is helpful to timely adjust the development strategy, further understand the distribution of remaining oil, and ultimately enhance the oil recovery.

### 2.2.7 Geological modeling methods (e.g., multi-point geostatistics)

The final result of reservoir description is a quantitative reservoir geological model as the basis of reservoir simulation, reservoir engineering, and oil recovery process (Yinan and Ziqi, 1996; Shenghe et al., 1999; Dianxing et al., 2015). At present, the method of facies-controlled modeling is often used to build the reservoir attribute model, while the method of variogram fitting is used in the process of sedimentary microfacies

**Table 2.2** Reservoir physical properties before and after steam flooding (Zhizhang and Jinhua, 1999).

| Block | Horizon | Sedimentary microfacies | Porosity % | | Horizontal permeability $10^{-3}\,\mu m^2$ | | Saturation % | | Remarks |
|---|---|---|---|---|---|---|---|---|---|
| | | | Before steam flooding | After steam flooding | Before steam flooding | After steam flooding | Before steam flooding | After steam flooding | |
| $Jiu_1^{\ 2}$ | $G_2^{\ 2-1}$ | Cross flow zone | 31.9 | 26.4 | 2300 | 263 | 77.8 | 45.5 | Corresponding to Well J275 |
| | $G_2^{\ 2-2}$ | Main channel | 31.2 | 25.5 | 1900 | 443 | 77.0 | 46.9 | |
| $Jiu_3$ | $G_2^{\ 2-1}$ | Main channel | 31.1 | 26.9 | 2460 | 56 | 69.1 | 50.0 | Corresponding to Well J280 |
| | $G_2^{\ 2-2}$ | Secondary channel | 30.7 | 26.5 | 2230 | 86 | 67.9 | 44.8 | |
| $Jiu_6$ | $G_2^{\ 2-1}$ | Cross flow zone | 30.6 | 26.1 | 2300 | 416 | 76.4 | 38.5* | Corresponding to Well J279 |
| | $G_2^{\ 2-2}$ | Secondary channel | 30.4 | 25.8 | 2190 | / | 69.2 | 41.0* | |
| $Jiu_1^{\ 1}$ | $G_2^{\ 2-1}$ | Main channel | 31.4 | 31.9 | 2722 | 1127 | 78.6 | 42.0 | Corresponding to Well J290 |
| | $G_2^{\ 2-2}$ | Main channel | 31.2 | 32.7 | 2533 | 1098 | 75.0 | 44.0 | |
| | $G_2^{\ 2-3}$ | Secondary channel | 30.5 | 32.1 | 2130 | 1423 | / | / | |

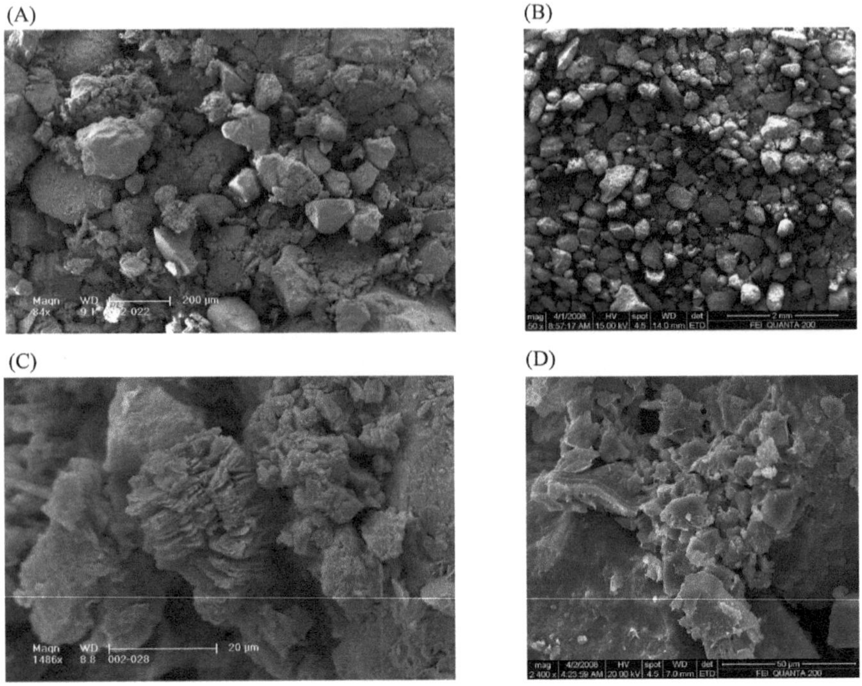


**Figure 2.10** SEM images showing the variation of heavy oil thermal recovery reservoir in Yulou reservoir in an area of Western Sag, the Liaohe Basin. (A) Well G41, Lianhua reservoir in the third member of Shahejie Formation, after the steam flooding, sandstone, loose sample, 751.32 m; (B) Well J2, Yulou reservoir in Sha 1 Member, before the steam flooding (steam stimulation), sandstone, loose sample, 989.18 m; (C) Well G41, Lianhua reservoir in the third member of Shahejie Formation, after the steam flooding, sandstone, clay minerals like kaolinite, 753.1 m; (D) Well J2, Yulou reservoir in Sha 1 Member, before the steam flooding (steam stimulation), sandstone, conversion from kaolinite to illite, 997.02 m.

modeling. We adopted the method of variogram fitting in the geological modeling of heavy oil thermal recovery reservoirs in the Western Sag, the Liaohe Basin (Fig. 2.11). Although this method is most widely used now, the biggest problem is that the geological genesis mechanism is not fully reflected in the modeling process, so the accuracy of the model is greatly affected. The application of multi-point geostatistics in stochastic modeling began in 1992 (Shenghe and Kewen, 2005). Its basic tool is the training image, which is equivalent to the variogram in traditional geostatistics. As far as sedimentary facies modeling is concerned, the training image is equivalent to a quantitative facies model, which is essentially a digital prior model containing facies contact that is believed by the modeler existing in

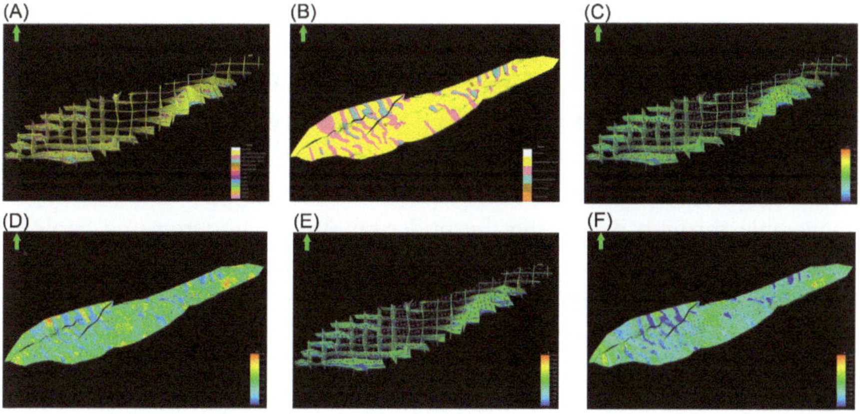


**Figure 2.11** Geological models of Yulou reservoir in an area of Western Sag, the Liaohe Basin. (A) Sedimentary microfacies model (fence diagram); (B) Sedimentary microfacies model ($yl_1^{1a}$); (C) Porosity model (fence diagram); (D) Porosity model ($yl_1^{1a}$); (E) Permeability model (fence diagram); (F) Permeability model ($yl_1^{1a}$).

the actual reservoir. Geologists need to capture the facies contact from the training images and correspond it to specific reservoir data (Jiahua and Tao, 2013). The method of multi-point geostatistics is advantageous in view of the fact that different types of data, including log data, the vertical proportional curve of sedimentary facies, and probability distribution of sedimentary facies (mostly extracted from seismic data) can be integrated into the reservoir modeling process through training images. The use of training images makes the geological model more in line with the actual underground geology. Of course, the method of multi-point geostatistics also suffers from some problems. It is still in the exploratory stage, but it will certainly become a leading method of geological modeling in fine reservoir description along with the deepening of research and the advancement of technology.

## 2.2.8 Multi-information remaining oil description

Most of the oilfields in the middle–late development stage have entered the high water-cut stage. On the one hand, additional reserves are more and more difficult to find, and the exploration degree is high. The newly discovered oilfields are generally smaller in scale; a large proportion of reserves additions are endowed with low-permeability reservoirs, and the additional and remaining reserves are difficult to produce. On the other hand, the water flooding oilfields in China suffer prominent issues of

"3H2L" (high comprehensive water-cut, high recovery percent, high oil recovery rate, low reserves-production ratio, and low recovery factor), and a large quantity of oil remains unrecoverable. Therefore, the study of remaining oil distribution for enhanced oil recovery has always been a core in oilfield development geologists' and reservoir engineers' work efforts (Jun and Jinliang, 2003). The remaining oil is generally considered to be recoverable through deepening the understanding of underground geological bodies and improving the level of production technology (Ping et al., 2004). Lin Chengyan (2000) summarized three foci in the studies on remaining oil distribution at home and abroad: (1) description of remaining oil distribution, (2) research on remaining oil saturation measurement and monitoring technology, and (3) research on remaining oil potential tapping technology. Now, the first is prominent in China, while the second and third ones are fruitful abroad. Guo Ping et al. (2004) systematically collected the research methods of remaining oil, involving microstructural interpretation, sand body sedimentary microfacies analysis, sequence stratigraphy, original oil in place abundance, dynamic analysis, numerical simulation, laboratory experiment, logging, and tracer. Meanwhile, the distribution pattern of the remaining oil is studied (Fig. 2.12). At present, the data for remaining oil research include geology, well logging, core, mud logging, analysis and testing, dynamic monitoring, and production performance. Based on the analysis results of microstructure, sand bed, sedimentary microfacies, and reservoir architecture, the methods of comprehensive geological analysis, data analysis of sealed coring wells and inspection wells, swept layer logging

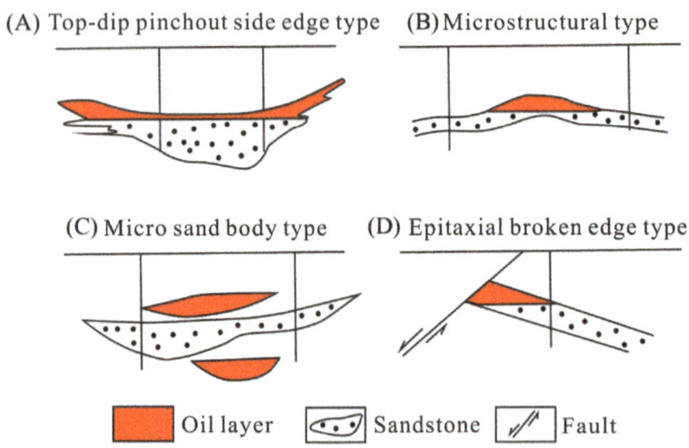


**Figure 2.12** Types of remaining oil in fluvial reservoirs (Guo Ping et al., 2004).

interpretation, reservoir numerical simulation, dynamic monitoring analysis, and reservoir engineering analysis are combined and cross-validated to ascertain the types and distribution of remaining oil. In production practice, the remaining oil characterization mainly depends on numerical simulation, which is a quantitative remaining oil description method but with less credibility of missing accuracy. On the one hand, the accuracy of numerical simulation is affected by the accuracy of geological modeling, where the inter-well sand prediction is not convincing on accuracy. On the other hand, the accuracy of numerical simulation is greatly affected by the artificial factors in parameter adjustment and some engineering factors in the production process. The book holds that the two points should be emphasized in the remaining oil description. First, the geological genetic type (tectonism, deposition, or diagenesis) of the remaining oil should be determined to ensure the accuracy of the remaining oil characterization. Second, dynamic monitoring data and production performance data should be fully integrated for analysis. For oilfields in the middle–late development stage, a large amount of dynamic data has been accumulated, which contains abundant information on underground fluids. The researchers should seriously consider how to combine these data to extract effective information reflecting the distribution law of remaining oil.

The above is the summary of eight key aspects of fine reservoir description in the middle–late development stage of oilfields. This summary can be conducive to oilfield development under the current background of low oil prices. On the one hand, oil companies have greatly cut their research projects in both numbers and funds for the purpose of "reducing costs and increasing efficiency." Accordingly, the arrangement of fine reservoir description projects has to be adjusted. Fine reservoir description involves all aspects of development geology study; thus, appropriate research contents and countermeasures need to be put forward for key bottlenecks in development and production practice so that the limited funds can be utilized effectively and efficiently. On the other hand, the fine reservoir description in the middle–late development stage should take into full account the certain reference results accumulated through the research in the exploration, appraisal, and early development stages, with the support of relevant research depending on current production practice. Therefore, it is particularly important to sort out and summarize the contents of fine reservoir description in the middle–late development stage. With the progress of oilfield development research, the problems encountered will change to some extent, and some new situations will emerge. At the same time, with the advancement of related methods and techniques, the contents of fine reservoir description will change. Nevertheless, there is no doubt that the ultimate goal of fine reservoir description is to solve the practical problems encountered in production practice so as to serve the target of enhancing oil recovery.

# CHAPTER 3

# Techniques of fine reservoir description

Fine reservoir characterization involves many aspects of development due diligence, such as structure, reservoir, and oil-gas-water relationship and, so, its methodology is diverse. This chapter introduces nine unique methods/technologies for fine reservoir description, including fine stratigraphic classification and correlation based on high-resolution sequence stratigraphy, multiinformation fracture characterization of the volcanic reservoir, reservoir architecture characterization based on sedimentary microfacies classification, genetic classification and quantitative characterization of reservoir pore structure, research on reservoir heterogeneity, comprehensive quantitative reservoir evaluation based on geological genesis analysis, classification of reservoir flow units, multipoint geostatistical modeling, and the remaining oil characterization.

## 3.1 Fine stratigraphic classification and correlation based on high-resolution sequence stratigraphy

Fine stratigraphic classification and correlation is one of the most basic and important aspects of fine reservoir characterization, which has always been the focus of oil and gas field development research efforts (Xiurong et al., 2001; XinTao and Pingping, 2007; Youliang and Liqiang, 2007; Huanqing and Xiaomin, 2008; Ailin, 2010). As the oil and gas fields gradually enter into the middle–late development stage, the implementation of (EOR) measures like the separated-layer water (gas) injection and the remaining oil potential tapping triggers a more immediate requirement for the precision of stratigraphic classification and correlation, which has turned from sand group to layer and even to bed. Due to the control of structure and sedimentary environment, the traditional "cyclic correlation, hierarchical control" method is no longer satisfactory to the production practice. The target reservoir in the study area is composed of fan delta front deposits, with the characteristics of near-source, short flow distance, multiple sand bodies, and small scale (Zaixing, 2003). On the one hand,

*Fine Reservoir Description*
DOI: https://doi.org/10.1016/B978-0-323-95401-3.00003-2

the frequent diversion of underwater distributary channels, short lateral extension distance, and vertical superimposition makes it very challenging to finely classify and correlate the strata. On the other hand, the tectonic activities have led to a well-developed synsedimentary fault system. The stratum thickness is greatly variable in the hanging and foot walls of the faults, and the strata are broken in the vicinity of faults impeding the tracing and correlation of boundary for stratigraphic division. All of these situations immensely challenge the fine stratigraphic classification and correlation. Under this circumstance, more sophisticated and scientific theories and techniques are required to guide the stratigraphic classification so as to improve the accuracy of stratigraphic correlation. The high-resolution sequence stratigraphy is just such an optimal choice.

The stratigraphic classification and correlation based on high-resolution sequence stratigraphy is a technique of stratigraphic correlation (including the evaluation and prediction of source bed, reservoir bed, and seal bed) within the genetic stratigraphic framework: namely, a high-precision stratigraphic correlation framework at the scale of the region, oilfield, or even the reservoir that is established using the fine stratigraphic classification and correlation technique according to the theory of high-resolution sequence stratigraphy and based on the outcrop, logging, core, and 3D high-resolution seismic reflection data (Hongwen et al., 2002; Yi et al., 2008; Rongcai et al., 2009; Rongcai, 2010). The stratigraphic correlation based on high-resolution sequence stratigraphy can realize the division of sequences at different orders, exacting the meeting of the requirement in the geological study for the precision of stratigraphic classification. The large-scale sequence hierarchy is well-consistent with the traditional stratigraphic boundary and classification system. The small-scale sequence hierarchy (four-order, five-order, etc.) can procure a fine stratigraphic classification at the level of the layer or even the bed, satisfying the fine stratigraphic classification in the middle—late development stage of oil and gas fields. Here, taking the Yulou reservoir in an area of Western Sag, the Liaohe Basin, as an example, the application of high-resolution sequence stratigraphy in fine stratigraphic classification and correlation is presented in detail.

### 3.1.1 Geology of Yulou reservoir

The study area, geographically in Linghai City of Liaoning Province and structurally in the Western Sag, the Liaohe Basin (Fig. 3.1), is developed

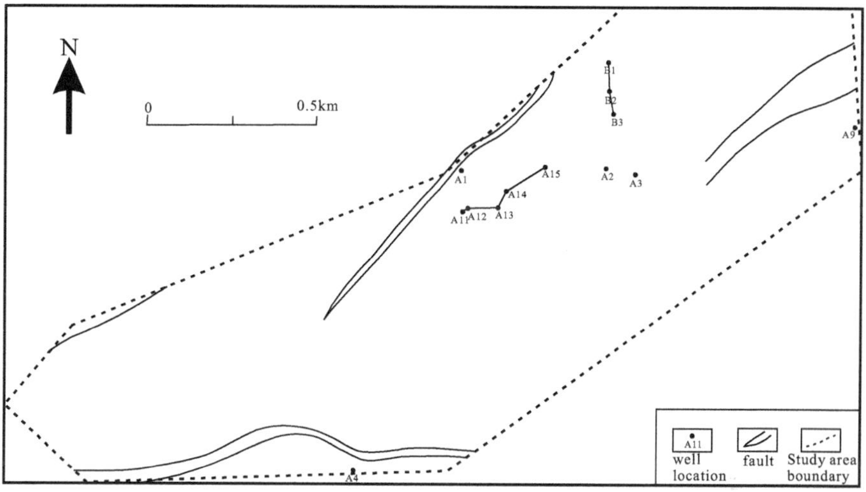


**Figure 3.1** Location of the study area in Western Sag, the Liaohe Basin (Well A1, Well A2, Well A3, Well A4, and Well A9 are critical wells with core data). *Note*: A1, A2, A3, A4, and A9 are key wells with core data.

in the Yulou and Xinglongtai reservoirs. The Yulou reservoir in the first member of Shahejie Formation of Paleogene is the target in this study. It presents as a monoclinic structure dipping southeastward with a dip angle of 2°–10°. The reservoir is classified as clastic sediment of fan delta front subfacies and, mainly, consists of thick inequigranular sandstone and medium to fine sandstone. Two sets of layer series are separated vertically for reservoir development with steam stimulation. Now, in the middle–late development stage, the production performance is getting worse. It is of great necessity to adjust the development mode. The Yulou reservoir is vertically divided into two pay zones, I and II, which include 12 layers. The frequent diversion of underwater distributary channel in each layer leads to the longitudinal superimposition between the channel skeleton sand bodies and interchannel sand bodies, on the one hand, and the combination of lateral bands, on the other, making the reservoir heterogeneous laterally and vertically given which the results follow in non-uniform producing in lateral and vertical directions. Ultimately, the promotion of steam flooding in this block is restricted. Considering the problems in the development and production, we propose to establish a high-resolution isochronous stratigraphic framework and to accurately characterize the single sand bodies by fine stratigraphic classification and correlation, thereby providing a reliable geological basis for adjusting the development mode and for tapping the potential of the remaining oil.

### 3.1.2 Sedimentary genesis as guidance

In China, the stratum with continental sedimentary genesis plays a dominant role (Yinan et al., 1997a,b). Sedimentary products and properties are dependent on their sedimentary environment, and they can be fully reflected in the strata. A preliminary understanding of stratum development pattern and distribution in space is indispensable for carrying out the stratigraphic classification and correlation of the area under study. For this purpose, it is necessary to identify the types of sedimentary facies or even sedimentary microfacies of the target interval. For an alluvial fan near the provenance, the thickness of the sedimentary stratum can vary greatly over a short distance due to the rapid deposition of source materials in the piedmont after rushing out of the mountain pass. The differential compaction during diagenesis in the later stage further aggravates the variation of stratum thickness. This should be considered in the stratigraphic classification and correlation to be carried out. For the fluvial deposits in the plain area where the terrain is flat, the sedimentary facies change slowly, the deposition is steady, and the stratum thickness varies slightly in space. Therefore the contour correlation and other means can be used in the stratigraphic classification and correlation being done of the region. The target interval in our study is classified as fan delta front deposits with frequent diversion of underwater distributary channel and the serious truncation and reworking to the early channel deposits by the later channel deposits, which causes great stratum variation in space. In addition to the yI top, yII bottom, and the mudstone between yI and yII, the rest mudstones show unstable distributions. The well-developed structural fault system and stratum breaking near the faults cause great challenges to be overcome before the stratigraphic classification and correlation of the region can be done. With the support of the classical fan delta front sedimentary model (Galloway and Hobday, 1983) (Fig. 3.2), we recognized the great stratum thickness variation in space. Thus the high-resolution sequence stratigraphy was adopted to establish a high-precision isochronous stratigraphic framework for the target interval.

### 3.1.3 Satisfaction to the production practice

For the stratigraphic classification and correlation based on high-resolution sequence stratigraphy of the region under study, the first step is to determine the scheme and hierarchy of stratigraphic classification and correlation given as dependent on the production practice. The ultimate goal of

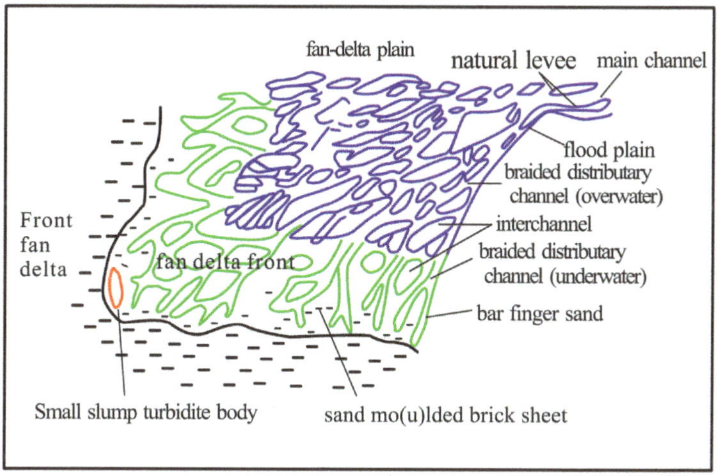


**Figure 3.2** Sedimentary model of fan delta (after Galloway and Hobday, 1983). *From Galloway, W.E., Hobday, D. K.,1983. Terrigenous Clastic Depositional Systems-Applications to Petroleum, Coal, and Uranium Exploration. New York: Spring-Verlag, New York, Inc.*

isochronous stratigraphic classification and correlation is to meet the requirements of practical production. The hierarchy of stratigraphic classification and correlation varies in different development stages of oil and gas fields. In the early development stage, the classification at the level of pay zone or layer is acceptable. In the middle–late development stage, and especially with the implementation of separated-layer water (or steam) flooding, plugging of dominant flow paths with water (or steam) channeling, and the remaining oil recovery, the classification at the level of bed corresponding to the single sand body becomes necessary. Thus the role of high-resolution sequence stratigraphy is gradually highlighted. To provide a geological basis for the fine sand body characterization and thereby support the development adjustment, we can mobilize the high-resolution sequence stratigraphy theory and techniques to divide the strata to the level of the fifth-order depositional cycle corresponding to the short-term base-level cycle, namely, the bed boundary. However, traditional stratigraphic classification and correlation methods do not work in this aspect. For the target interval in the study area, as the heavy oil thermal recovery by steam stimulation progresses into the middle–late stage, both production and pressure begin to decline. It is urgent to convert the development mode from steam stimulation to steam flooding to enhance the heavy oil recovery and to tap the potential of remaining oil. Therefore the layer-level classification currently being used in the oilfield should be broken down to the bed level (corresponding to the fifth-order depositional cycle or the fifth-order sequence) to characterize the single sand body. Specifically, the regional marker bed is firstly determined based on the seismic data; then, through seismic layer velocity calibration and synthetic seismogram making, a large-scale isochronous stratigraphic framework is established after verification and modification with logging–seismic data. Within the large-scale stratigraphic framework, depending on the well data, including log responses, core, depositional cycles, and fluid properties, the strata are divided by identifying different orders of base-level cycles on critical wells; then, extending to the whole area by virtue of skeleton section network, the high-precision isochronous stratigraphic framework of the whole area is established through fine stratigraphic classification and correlation at the level of bed (Fig. 3.3), which can provide a reliable geological basis for sand body characterization.

### 3.1.4 Principles for bed classification and correlation

The fine stratigraphic classification in this study is set to bed level. By definition, the "bed" is a relatively independent oil- or gas-storing sand bed, above and under which there are barriers and within which there is an isolated fluid flow unit (Yinan et al., 1997a,b). The scheme of fine

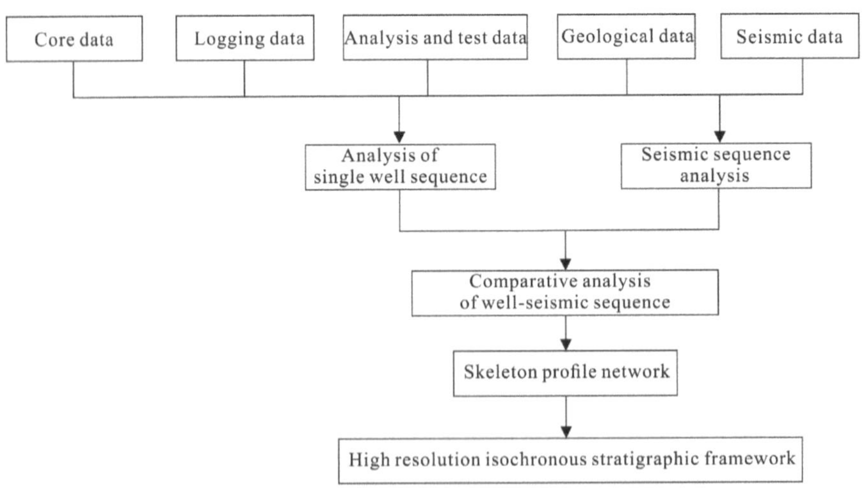


**Figure 3.3** Process of stratigraphic classification and correlation based on high-resolution sequence stratigraphy.

stratigraphic classification and correlation based on high-resolution sequence stratigraphy is also one of the cores. This book considers that the bed-level classification should meet the following principles:

1. Only one sand body is developed in more than 50% of the beds divided.
2. The sand thickness in a bed should not exceed the maximum thickness of the single-stage channel sand body to ensure that the vertically superimposed sand bodies can be separated.
3. The well pattern should be dense enough in the bed classification to ensure that the well spacing is close to or less than the single-channel width used to separate different single-stage channels.
4. The bed boundary generally corresponds to the maximum value of the conductivity log, indicating the position of the flooding surface.
5. The geological time of the bed is approximately 0.03 Ma.
6. The bed roughly corresponds to the short-term base-level cycle in the high-resolution sequence stratigraphy.
7. The bed basically corresponds to the fifth-order sequence in the classical sequence stratigraphy.
8. For the noncoring wells, most bed boundaries can be divided by referring to the short-term base-level cycle response model of critical wells (the division of short-term base-level cycle is feasible for noncoring wells).

It is worth mentioning that the bed classification is not a simple process. First, based on the log morphology, the depositional cycle, and other information, the primary bed classification scheme is determined. Then, the scheme is extended to the wells on the skeleton section network through critical wells and, then, to the skeleton section network to obtain the final scheme. Too rough a classification is incapable of development adjustment. Too fine a classification is difficult to operate on-site and random and also generates unnecessary workload. In short, the bed classification should be closely integrated with the reservoir development adjustment. Considering the production practice in the study area, we adopt the widely used classification system in this study, including long-term base-level cycle, medium-term base-level cycle, and short-term base-level cycle. The Yulou reservoir in the first member of Shahejie Formation of Paleogene roughly corresponds to 1.2 Ma (Rongcai et al., 1999). It is subdivided into two long-term base-level cycles, namely, yI and yII. These two long-term base-level cycles are subdivided into 12 medium-term base-level cycles ($yI_1^1$, $yI_1^2$, $yI_2^3$, $yI_2^4$, $yI_3^5$, $yI_3^6$, $yII_1^1$, $yII_1^2$, $yII_2^3$, $yII_2^4$, $yII_3^5$ and $yII_3^6$), which are further subdivided into 29 beds respectively ($yI_1^{1a}$, $yI_1^{1b}$, $yI_1^{2a}$, $yI_1^{2b}$, $yI_1^{2c}$, $yI_2^{3a}$, $yI_2^{3b}$, $yI_2^{3c}$, $yI_2^{4a}$, $yI_2^{4b}$,

$yI_2^{4c}$, $yI_3^{5a}$, $yI_3^{5b}$, $yI_3^{5c}$, $yI_3^{6a}$, $yI_3^{6b}$, $yI_3^{6c}$, $yII_1^{1a}$, $yII_1^{1b}$, $yII_1^{2a}$, $yII_1^{2b}$, $yII_1^{3a}$, $yII_1^{3b}$, $yII_2^{4a}$, $yII_2^{4b}$, $yII_3^{5a}$, $yII_3^{5b}$, $yII_3^{6a}$ and $yII_3^{6b}$), corresponding to 29 short-term base-level cycles.

### 3.1.5 Establishment of the large-scale isochronous stratigraphic framework with logging and seismic data

Log data and seismic data are the most fundamental and crucial data in stratigraphic classification and correlation of any region. The log data can provide relatively accurate information about vertical stratum development at the well site, and the seismic data can characterize the macroscopic stratum development. Due to limited log data, the research efforts related to oil and gas exploration pay more attention to seismic data. With the progress of exploration and development, the availability of more and more log data and the more immediate requirement for finer activities in oil and gas field development, the significance of log data has become increasingly prominent. The logging and seismic data should be fully combined to establish a high-precision isochronous stratigraphic framework based on high-resolution sequence stratigraphy for oil and gas fields in the middle—late development stage. In this process, seismic data play an extremely important role. The tracing and identification of regional marker bed on the seismic profile can help avoid the common "diachroneity" mistake during the fine stratigraphic classification and correlation being carried out for any region. In addition, the regional marker bed can be calibrated on the well through stratum velocity calibration or synthetic seismogram calibration so that the stratigraphic correlation is calibrated with logging and seismic data, given that the data calibration needs to act as certain constraints for the stratigraphic classification and correlation on the well (Fig. 3.4).

First of all, the regional marker bed is identified using seismic data. There are three sets of regional marker beds in the Yulou reservoir, that is, y_top, y_bottom, and yI—yII boundary. The y_top is a set of mudstone with a strong reflection on the seismic profile, which shows a distinct boundary with the basalt at the bottom of the overlying Dongying Formation. Both the y_bottom and the yI—yII boundary are mudstones in stable regional distribution, showing relatively strong reflection on the seismic profile, which is lower than that of the y_top. These three sets of regional marker beds can also be reflected on wells. The y_top (the bottom of Dongying Formation) is observed as a set of dark gray basalt with low induction, low interval transit time, and straight spontaneous

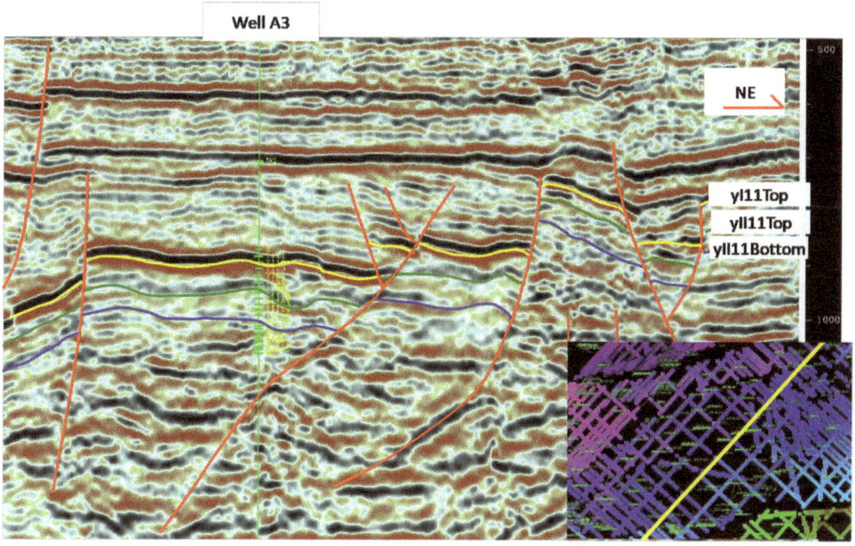


**Figure 3.4** Establishment of large-scale sequence stratigraphic framework with logging–seismic data in an area of Western Sag, the Liaohe Basin.

potential. The yI–yII boundary is a set of "bell-shaped" gray mudstone, which is characterized by low induction, high interval transit time, and straight spontaneous potential, with the induction and resistivity responses usually as bell-shaped. The y_bottom is a set of "funnel" mudstone, which serves as the boundary between the Yulou reservoir and the Xinglongtai reservoir and is characterized by ultra-low induction with funnel-shape, high interval transit time, and straight spontaneous potential. The marker bed identification can be extended to the well through seismic velocity calibration, and finally, the large-scale isochronous stratigraphic framework with logging–seismic data is established after the verification and modification of marker bed identified on the well. Under this framework, the target interval can be subdivided to bed level. In the logging–seismic sequence division, the logging-based results may be inconsistent with the seismic-based results, which is believed to have resulted from the differences in data accuracy and division basis (Huanqing et al., 2009). The seismic-based sequence division is mainly based on unconformity. The logging-based sequence division, limited by the accuracy of the fossil zone, is mainly based on the lithology and electric associations. However, there is often no evident change in lithoelectric properties atop and beneath this local unconformity, so the logging-based division by lithoelectric properties may usually miss the local unconformity. The seismic sequence is a chronostratigraphic unit, while the lithoelectric sequence is generally a lithoelectric unit, which shows incompatible properties. Therefore the seismic-based division should be the preferential treatment in case of unsolvable contradiction between two-division systems.

### 3.1.6 Combination of high-resolution sequence stratigraphy and traditional method for stratigraphic classification and correlation

It should be noted that the high-resolution sequence stratigraphy is not contradictory to the traditional "cyclic correlation, hierarchical control" method for stratigraphic classification and correlation. The former is an improved stratigraphic classification and correlation system on the basis of the latter, and it highlights the significance of stratigraphic genesis, thus providing a more reliable and rational stratigraphic classification and correlation. The marker bed correlation, depositional cycle pattern, logging response, reservoir fluid property variation, and other aspects in the traditional method can be closely integrated with the high-resolution sequence stratigraphy for mutual validation and supplementation so as to provide a finer stratigraphic classification and correlation (Figure 2—5, Fig. 3.6). For example, stable mudstones are developed widely at y_top, y-bottom, and yI—yII, which can be regarded as the regional marker beds. These marker beds provide a reliable foundation for the establishment of a large-scale isochronous stratigraphic framework and the fine stratigraphic classification and correlation based on high-resolution sequence stratigraphy. Similarly, reservoir fluid property variation can also present favorable references for the stratigraphic classification and correlation of noncoring wells. In general, fine stratigraphic classification and correlation in the development stage is considered as a multiinformation activity, and it requires the close combination of high-resolution sequence stratigraphy and the traditional method.

### 3.1.7 High-resolution sequence stratigraphy analysis

#### *3.1.7.1 Single-well high-resolution sequence stratigraphy analysis*

Single-well high-resolution sequence stratigraphy analysis is critical for bed classification and correlation (Fig. 3.5). Seismic data is incapable of bed classification due to the limitation of acquisition accuracy. The high-resolution sequence stratigraphy analysis is essentially the identification and division of different orders of base-level cycles, where the

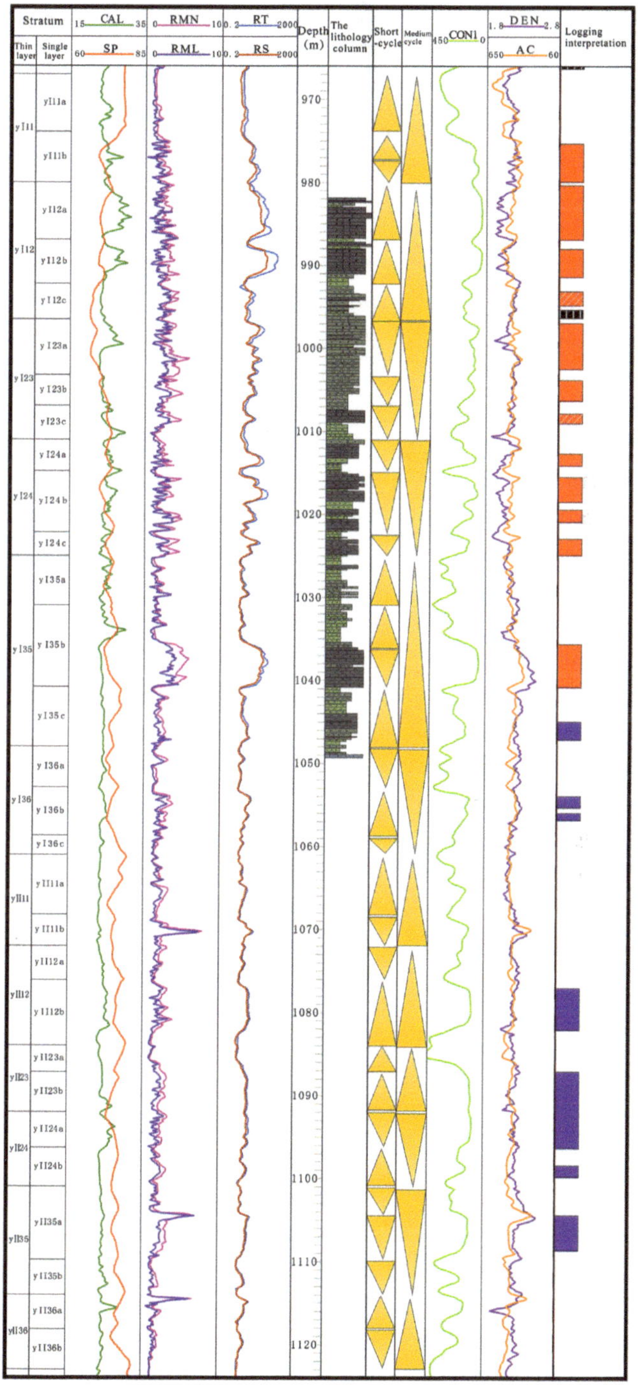

**Figure 3.5** Short-term base-level cycle response model of Well W2 in an area of Western Sag, the Liaohe Basin.

establishment of a base-level cycle response model for critical wells is a core. Currently, there are many classification schemes for base-level cycles (Hongwen et al., 2002; Rongcai, 2010). Of the nearly 400 wells in the study area, the conductivity log is mainly adopted to find the turnaround position of the short-term base-level cycle, namely, the position of the flooding surface. The base level is a potential energy surface that reflects the imbalance degree between the Earth's surface and the process to strive for a balanced surface. A genetic sequence is a stratigraphic unit with sediment aggradation/progradation during a base-level cycle with increasing or decreasing accommodation space. The semi-cycle boundary of a genetic sequence is usually located in the turnaround position with the base level rise to fall or fall to rise. The stratigraphic cycles resulting from all the base-level cycles, despite scales, are considered as the time stratigraphic unit. All the stratigraphic cycles are composed of the stratigraphic records that accumulate in an associated sedimentary environment during the base-level cycle variation. Due to the denudation resulting from the base-level cycle movement under the surface, the whole time of the base-level cycle consists of stratigraphic record (rock) and sedimentary discontinuity. The identification and division of multi-order base-level cycles are fundamental for the establishment of a high-resolution sequence stratigraphic framework (Rongcai, 2010).

In this study, what it means is to identify the flooding surface position through the conductivity log. In the Western Sag, the Liaohe Basin, due to the influence of radioactive materials in the parent rock of provenance, the natural gamma log that is sensitive to sand content is limited in the identification of sandy mudstone and the cycles resulted from sand content variation in this area; instead, induction log is usually the required logging sequence (Rongcai, 2010). Of the nearly 400 wells in the study area, the conductivity log is mainly adopted to find the turnaround position of the short-term base-level cycle, namely, the position of the flooding surface. In practice, the conductivity log is firstly used to identify the flooding surfaces at different orders (especially the turnaround position of the short-term base level, namely, the fifth-order sequence boundary). In addition, the spontaneous potential, resistivity, interval transit time, and density logs (as well as the patterns of different vertical cycle associations of core samples for the coring well) are referred to establish the response models of medium-term and short-term base-level cycles (Fig. 3.5) for critical wells, thereby to divide base-level cycles (especially short-term base-level cycle). Then the division is extended to the whole area through the skeleton section network, and finally, the bed classification and correlation is realized. The vertical cycle associations of core samples should also be considered for the short-term base-level cycle division on critical wells, so that the accuracy of base-level cycle identification can be ensured and the systematic errors of logs resulting from physical and human factors can be eliminated. The establishment of response models of base-level cycles at different orders for critical wells is crucial for the successful application of high-resolution sequence stratigraphy in the stratigraphic classification and correlation during oil and gas field development. To finely classify the strata based on high-resolution sequence stratigraphy, we should first determine a fine stratigraphic division scheme, with the short-term base-level cycle response model established for the critical well with coring data; then, the division is extended to the whole study area through the skeleton section network (Figs. 3.5 and 3.6).

### 3.1.7.2 Sectional high-resolution sequence stratigraphy analysis

The cross-section high-resolution sequence stratigraphy analysis mainly relies on the responses of spontaneous potential, resistivity, conductivity, and interval transit time logs, as well as the depositional cycle association pattern, stratum thickness, and properties of the oil, water, and other fluids. The fine stratigraphic classification for critical wells can be extended to the whole study area through the skeleton section network, and then the

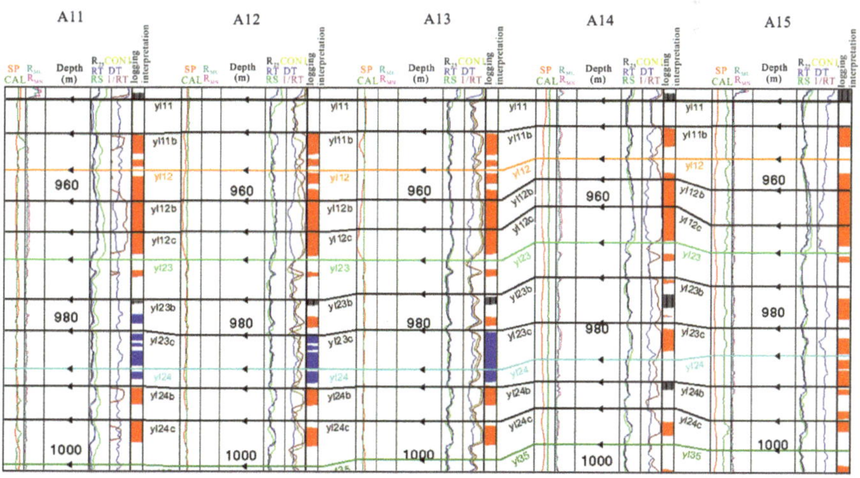


**Figure 3.6** Bed-level high-resolution sequence stratigraphy analysis in an area of Western Sag, the Liaohe Basin (the section position is shown in Fig. 3.1).

high-resolution isochronous stratigraphic framework for the whole study area is established (Fig. 3.6). The skeleton section can be selected in the directions: roughly parallel or perpendicular to the provenance. Critical wells should be above the skeleton section, and the skeleton section should bypass faults as far as possible to improve the stratigraphic correlation accuracy and avoid the influence of stratum leap resulting from a fault.

#### 3.1.7.3 Results of high-resolution sequence stratigraphy analysis

The Yulou reservoir is divided into 29 beds ($yI_1^{1a}$, $yI_1^{1b}$, $yI_1^{2a}$, $yI_1^{2b}$, $yI_1^{2a}$, $yI_2^{3a}$, $yI_2^{3b}$, $yI_2^{3c}$, $yI_2^{4a}$, $yI_2^{4b}$, $yI_2^{4c}$, $yI_3^{5a}$, $yI_3^{5b}$, $yI_3^{5c}$, $yI_3^{6a}$, $yI_3^{6b}$, $yI_3^{6c}$, $yII_1^{1a}$, $yII_1^{1b}$, $yII_1^{2a}$, $yII_1^{2b}$, $yII_1^{3a}$, $yII_1^{3b}$, $yII_2^{4a}$, $yII_2^{4b}$, $yII_3^{5a}$, $yII_3^{5b}$, $yII_3^{6a}$ and $yII_3^{6b}$), corresponding to 29 short-term base-level cycles (Fig. 3.6). According to the lateral distributions of $yI_2^{4a}$, $yI_2^{4b}$, $yI_2^{4c}$, $yI_3^{5a}$, $yI_3^{5b}$ and $yI_3^{5c}$ (Fig. 3.7), the areas with shifty bed thickness generally show a banded shape with NW-SE direction, which indicates the NW provenance direction for the target interval in this area. The rest beds show similar patterns. In addition, these stratum bands with shifty thickness are featured by multiple intersections and separations, which is consistent with the characterization of braided-river delta front deposit and frequent diversion of underwater distributary channel. The high-resolution sequence stratigraphy exhibits some

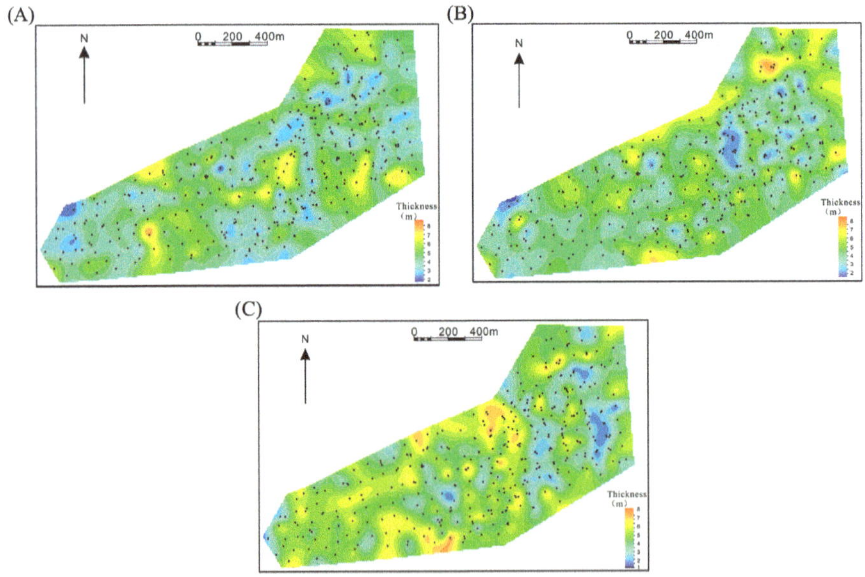


**Figure 3.7** Lateral distribution of stratigraphic thickness in an area of Western Sag, the Liaohe Basin. (A) $YI_2^{4a}$, (B) $YI_2^{4b}$, (C) $YI_2^{4c}$, (D) $YI_3^{5a}$, (E) $YI_3^{5b}$, (F) $YI_3^{5c}$

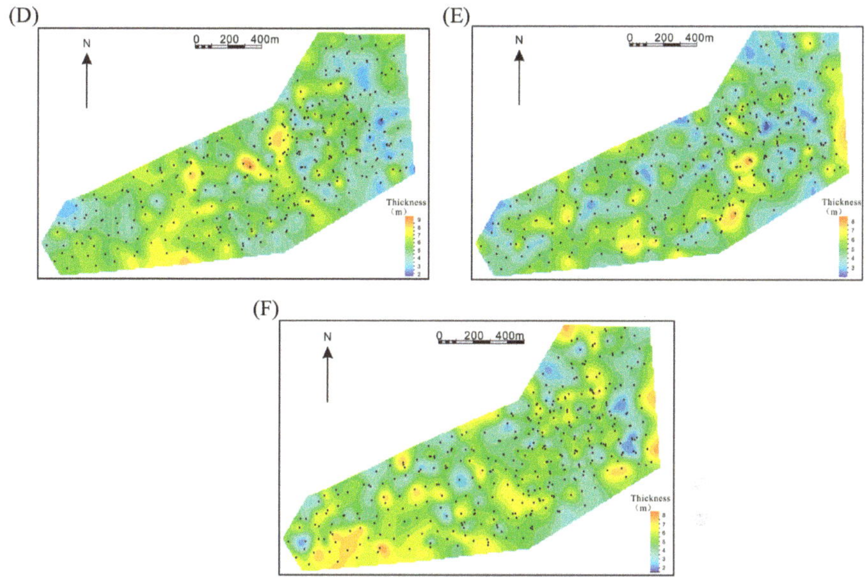


**Figure 3.7** (*Continued*)

unique advantages to the traditional method in establishing a high-precision isochronous sequence stratigraphic framework. It takes the stratum sedimentary genesis into account and determines the boundary based on the identification of base-level cycle with accumulation pattern. Moreover, it reflects the isochronous correlation principle in the stratigraphic correlation since the turnaround position of the base-level cycle, that is, the position at which the base-level cycle turns from fall to rise or from rise to fall can be regarded as the preferred position for time stratigraphic correlation.

#### *3.1.7.4 Significance of high-resolution sequence stratigraphy analysis for reservoir development*

Hydrocarbons are stored in the underground reservoir, and the single sand body is considered as the basic unit for hydrocarbon flow. Therefore the internal heterogeneity of sand body is the key geological factor for oil displacement performance (direction, lateral, and vertical sweep pattern, displacement efficiency, etc.) and accumulation and distribution of the remaining oil, and it is also one of the core contents of development geology (Shizhong et al., 2008). A single sand body refers to a geological sand body that is continuous in its lateral and vertical directions but separated from the overlying and underlying sand bodies by mudstone or impermeable baffles (Zhang et al., 2008). The bed-level high-resolution isochronous stratigraphic framework established by

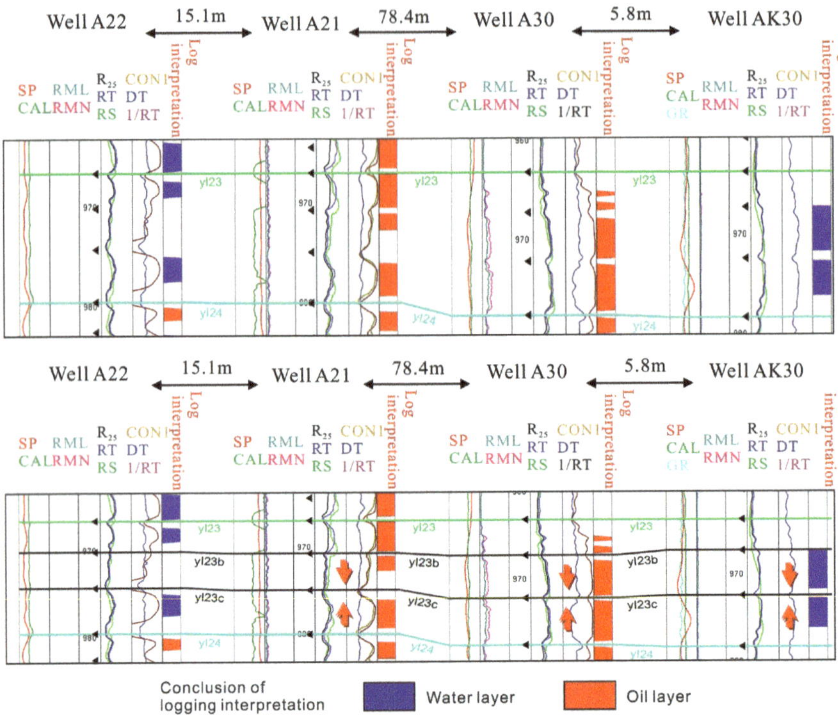


**Figure 3.8** Comparison between layer-level framework and bed-level framework for an area of Western Sag, the Liaohe Basin.

high-resolution sequence stratigraphy is of great practical significance for reservoir development adjustment. The sand body in the bed corresponds to a single sand body. According to the comparison between the layer-level framework and the bed-level framework (Fig. 3.8), it can be seen that the sand bodies superimposed in different sedimentary periods are divided in the bed-level framework, which can provide a reliable geological reference for single sand body fine characterization and reservoir development adjustment. The single sand body identification and characterization within the bed-level high-resolution isochronous stratigraphic framework are qualified for the fine stratigraphic classification and correlation in the middle—late development stage, greatly supporting the tapping of remaining oil and enhancing oil recovery.

## 3.2 Multiinformation fracture characterization of volcanic reservoir

A fracture refers to the discontinuous surface generated by rock breaking, and it is a result when the rock is broken under force. The fracture is an important pore space in hydrocarbon reservoirs (especially fractured reservoirs) and also a favorable flow path. Many large and extra-large oil and gas fields in the world are classified as fractured reservoirs. A systematic study of the fracture types, properties, patterns, and distribution is of great significance for the exploration and development of fractured reservoirs like the volcanic reservoir (Yuanyan et al., 2005). Through literature review, the following methods to study volcanic fractures can be concluded:

1. Conductivity logging, acoustic logging, radioactive logging, formation dip logging, and FMI imaging logging techniques are combined to identify and predict fractures (Pan et al., 2002; Yongjun et al., 2007a, b; Zhidi et al., 2008).
2. Comprehensive probability, finite deformation, and other mathematical methods are used to analyze the fracture development patterns (Junsheng et al., 2003; Xiaoyan et al., 2009).
3. The fracture analysis with flexural strength (the ratio of P-wave velocity) and other methods based on seismic data and attributes are used to study the fissure development level and patterns in volcanic rocks (Kaixuan et al., 2007; Fenglian et al., 2007).
4. The numerical simulation technique, together with reservoir geology, development, and production performance data, is used to predict fractures (Zhengming et al., 2010).
5. Key fracture characterization techniques are developed. Zhizhang et al. (1999a,b) proposed a set of fracture characterization techniques when describing the fractures in Huoshaoshan Oilfield in the Junggar Basin, which includes the field similar outcrop investigation and core observation, fracture prediction through numerical simulation of tectonic stress field, laboratory analysis and test, comprehensive analysis of sedimentary microfacies, critical well analysis and multi-well evaluation, reservoir dynamic analysis, drilling engineering analysis, multivariate statistical analysis, neural network simulation and prediction, fractal prediction, geostatistical prediction, geological analysis of fluid flow, and fractured reservoir modeling.
6. Systematic isothermal remanent magnetization and thermal demagnetization analysis is carried out to understand the fracture development (Fengqi et al., 2007).
7. The genetic mechanism of fractures is analyzed to obtain the relationship between fracture and reservoir and, thus, guide the exploration and development (Chunlin et al., 2004; Li et al., 2003). Huteau et al. (2001) carried out research on the long, narrow fractures in the Upper

Cretaceous Castillo tuff in the San Jorge Basin, Argentina, and found that the hydraulic fracturing performance is poor, which is mainly attributed to the premature water injection in the multifracture system, to the hydraulic fractures parallel to the original fractures, or to the poor connectivity between induced fractures and wellbore.

In this book, the fractures in target interval in the study area are characterized in detail by means of core fracture visual observation, microscopic fracture characterization with thin-section, imaging logging response analysis, and macroscopic fracture prediction with seismic data, and the fracture genesis is analyzed, in order to provide a geological basis for the efficient development of volcanic gas reservoir (Huanqing, 2016). Although the volcanic gas reservoir is exemplified, the fracture characterization technique is reproducible for oil reservoirs and can also be used as a reference for fracture research in fine reservoir characterization.

### 3.2.1 Geology of the volcanic gas reservoir in Ying 1 Member in Xudong area, the Songliao Basin

The Xushen Gas Field in the Songliao Basin is located in Daqing-Anda of Heilongjiang Province, and it's about 45 km long from north to south and about 10 km wide from east to west. Structurally, the gas field lies in the Xujiaweizi Faulted Depression in northern Songliao Basin, and it was formed in the Late Jurassic—Early Cretaceous period. The strata include (from bottom to top): Huoshiling Formation, Shahezi Formation, Yingcheng Formation, Denglouku Formation, and the first and second members of Quantou Formation. The frequent volcanic eruption activities led to a large number of volcanic products in the Yingcheng Formation. The volcanic reservoirs dominated by acidic eruptive rocks are mainly distributed in the first and third members of the Yingcheng Formation of the Lower Cretaceous. At present, there are 69 wells in this area, including 38 wells with commercial gas flow obtained, and the scale of gas reserves is confirmed to reach 100 billion cubic meters, of which 89.8% is endowed by volcanic gas reservoirs, recording as the major prospect of natural gas development in Daqing Oilfield (Wu et al., 2002; Yingnan and Yuqing, 2009). The study area, Xudong, is located in the central part of the Xujiaweizi Faulted Depression, where Well XS1 was drilled to discover the Xushen Gas Field. Xudong has become one of the most important gas prospects in Xushen Gas Field (Fig. 3.9). The analysis of fractures in Xudong is practically significant for the efficient development of volcanic gas reservoirs in Xushen Gas Field and also presents a

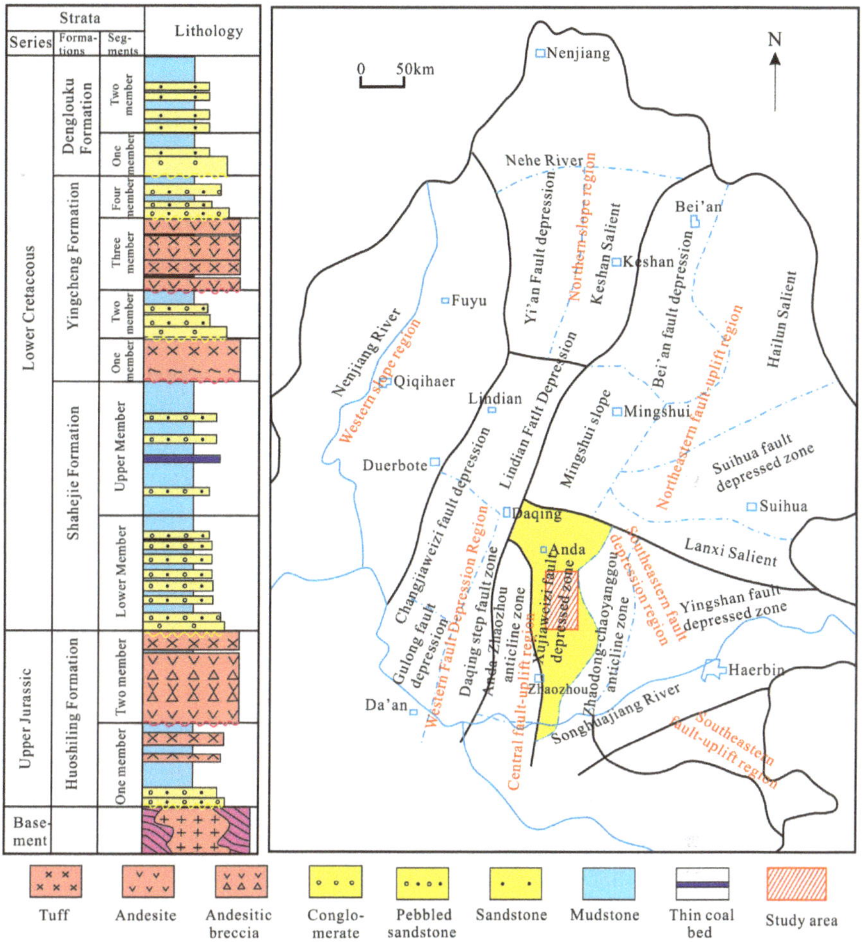


**Figure 3.9** Stratigraphic classification and geographical location of Xudong area of the Songliao Basin (after Heyong et al., 2002). *From Yingnan, W, Yuqing, X., 2009. Study on the characteristics of volcanic lithology, facies, and pore structure at the First Member of Yingcheng Formation in Xingcheng Area, Songliao Basin. China Petroleum Exploration, 1:24–29.*

reference for the development of volcanic reservoirs in the Songliao Basin and other basins in China. The target interval in the Xudong area is the volcanic rocks in Ying 1 Member of Cretaceous.

### 3.2.2 The approach of fracture characterization

Chen et al. (2011a,b,c) adopted drilling cores, thin-sections, FMI (whole-hole formation microresistivity scanning imaging logging), and conventional log data to identify fractures before fracture classification and genesis analysis. On this basis, the structural interpretation with seismic data and fault coherence cube analysis was applied to characterize the lateral development patterns of fractures in volcanic gas reservoirs. In addition, the basin structural history, burial diagenetic history, and other aspects of fault development were taken into consideration to identify the fault development history and its effects on fracture development. The fracture development period was defined to determine its evolution patterns and the spatial distribution of poro-permeability. The effects and contribution of different periods of fractures to the storage and preservation capacity of the gas reservoir were investigated, and the reservoir development geologic features of these fractures in different regions were analyzed so that the fracture evaluation could be ensured as more pertinent and referential for gas reservoir development. In this study, in addition to geological data, the seismic data were also integrated. With the support of the computer workstation and Petrel software, ant tracking of fractures was conducted to understand the development of fractures, deriving a finer characterization than the conventional fracture coherence cube analysis.

### 3.2.3 Classification of fractures in the target interval

According to the genesis, the fractures in the volcanic gas reservoir in Ying 1 Member in Xudong area, the Songliao Basin can be classified into: structural fracture, condensing shrinkage fracture, explosive fracture, dissolved fracture, stylolitic fracture, and weathering fracture (Table 3.1). Structural fractures refer to the fractures generated or associated with local tectonism, and they are mainly related to fault and fold. Their orientation, distribution, and formation are all related to the formation and growth of local structures. Structural fractures are usually oriented and exist in groups, with a long extension and deep cutting. They can connect other pores in spite of their small contribution to pore space, so they often serve as flow paths in volcanic reservoirs, which significantly improves the rock storage capacity. The structural fractures are generally 0.01–0.1 mm wide or narrower occasionally. Some wider structural fractures are partially or fully filled with calcite or quartz. The structural fractures in the target interval in the study area are mainly related to faults. Condensing shrinkage fractures are diagenetic fractures generated when the magma shrink during the condensation and solidification after they spill to the surface. They are mainly found in lava and pyroclastic lava (e.g., rhyolite and

**Table 3.1** Classification of fractures in a volcanic reservoir in Ying 1 Member in Xudong area, the Songliao Basin.

| Feature classification | Scale | Genesis | Development level | Common rocks | Data basis |
|---|---|---|---|---|---|
| Structural fractures | Large | Tectonism | Well | All rock types | Core photo, seismic data |
| Condensing shrinkage fractures | Small | Volcanic eruption, diagenesis | Local | Lava and pyroclastic lava | Microscopic thin-section |
| Explosive fractures | Small | Volcanic eruption | Local | Tuff lava, crystal tuff | Microscopic thin-section |
| Dissolved fractures | Small to medium | Diagenesis | Well | All lithologies | Microscopic thin-section |
| Stylolitic fractures | Small | Diagenesis | Poor | Tuff and volcanic breccia | Microscopic thin-section |
| Weathering fractures | Unequal | Diagenesis | Poor | Volcanic breccia | Core photo |

breccia lava) and sometimes in common pyroclastic rocks. Explosive fractures are formed by detrital materials carried by magma in the process of explosion with a volcano's eruption. The explosive fractures are common in various volcanic rocks, especially tuff lava and crystal tuff. Dissolved fractures are derived from preexisting fractures due to dissolution. The matrix in the rhyolite volcanic breccia is corroded to form reticular fractures, and the intergranular space of volcanic breccia is dissolved to form secondary fractures. These fractures are wide and effective but less in number. It is found that only 10% of the structural fractures have been dissolved. Stylolitic fractures are unique for their jagged shapes. The stylolitic fractures in the target interval in the study area often cut the plaque and matrix of lava or the volcanic debris in volcanoclastic rocks. These fractures are mostly filled or partially filled with iron or mud but rarely unfilled. The stylolitic fractures are occasionally seen in tuff and volcanic breccia and have not been observed in other types of volcanic rocks. Weathering fractures are associated with various mechanical and chemical weathering processes (e.g., freezing-thawing cycle, small-scale rock disintegration, mineral alteration, and diagenesis) and downwasting on or near the surface. Condensing shrinkage fractures can also be observed in mudstone (Yuanyan et al., 2005), while explosive fractures are unique in volcanic rocks. Other fractures can be seen in clastic rocks and carbonate rocks. The fracture classification in this book considers the features of volcanic rocks, so it is more comprehensive and integral than other general genesis-based classifications.

## 3.2.4 Multiinformation fracture characterization of volcanic reservoir

### 3.2.4.1 Outcrop

To generally and intuitively understand the presence of fractures in the target interval in the study area, the data of some outcrops with fractures in the Yingcheng Formation of the Songliao Basin were acquired (Fig. 3.10) (Pujun et al., 2008). These outcrops can intuitively exhibit the fracture scale, density, and orientation in the Yingcheng volcanic reservoir in the Songliao Basin, thereby providing a reference for analyzing the fractures in a volcanic reservoir in the target interval.

### 3.2.4.2 Core

Core observation is a direct means to understand reservoir fractures. The cores, taken from a volcanic reservoir in Xudong area of the Songliao Basin as examples, exhibit extensive vertical structural fractures and high-angle structural fractures, which vary in presence levels from region to region and from lithology to lithology (Fig. 3.11). The fractures observed in the core samples are usually relatively large, with extensions greater

(A) (B)

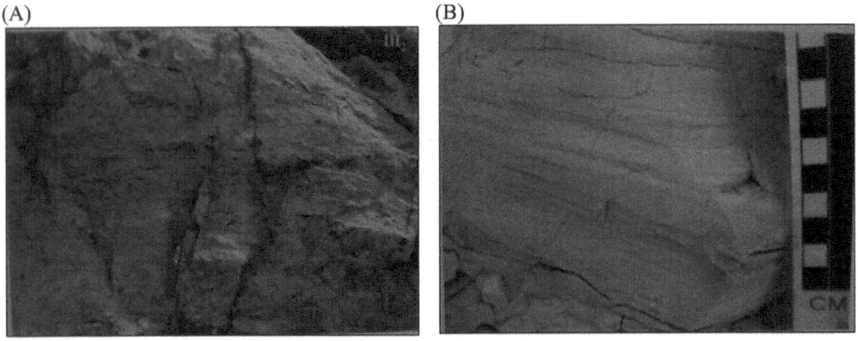


**Figure 3.10** Outcrops with fractures in the Yingcheng volcanic reservoirs, the Songliao Basin. (A) Rhyolite, with vertical structural fractures; (B) Rhyolite tuff, with fracture network, Shanghewan of Jiutai, Jilin Province. *From Pujun, W., Zhiqiang, F., et al., 2008. Basin Volcanic Lithology, Lithofacies, Reservoir, Gas Reservoir, Exploration. Beijing: Science Press.*

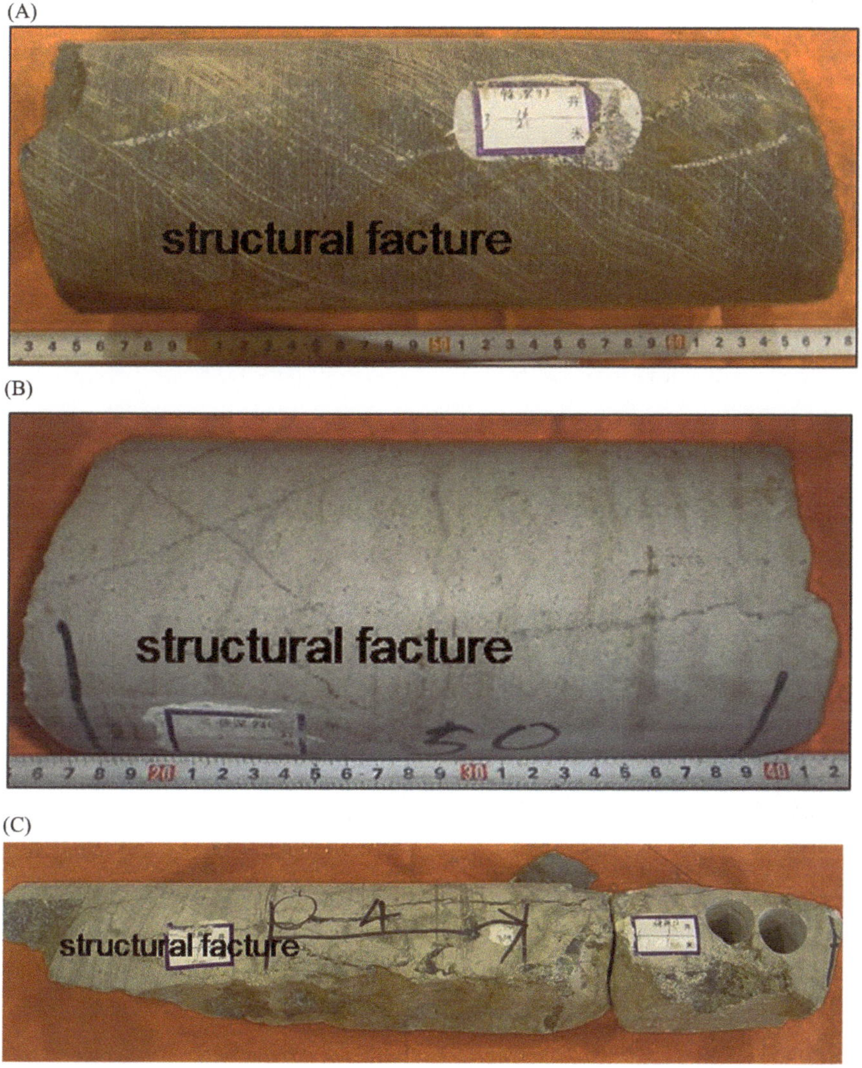


**Figure 3.11** Fractures in a core taken from a volcanic reservoir in Ying 1 Member in Xudong area, the Songliao Basin. (A) High-angle structural fractures filled with calcite vein, Well XS17, 3649.89–3650.15 m; (B) High-angle structural fractures filled with calcite vein, Well XS231, 3760–3760.23 m; (C) Vertical structural fractures, Well XS12, 3730.06–3730.57 m; (D) High-angle structural fractures and vertical structural fractures, Well XS21, 3656.24–3656.58 m; (E) High-angle structural fracture network, Well XS12, 3667.83–3668.24 m.

(D)

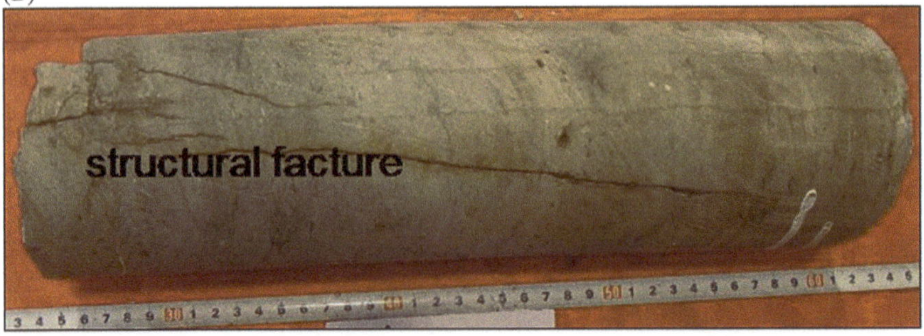


(E)

**Figure 3.11** (*Continued*)

than 0.5 m. Locally, multiple small fractures (with an extension less than 0.1 m) create a three-dimensional fracture network.

The fracture development periods can be qualitatively divided according to the cross-cutting relationship and filling of structural fractures in the cores. The structural fractures in the study area are divided into three periods: (1) low-angle structural fractures cut by high-angle structural fractures and filled, with the early structural fractures usually cut by the late structural fractures; (2) high-angle structural fractures cutting through low-angle structural fractures and filled, which are filled with calcite vein or intruded by magma, so that their development period is generally earlier than that of open or semi-open high-angle structural fractures; and (3) open high-angle structural fractures cutting through low-angle structural fractures. In general, the early fractures are filled with calcite vein, with large width (>0.01 m) and long extension; the late fractures are usually in an open state, with long extension (>0.5 m) generally, and about 1/3 of these fractures are concentrated and small in scale to create fracture network. Analysis shows that the fractures in the latter two periods have the most influence on reservoir properties.

**Figure 3.12** Microscopic thin-section showing the fractures in a volcanic reservoir in Ying 1 Member, Xudong area, the Songliao Basin. (A) Crystal explosive fractures, Well XS42, 3702.06 m, ×10, (-); (B) Dissolved fractures, Well XS21, 3658.66 m, ×10, (-); (C) Dissolved fractures partially filled with quartz, Well XS23, 3899.47 m, ×40, (-); (D) Fractures filled with chlorite, Well XS12, 3731.47 m, ×10, (-); (E) Characteristics of fractures in different periods, Well XS12, 3611.07 m, ×10, (-) (Numbers 1, 2 and 3 represent fractures formed in different geological historical periods with chronological order.); (F) Condensing shrinkage fractures in volcanic breccia, Well XS401, 4176.48 m, ×10, (-); (G) Stylolitic fractures, Well XS12, 3731.47 m, ×40, (-); (H) Connected fracture network, Well XS21, 3731.17 m, ×40, (-).

### *3.2.4.3 Microscopic thin section*

Microscopic thin-section observation is one of the most direct means to understand fractures, and it can allow the researchers to thoroughly recognize the microscopic development of fractures (Fig. 3.12). The fractures in the target interval in the study area vary in geometry and scale. There are large-scale fractures with a long extension and also small-scale fractures resulting from late diagenesis. Locally, some well-developed fractures can create a three-dimensional fracture network (Fig. 3.12B), which can serve as a favorable pathway for hydrocarbon migration and transportation but

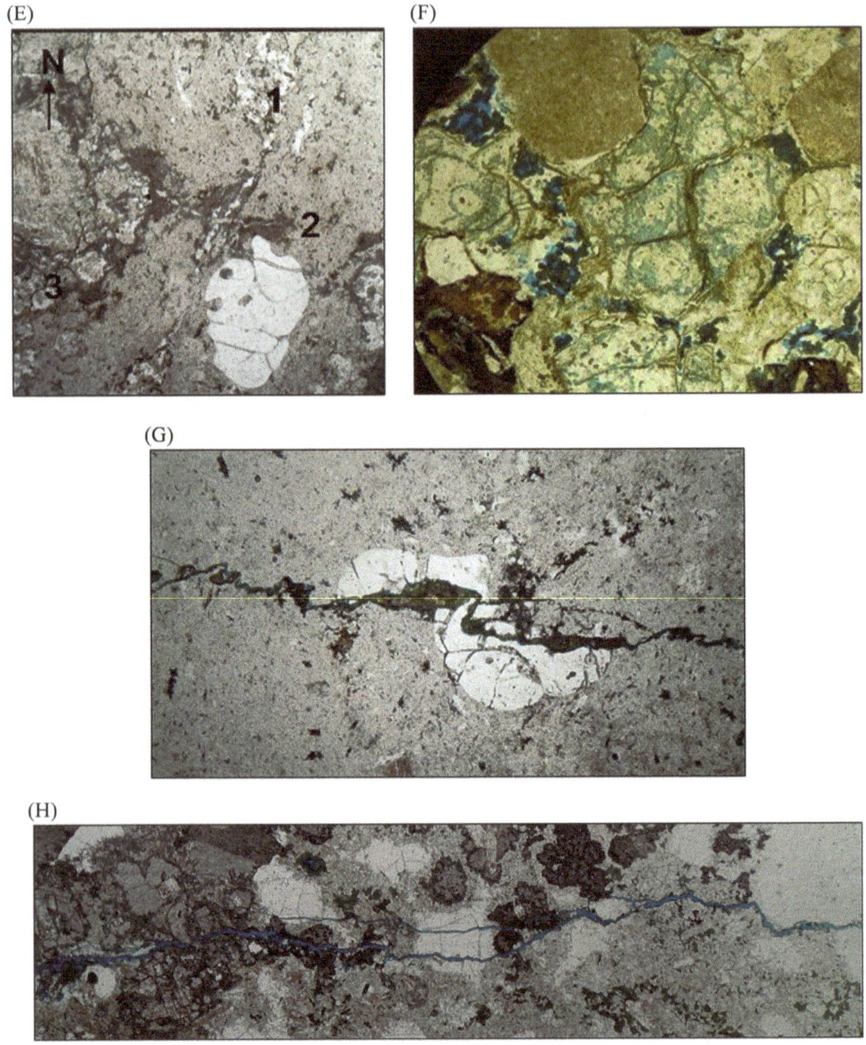


**Figure 3.12** *(Continued)*

may lead to water channeling, which will seriously influence the development performance.

Explosive fractures (Fig. 3.12A), dissolved fractures (Fig. 3.12B and C), structural fractures (Fig. 3.12D and E), condensing shrinkage fractures (Fig. 3.12F), and stylolitic fractures (Fig. 3.12G) show distinct characteristics on thin-sections (Table 3.1). Due to the limitations of data and observation scale, dissolved fractures are the most common ones under the microscope, followed by explosive fractures, while other types of fractures are rarely observed. The fracture development periods can also be judged through thin-section study (Fig. 3.12E), which shows three periods (the orientations of fractures in different periods can refer to the orientation planar graph), agreeing with the core observation.

### 3.2.4.4 Logging

There are many logging techniques available for fracture characterization. The FMI imaging logging acquires the borehole wall image by scanning the well axis with the current beam and acoustic beam (Zhizhang et al., 1999a,b), which is effective in reservoir fracture characterization. In this study, in addition to the conventional logging interpretation which is used to analyze the relationship between fractures and reservoir lithofacies, the imaging logging and core correlation were combined to finely characterize the fractures in the target interval (Fig. 3.13). The vertical, high-angle, low-angle, and horizontal fractures can be evidently reflected on the logging images, with the third-period fractures in predominance. Moreover, many microfractures are intuitively observed on the FMI images, showing width of 1–40 μm.

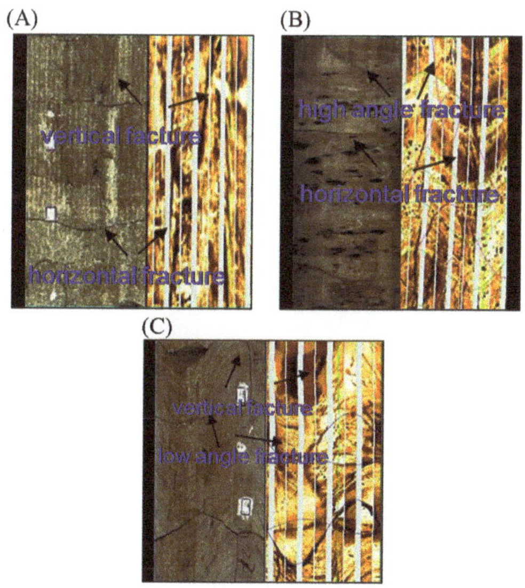


**Figure 3.13** Imaging logging responses of fractures in a volcanic reservoir in Ying 1 Member in Xudong area, the Songliao Basin. (A) Horizontal and vertical fractures, Well XS13, 4130–4131.2 m; (B) High-angle and horizontal fractures, Well XS14, 4207.7–4209.3 m; (C) Low-angle and vertical fractures, Well XS21, 3732.4–3734.4 m.

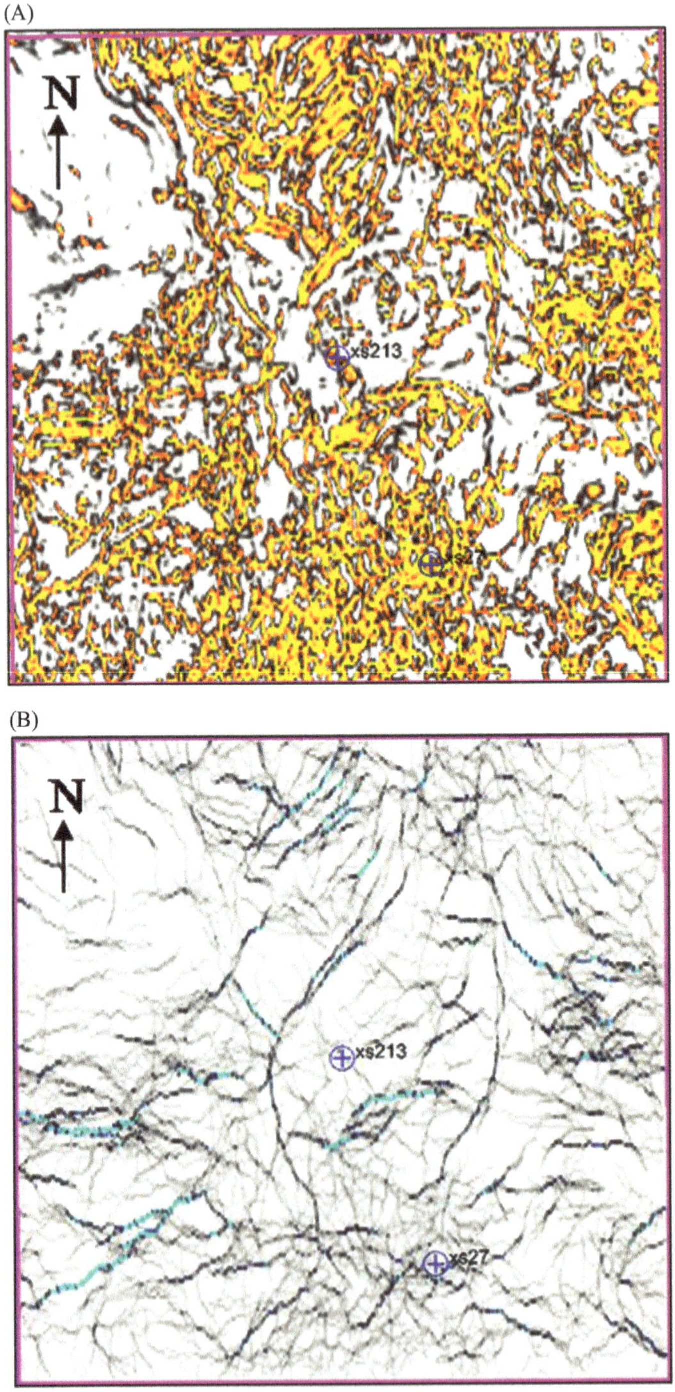


**Figure 3.14** Seismic characterization of faults in Ying 1 Member, Well XS27, Xudong area, the Songliao Basin. (A) Lateral fault by seismic coherence analysis; (B) Lateral fault by seismic ant tracking. Note: Dark lines are the lateral development positions of faults, and the location of Well XS27 is shown in Fig. 3.15.

#### 3.2.4.5 Seismic data

To finely characterize the lateral fractures, we carried out the fracture coherence analysis (especially in the well-free area). Fig. 3.14A illustrates the fracture development in Well XS27 in the Xudong area according to the coherence cube. It can be seen that the fractures are most developed in the northeastern and southwestern parts of the well site but less developed in the central part. Fig. 3.12B illustrates the results of ant tracking with Petrel software based on fracture coherence analysis. It can be concluded that the microfractures mainly develop along large-scale faults and varies in density from region to region. Comparison of two views in Fig. 3.14 indicates that the ant tracking provides a finer fracture characterization than the coherence analysis. The ant tracking can fully excavate the information in the seismic data volume by using a characteristic algorithm and displaying it on the plane. Thus it provides a very effective means for fine characterization of fractures in well-free areas.

Ant tracking is a new function of Petrel. It can finely characterize large-scale faults but also small-scale faults and fractures. It is found that the fractures are macroscopically most developed along the Xuzhong and Xudong faults and roughly strike in nearly the NS direction. In the northern and central parts of the study area, fractures are large in size but small in number, with the concentrated distribution. In the southern part, fractures are small in size but large in number, with the dispersed distribution. In view of the position, the faults are mainly developed in the western half of the northern and central Xudong area and across the southern Xudong area (Fig. 3.15). Macroscopically, the fractures develop in consistency with faults and mostly near the faults. Microscopically, there are many EW or nearly EW fractures in addition to the NS fractures near the fault zones. The former is connected with the latter, but the latter extends shorter laterally than the former. All these fractures in various directions constitute the three-dimensional fracture network in the Xudong area, which greatly affects the reservoir properties of the target interval in this area.

### 3.2.5 Fracture genesis and influencing factors

From the geological perspective, fractures are formed under various geological processes, such as local tectonism, regional stress, diagenetic contraction, unloading, weathering, and even sedimentation, and controlled by different factors in different regions (Yuanyan et al., 2005). It is found that the fractures in a volcanic reservoir in Ying 1 Member in the

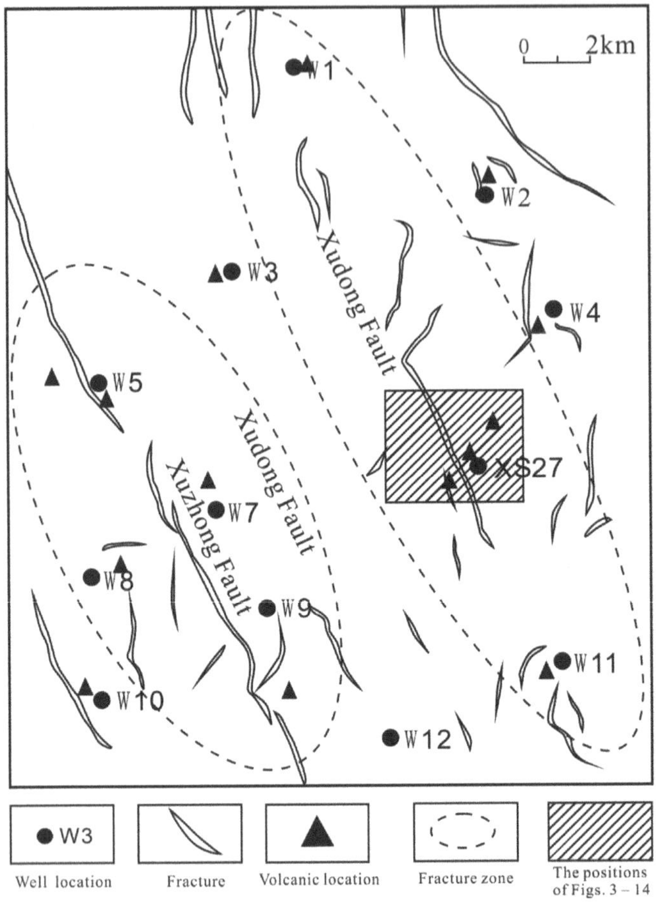


**Figure 3.15** Lateral faults in a volcanic reservoir in Ying 1 Member in Xudong area, the Songliao Basin.

Xudong area, the Songliao Basin, were formed under the control of tectonism, volcanic lithology, volcanic rock body, volcanic facies, diagenesis, and other factors. According to the macroscopic fracture distribution in space and the microscopic thin-section data, the dominant factors include tectonism and diagenesis, and the secondary factors include volcanic lithology, volcanic facies, and volcanic rock body.

### *3.2.5.1 Tectonism*

Tectonism is one of the main controlling factors for volcanic reservoir development, so is it for fractures. Tectonism mainly controls the fracture creation and development through the faulting activity, and the paleotopography resulted from tectonic uplifting and compaction, with the faulting activity in dominance. In Xushen Gas Field, the faulting activity occurred in stages. Accordingly, three fault systems can be derived by stages, including early fault (Huoshanling Formation—Shahezi Formation sedimentary period), inherited fault (Huoshanling Formation—Yingcheng Formation sedimentary period), and late fault (after Denglouku Formation sedimentary period). In space, there are two major fault systems: Xuzhong and Xudong (Fig. 3.15). Previous theoretical studies and field observations (Yuanyan et al., 2005) indicate that fault and fracture possess the consistent formation mechanism, and a fracture is the prototype of a fault. Generally, there are always fractures associated in the vicinity of the fault, and both fracture and fault are developed in a consistent stress field.

Laterally, fault and fracture are well-developed near the Xuzhong and Xudong fault systems and less developed far away from the two systems. Vertically, fault and fracture are better developed in the first member of the Yingcheng Formation than that in the fourth member. In Xudong, both volcanic activity and tectonism exist, and there are multiple faulted anticlines and faulted blocks. The present-day structure is generally low in the central part and high in the periphery. As the major part, the Xudong slope zone is high in the east and low in the west. This kind of topography controls the volcanic rock body distribution and, thereby, the fracture development.

The fissure—central eruption dominantly occurred in the target interval. Thus the fractures are better developed in the high structural position near the crater than in the low structural position far away from the crater. In this study, the crater was identified using seismic data. Usually, the crater is reflected as an inverted-cone shape on a seismic profile, with chaotic reflection in the cone. In addition, the lithoelectric logging responses of the well penetrating the crater (if any) can be considered. The lateral fault development map (Fig. 3.15) illustrates that the crater is mostly located in the region with faults developed, where fractures are also observed. It is worth mentioning that Figs. 3.15 and 3.14 seem to be somewhat inconsistent because the latter with higher precision is a partial view of the former.

#### *3.2.5.2 Volcanic lithology*

The volcanic reservoir in Ying 1 Member, Xudong area, is formed as a result of multiple phases of volcanic eruptions and contains diverse

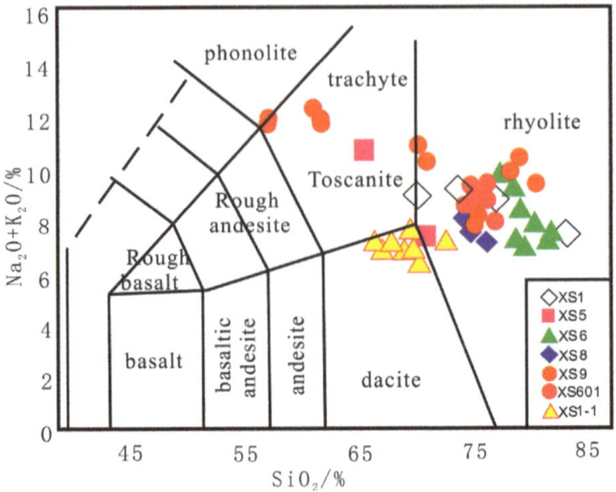


**Figure 3.16** Total alkali-silica map of volcanic rock in Xudong area of the Songliao Basin (TAS chart). *From Yongjun, W., Lin, Y., Qiquan, R., et al., 2007a. Lithologic identification of deep volcanic rock gas reservoir in Xingcheng gas field. Journal of Southwest Petroleum University 29(2): 78–81. Yongjun, W., Yongle, H., Qiquan, R., et al., 2007b. Fracture development evaluation for deep volcanic gas reservoir. Natural Gas Industry 27(8): 31–34.*

rock types. Test data, TAS (total alkali silica) chart, and microscopic observation (Fig. 3.16) demonstrate that there are two rock types (volcanic lava and pyroclastic rock) with 17 lithologies in the coring interval. Volcanic lava varies in a wide range, including acid rock, intermediate-acid rock, intermediate rock, and intermediate-basic rock, with acid rock in dominance. The rock types include breccia lava, tuff lava, welded breccia, basaltic breccia lava, rhyolite, tuff, tuffite, volcanic breccia, basalt, ignimbrite, sedimentary volcanic breccia, conglomerate, crystal tuff and basaltic volcanic breccia, tuff breccia, siltstone, fine sandstone, etc. The most developed types are rhyolite, breccia lava, and tuff lava.

From the lithological perspective, rhyolite, ignimbrite, and tuff with dense lithology contain well-developed fractures, while the rest of the volcanic rock types have fewer fractures. Dissolved fractures are most developed in volcanic breccia, welded breccia, and breccia lava. The target interval is dominated by acid rocks, such as rhyolite, ignimbrite, and crystal tuff, which are mostly continuous, so the structural fractures therein usually extend long (>0.5 m). This has been demonstrated from fractures in the core observation.

### 3.2.5.3 Volcanic rock body

There are multiple volcanic rock bodies in Ying 1 Member, Xudong area. These volcanic rock bodies of different eruption cycles are superimposed in space to comprise the target volcanic interval. Along the interfaces of these volcanic rock bodies, there are often low-angle or horizontal fractures. The scales and superimposition patterns of volcanic rock bodies partly control the fracture angle and extension. Volcanic rock bodies can be preliminarily traced on the seismic profile by the crater identification and by the continuity and strength of seismic events of volcanic rock bodies of different phases on the seismic profile. In addition, the lithoelectric logging responses of volcanic rock bodies in the well penetrating the rock bodies (if any) can be considered. The seismic profile illustrates that high-angle fractures with large vertical extension are usually developed near the interface of volcanic rock body near the crater, while horizontal and low-angle fractures with relatively large lateral extension are developed on the boundary of volcanic rock body far away from the crater, and where these fractures show banded shape with chaotic internal reflections on seismic profile; fractures can also be observed near the faults with complete disconnection of the seismic event (Fig. 3.17). During the identification of fractures with seismic profiles, the above characteristics should be repeatedly compared and validated on multiple adjacent or perpendicular intersecting profiles. The fracture scale and angle are also partly dependent on the topographic condition. According to the lateral fracture distribution, the fractures mainly exist in the region near the crater within the volcanic rock body, and their presence gradually weakens with the increase of distance to the crater.

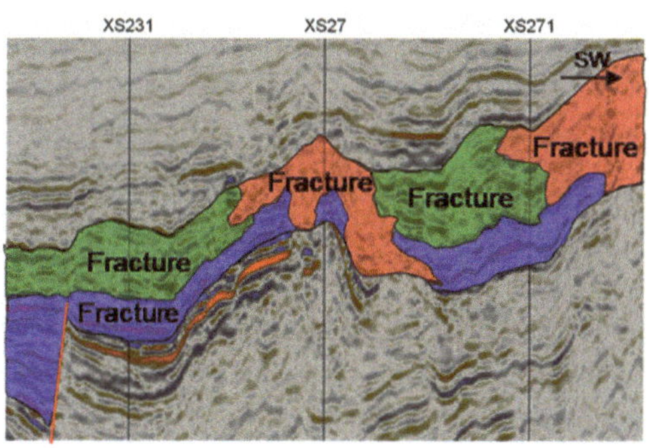


**Figure 3.17** Logging–seismic identification of volcanic rock bodies in Ying 1 Member in Xudong area, the Songliao Basin. *Note*: Different colors represent different volcanic rock bodies.

**Table 3.2** Classification of volcanic facies in Ying 1 Member in Xudong area, the Songliao Basin.

| Facies | Thickness proportion in single well (%) | Subfacies |
|---|---|---|
| Volcanic conduit facies | 4.55 | Volcanic neck subfacies, sub-volcanic subfacies, cryptoexplosion breccia subfacies |
| Extrusive facies | 1.35 | Inner-zone subfacies, meso-zone subfacies, outer-zone subfacies |
| Explosive facies | 44.71 | Splashing subfacies, hot debris flow subfacies, thermal base-surge subfacies, fallout subfacies |
| Effusive facies | 43.40 | Top subfacies, upper subfacies, middle subfacies, lower subfacies |
| Volcanogenic sedimentary facies | 5.99 | External debris-bearing subfacies, rehandling subfacies |

### 3.2.5.4 *Volcanic facies*

The volcanic rocks in the target interval can be divided into five volcanic facies and 16 subfacies (Table 3.2). The volcanic conduit facies exhibits high value and jagged shape on logs, with a thickness of around 30 m. The extrusive facies usually develops near the volcanic conduit facies and shows medium to low logging value, with a thickness of less than 20 m. The explosive facies presents jagged shapes on logs with medium to low resistivity. The effusive facies is characterized by the thick layer, slightly jagged shape, and medium to high resistivity. The volcanogenic sedimentary facies usually shows rhythmic log pattern with unequal thickness. According to the relationship between the fracture porosity and permeability from conventional logging interpretation and the volcanic facies, the volcanogenic sedimentary facies shows the best fracture properties while the extrusive facies indicates the worst fracture properties; the rest of volcanic facies have fracture properties between the former two facies (Fig. 3.18). The volcanogenic sedimentary facies is generally composed of various volcanic clastic rocks with poorly cemented grains, leading to the probability of fracture development. According to the lateral superimposition of faults and fractures and the comparison of distributions between crater and faults, the fissure—central eruption is dominant in the study area. Therefore fractures are relatively developed in the regions with volcanic conduit facies and

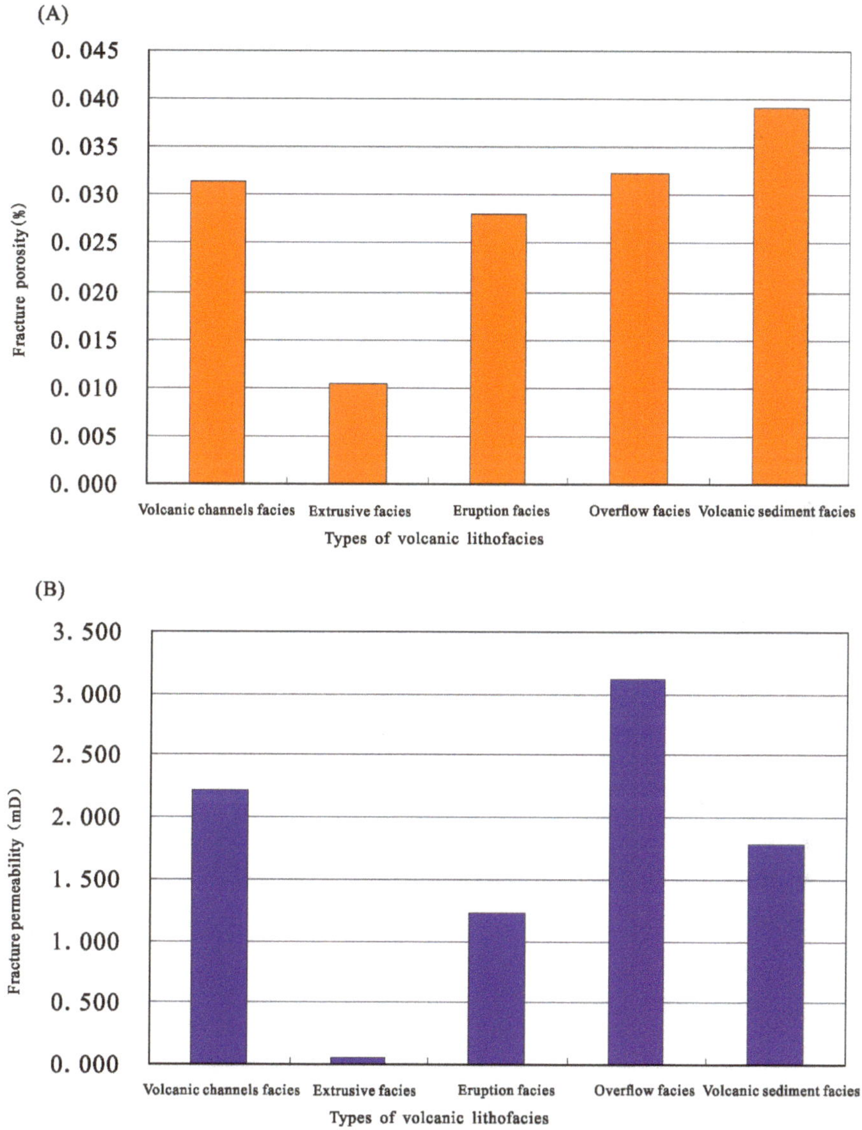


**Figure 3.18** Relationship between reservoir fracture properties and volcanic facies based on single-well logging interpretation in Ying 1 Member in Xudong area, the Songliao Basin. (A) Relationship between porosity and volcanic facies; (B) Relationship between permeability and volcanic facies.

explosive facies but less developed in the regions with effusive facies and volcanogenic sedimentary facies far away from the crater. Lithology is the major factor leading to a higher fracture development level of volcanogenic sedimentary facies than that of effusive facies. The fracture development index is a parameter to reflect the fracture development level by fracture storage-permeability properties, which is defined as (Jianchao et al., 2001):

$$F = 100 f_f K_f h$$

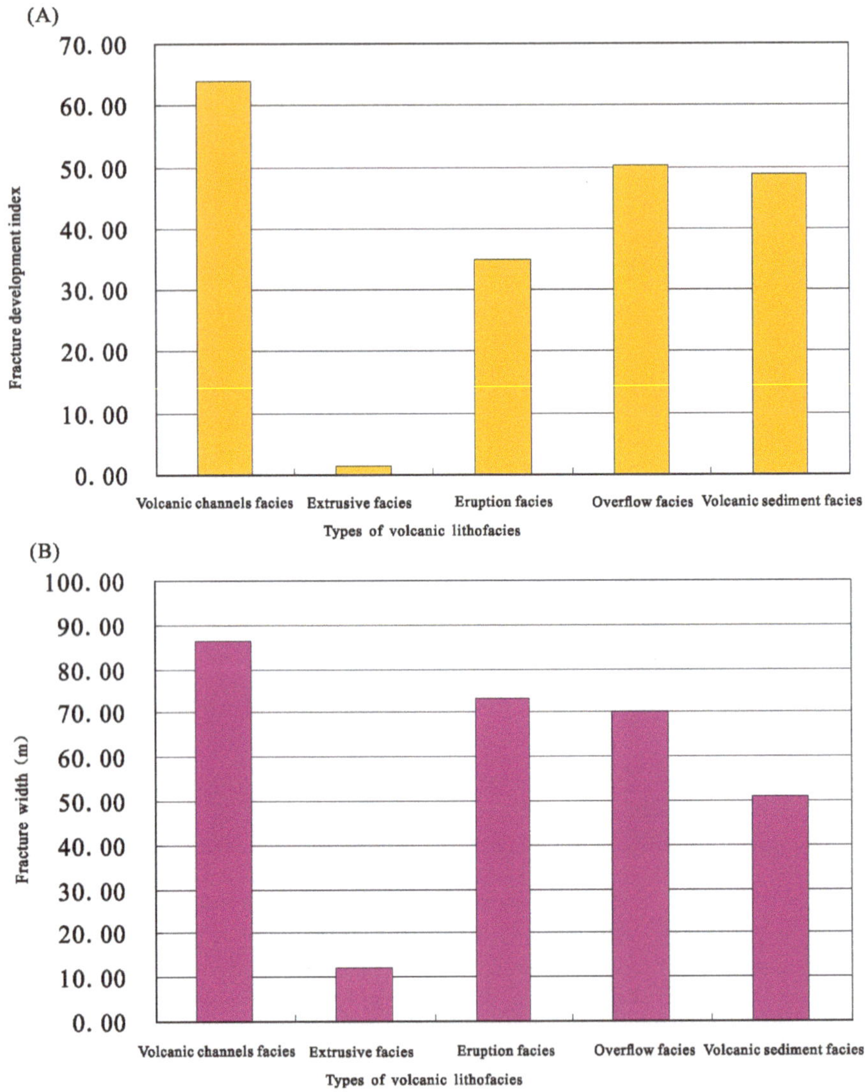


**Figure 3.19** Relationship between the reservoir fracture development level and volcanic facies in Ying 1 Member in Xudong area, the Songliao Basin. (A) Relationship between fracture development index and volcanic facies; (B) Relationship between fracture width and volcanic facies.

where $\phi_f$ is the fracture porosity, %; $K_f$ is the fracture permeability, mD; h is the reservoir thickness, m.

According to the histograms (Fig. 3.19) of fracture development index and width, volcanic conduit facies shows the most developed fractures in all types of volcanic facies. The fracture development index can be obtained by the fracture porosity and fracture permeability, and the fracture width can be determined through FMI imaging logging. The equation to calculate the fracture width is written as:

$$\varepsilon = aAR_{xo}^{b}R_{m}^{1-b}$$

where *a* and *b* are the constants associated with the device, and b approaches zero; A is the abnormal conductivity area resulting from fracture, $mm^2$; $R_{xo}$ and $R_m$ are the resistivity of intrusion zone and drilling fluid, $\Omega \cdot m$. These data can be acquired from log data for the purpose of calculating the fracture width.

### *3.2.5.5 Diagenesis*

From the perspective of reservoir characterization, the diagenesis of volcanic rock is defined as the sum of various processes experienced before the molten lava (the product of volcanic eruption) and/or pyroclastic materials transit into rocks until growing to metamorphic rocks or weathering rocks. The effects of diagenesis on fracture development can be obtained by the observation of microscopic thin-section. Such effects can be analyzed from the two aspects of diagenesis: positive processes and negative processes. The positive processes include condensing/cooling shrinkage, dissolution, and weathering; and the negative processes include compaction and filling (Fig. 3.12). Condensing/cooling shrinkage refers to the shrinkage of volcanic eruptive materials due to condensation/cooling. Dissolution is classified into four categories: partial intragranular dissolution (dissolution inside crystal fragment, dissolution inside volcanic breccia debris), complete particle dissolution (to form moldic pores, such as feldspar phenocryst or clastic dissolved moldic pores), matrix dissolution of pyromeride, and intergranular dissolution. Intergranular dissolution has greater effects on fracture development in the target interval than intragranular dissolution. Weathering is the process that the weathering fractures are created in rocks due to various mechanical or chemical reworking actions when the rocks are exposed to the surface or at the surface, which can improve the reservoir properties. Weathering has less effect on the fracture development in the target interval than condensing/cooling shrinkage and dissolution. Compaction refers to the process that the pyroclastic materials form rocks after compaction and consolidation. Microscopic observation shows that filling in the target interval includes albite filling, autogenous quartz filling, chlorite filling, carbonate filling, etc., of which the chlorite filling is considered as causing the greatest damage to fractures. Compared to filling, compaction has much less effect on fractures. Dissolution and filling are the most common processes.

The above genetic factors are believed to have interacted in the process by which the fractures were formed in the volcanic reservoir in Ying 1 Member in the Xudong area, the Songliao Basin. For example, the zone with structural fractures, where the rocks are fragmental and unconsolidated, is susceptible to dissolution to form dissolved fractures (Fig. 3.12E). Sometimes, a fracture might be generated under the control of several factors with different contributions. Fig. 3.12A illustrates the fractures resulting from the tectonism-derived fractures under the control of filling. Therefore the fracture genesis mechanism should be analyzed comprehensively.

## 3.3 Reservoir architecture characterization based on sedimentary microfacies classification

Reservoir architecture refers to the pattern, scale, direction, and overlapping relationship of different orders of reservoir architectural elements (Shenghe, 2010). It involves the reservoir interfaces at different orders and the geological bodies of different periods separated by these interfaces. The reservoir architecture can be qualitatively and quantitatively characterized through analysis of sedimentation, diagenesis, and intercalation. It is of great significance for predicting the favorable accumulations in the hydrocarbon exploration stage and tapping the potential of the remaining oil in the development stage (Chaoli et al., 2001; Ray and Chakraborty, 2002; Prather, 2003; Williams and Robertd, 2004; Dali et al., 2007; Huanqing and Xiaomin, 2008; Zhenlin et al., 2010; Ping et al., 2010; Xiangping, 2010; Olariu et al., 2011; Fenglan et al., 2011). Dali et al. (2007) analyzed the architecture of the meandering-river point-bar reservoir with the closer-well-spacing zone 11J11 in Gudao Oilfield of Shengli Oilfield as an example. Jiagen et al. (2008) established a 3D architecture model of channel bar based on the braided-river sand body of Neogene Guantao Formation in Kongdian Oilfield of Huanghua Depression. The previous studies on reservoir architecture usually focused on the meandering river and the braided river. Currently, the architecture of the alluvial fan reservoir has been rarely reported and doesn't follow any recognized architecture model and mature classification scheme (Dali et al., 2007; Xiangping, 2010; Fenglan et al., 2011). Zhenlin et al. (2010) dissected the architecture of the Lower Karamay Formation reservoir in Block 6-Mid of Karamay Oilfield, Xinjiang, and concluded that there is abundant remaining oil near the architecture boundaries; however, Yi's study did not explicitly provide the conglomerate reservoir architecture classification. The reservoir architectures of different orders and types are variable in development patterns due to their distinct physical properties. The architecture hierarchy determination and fine classification for the conglomerate reservoir in the high water-cut stage will be meaningful for the development adjustment and the remaining oil recovery. Shenghe et al. (2012) analyzed the internal architecture of the alluvial fan reservoir in the Lower Karamay Formation of Triassic in Karamay Oilfield, Xinjiang, and established the architecture model of the alluvial fan sedimentary reservoir; however, Wu did not deal with the lithoelectric properties and identification markers of architectures.

## 3.3.1 Current status

### *3.3.1.1 Contents*

#### 3.3.1.1.1 Data base

Reservoir architecture characterization is a systematic task involving multiple disciplines and diverse data. An abundant database is indispensable to the detailed reservoir architecture characterization. The necessary data include field outcrop, logging, seismic, drilling core, test, and production performance (Fig. 3.20). In practical application, logging and seismic data are the most direct options. Particularly, seismic data play an important role in the characterization of architecture boundaries and architectural elements defined by different orders of interfaces. Deptuck et al. (2003) used high-resolution multi-wave 2D and 3D seismic data to analyze the complex architectures of submarine channel and embankment reservoirs in the upper slope of the Niger Delta, Arabian Sea. McLaurin and Steel (2007) adopted the field outcrop data to analyze the architecture of the deformed thin fluvial sedimentary sandstone reservoir in Lower Castlegate, Book Cliffs, Utah, USA, and its genesis, and also made quantitative statistics on the scale of fluvial and bar sedimentary systems.

(A) (B) (C) (D)

**Figure 3.20** Modern sediment and field outcrop data for characterizing the architecture of the Lower Karamay Formation reservoir in an area at the northwestern margin of the Junggar Basin. (A) Modern sedimentary characteristics of alluvial fan; (B) Modern sedimentary characteristics of alluvial fan; (C) Field outcrop profile characteristics of alluvial fan reservoir configuration; (D) Field outcrop profile characteristics of alluvial fan reservoir configuration.

### 3.3.1.1.2 Architecture hierarchy

Architecture boundary refers to a set of bedding contacts with hierarchical sequences, which can be taken to divide the strata into genetically-related stratigraphic blocks ( Shenghe, 2010). Miall A. D. (1985, 1996) classified eight orders of interfaces in the architecture analysis, which together make up an interface hierarchy representing various time-limits; also, where different scales of sedimentary units are defined (Fig. 3.21). The first-order boundary is the boundary of cross-bedding, at which there is little or no erosion and which is invisible on the core and can generally be identified by the truncation and pinching-out of cross foreset. The second-order boundary is the interface of simple layer series, above and below which the lithofacies are different. The third-order boundary is a low-dip (generally <15°) erosion surface in a large bed form, which being a low-angle boundary truncates one or more underlying cross beddings, with a mudstone cover overlaid by intraclast boulder-clay, and with similar lithofacies

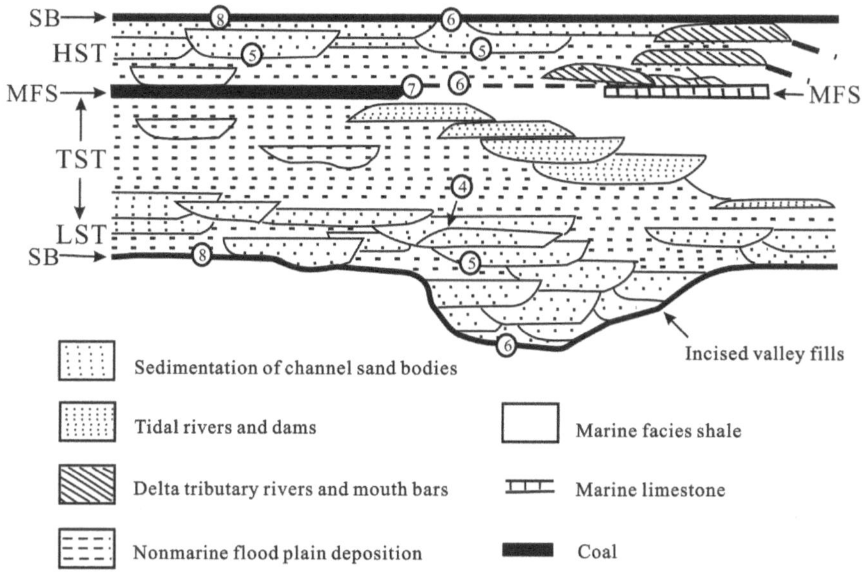


**Figure 3.21** Schematic diagram showing different orders of reservoir architecture boundaries (Miall et al., 1985, 1996; Huanqing and Xiaomin, 2008).

associations above and below. The fourth-order boundary is the boundary of a large bed form, such as the boundaries of channel bar and point bar, the scour surface (spillway channel) of the small-scale channel, and the crevasse splay boundary. The fifth-order boundary is a large-scale sand sheet, such as the interface of large channel and channel-fill complex, which is generally flat to slightly upward-concave and marked by cutting-filling topography and bottom-retaining gravel. The sixth-order boundary is the boundary of restricted channel group or incised valley, which is roughly equivalent to the interface of member or sub-member (mapped stratigraphic unit). The seventh-order boundary is the boundary of allogenetic cycle sedimentary deposit, such as the maximum flooding surface. The eighth-order boundary is a regional unconformity surface, which is equivalent to the third-order sequence boundary. The introduction of reservoir architecture boundary is to apply a set of bedding contacts with hierarchical sequences to divide the sand bodies into genetically-related stratigraphic blocks. The analysis of different orders of reservoir architectures can provide a further understanding of the sedimentary characteristics of target reservoir intervals (Huanqing and Xiaomin, 2008). In this study, we combined various data to classify the architecture boundaries of the alluvial fan reservoir in the Lower Karamay

Formation in an area at the northwestern margin of the Junggar Basin so as to identify various reservoir architectural elements.

### *3.3.1.2 Methods*

#### 3.3.1.2.1 Sequence stratigraphy

Sequence stratigraphic architecture characterization is also covered in reservoir architecture characterization. It is realized through analyzing the bedding contacts and the sedimentary systems in different strata by means of sequence stratigraphy. Hampson et al. (2004) used sedimentology to study the reservoir architecture in mature producing zones in Brent, North Sea, UK. In his study, the high-resolution sequence stratigraphy contributed to: (1) enhancing the precision of characterization in time and space inside and between sedimentary reservoirs; (2) improving two sets of regional reservoir sedimentary prediction models (Broom and Tarbert); and (3) identifying the factors controlling the reservoir architectures, such as fine structure and stratigraphy, which are related to the tectonic evolution of the Middle Jurassic in the North Sea. Gary J. Hampson's study helped enhance the exploration potential and ultimate oil and gas recovery in the area. Allen and Fielding (2007) conducted the sedimentology and stratigraphic architecture analysis for the coal-producing coastal alluvial plain deposits in the northwestern margin of the Late Permian Galilee Basin, Queensland, Australia, and he used the sequence stratigraphy to divide the whole sediment architecture into six genetic units separated by an unconformity. The sequence stratigraphy has been successfully applied in the clastic and carbonate reservoir architecture characterization, but it is insufficient to address key problems for some reservoirs, such as the volcanic reservoir.

#### 3.3.1.2.2 Sedimentology

Sedimentology is the most commonly used and also the widely used method in reservoir architecture characterization. In the sedimentologic architecture characterization, the spatial development scale and superimposition pattern of sedimentary subfacies and sedimentary microfacies are described to identify the corresponding reservoir architectural element and boundary. Jo and Chough (2001) analyzed the Cretaceous alluvial sequence architecture in the northwestern part of the Kyongsang Basin in southeastern Korea with sedimentology, focusing on thick sandstone, thin sandstone, and mudstone. This is the most effective method to characterize reservoir architecture. Even the widely used Miall classification of reservoir architecture has also been established on the basis of sedimentology. Nevertheless, the sedimentologic architecture characterization has weaknesses. For example, this method alone fails to thoroughly highlight the genetic patterns of the alluvial fan and other sedimentary reservoir architectures; instead, it will combine with other methods to obtain a fine and accurate characterization of reservoir architecture.

#### 3.3.1.2.3 Diagenesis

Reservoir architecture is a result of several geological processes. Diagenesis is the dominant process for some reservoirs. The diagenetic architecture characterization analyzes the influences of various processes of diagenesis on reservoir properties, especially the resulting variation of reservoir architecture. Weber and Ricken (2005) analyzed the relationship between quartz cementation and sedimentary architecture of the Triassic Solling Formation braided-river sedimentary system in the Reinhardswald Basin, Germany, by cathodoluminescence observation, rock inclusion analysis, sedimentary thermal simulation of continuous burial history, and other means, and proposed the concept of diagenetic architecture. This immature method is somewhat limited since strong diagenesis is not a feature in all reservoirs.

#### 3.3.1.2.4 Geostatistics

More and more researchers have introduced geostatistics into reservoir architecture characterization. Zhenlin et al. (2010) quantitatively calculated the width and thickness of different architectural elements in the alluvial fan conglomerate reservoirs in the Lower Karamay Formation in Block 6-Mid, Karamay Oilfield, Xinjiang, thereby providing a reliable basis for the remaining oil recovery and production adjustment. The geostatistical reservoir architecture characterization helps greatly to drive the process of quantitative reservoir architecture characterization and facilitates the application of characterization results in the well infilling and adjustment of mature oilfields. However, this method needs the support of a powerful database.

#### 3.3.1.2.5 Geological modeling

Geological modeling is increasingly applied to reservoir architecture characterization, especially in interwell sand body and fluid prediction. In the reservoir architecture characterization by geological modeling, a 3D geological model of reservoir architecture is established to display the reservoir architecture in three dimensions and predict the interwell reservoir architecture development. The 3D geological modeling of reservoir architecture is not substantially different from conventional modeling of sedimentary facies in view of the process and the procedures, but the former is more precise. Donselaar and Geel (2007) established a geological model of architecture for the tidal sedimentary facies of various lithofacies (mainly four lithofacies: tidal channel sandstone, sand-dominated internal tide beach multi-lithology association, mud-dominated internal tidal channel, and fresh-water mudstone) in the Holocene tidal basin, Netherlands. Sylvester et al. (2011) analyzed the stratigraphic architecture of submarine channel and embankment sedimentary systems by establishing a simple 2D model of channel trajectory, and they proved that this model could be used to characterize the incising, migrating, and depositing positions of a single channel during a period in a seemingly complex system. In China, many oilfields such as Dagang and Daqing have also conducted reservoir architecture modeling, with preliminary results achieved. Geological modeling can allow an intuitive and visual reservoir architecture characterization, and it can be feasibly applied to the practice of remaining oil recovery. However, it is difficult to realize due to the limitations of data and algorithms, and its verification is also a huge challenge.

#### 3.3.1.2.6 Numerical simulation

A further understanding of the architecture genesis is essential to reservoir architecture characterization. Numerical simulation provides an intuitive quantitative means for understanding the reservoir architecture genesis. The numerical simulation of the reservoir architecture development process can quantitatively reproduce the spatial evolution of reservoir architectures in different sedimentary stages. Gouw (2007) analyzed the alluvial architectures of river-delta strata and established a simple alluvial architecture model, which was used to simulate the proportion and distribution of channel coarse-grained deposits and floodplain fine-grained deposits. Hill and Griffiths (2008) established a stochastic sedimentary continuous model based on the reservoir architecture identification, which was used for 2D numerical simulation for the Shrubs-canyon outcrop section and Bengal fan seismic facies. Numerical simulation can quantitatively characterize the reservoir architecture development pattern, but it must be bolstered by a relatively reliable geological model.

#### 3.3.1.2.7 Other new techniques/methods

With the advancement of related disciplines, more and more qualitative and quantitative new methods are applied to reservoir architecture characterization. Skelly et al. (2003) used the ground penetrating radar (GPR) to analyze the architecture of aggrading shallow sandbed braided-river channel deposits downstream of the Niobrara River in northeastern Nebraska, USA. The radar facies identification, based on the GPR data, could reproduce the architectural elements, such as channel bar complex, large and small river beds (including 2D and 3D sand dunes), and channel. Baas et al. (2004) conducted the flume experiment to analyze the riverbed geometry, configuration, and composition with a channel analog of gradually weakening hyperpycnal flow, and thereby presented the sedimentary architecture, and flow properties and deposition patterns of hyperpycnal flow. Crawford et al. (2010) used the aeromagnetic data to analyze the evolution of underground architecture in the Paleoproterozoic Wernicke Formation, Yukon, Canada, when the geological and geophysical data were also used. All these new techniques/methods enable the reservoir architecture characterization to be increasingly quantitative, systematic, and accurate. Of course, these techniques/methods are not omnipotent. At any stage, on a solid geological basis, an appropriate technique/method should be selected depending on the specific object.

Reservoir architecture characterization is a systematic task involving multiple aspects. An ideal characterization should combine the abovementioned techniques/methods with the actual geological conditions and data to procure the architecture characterization for the target interval.

### *3.3.1.3 Problems and trend*

#### 3.3.1.3.1 Problems

There are the following problems in the reservoir architecture characterization.

1. There is no consistent understanding of the concept and connotation of reservoir architecture. Most researchers believe that the reservoir architecture is the internal architecture of the reservoir, while a considerable number of researchers hold that the reservoir architecture also includes the sequence stratigraphic architecture of the reservoir. This is mainly dependent on the research purpose and relevant stage (exploration or development). In the exploration stage, the researchers may focus on large-scale reservoir architecture laterally, such as ranges of

the basin, sag, and zone, and possibly on, at most, the third-order sequence vertically. The case is not the same in the development stage. In this stage, laterally, the research may target a reservoir, or a sedimentary body, or even a well cluster; vertically, the research may deal with an oil pay zone or sand group, and even a single sand body and, correspondingly, the reservoir architecture characterization may be fine as to the third-order or fourth-order. In this study, we divided the architectures of the alluvial fan glutenite reservoir in the Lower Karamay Formation in an area at the northwestern margin of the Junggar Basin into four orders. This classification can meet the practical production requirements and also correspond to the architecture identification precision that can be achieved with log data (Fig. 3.22). In particular, the reservoir architecture characterization is to meet the actual requirement and goal rather than describe as fine as possible.

2. There is a lack of analogy and applicability studies on the architecture patterns and models derived from the modern sediment data and the field outcrop data and the underground reservoir architectures. Currently, the reservoir architecture characterizations are mostly based on modern sediment data and field outcrop data. However, these data are not completely consistent with the underground reservoir geology due to multiple geological factors such as sedimentation, tectonism, and diagenesis. The understanding and research results (e.g., the reservoir architecture patterns of different sedimentary facies) derived from modern sediment and field outcrop data must be treated by analogy and revision before they are used practically for analysis of underground reservoir architectures so as to avoid fatal errors.
3. The architecture patterns of reservoirs of different sedimentary genesis need to be improved. The architecture pattern of the meandering-river reservoir has been fully ascertained, but the architecture patterns of the delta, the alluvial fan, and some other sedimentary systems are not known thoroughly and scientifically (Chaoli et al., 2001; Dali et al., 2007; Zhenlin et al., 2010; Zi et al., 2011). The summary and refinement of reservoir architecture patterns play an important role in predicting the favorable development zones and for tapping the potential of the remaining oil, and it is a vital aspect for reservoir architecture characterization. McHargue et al. (2011) analyzed the architecture of continental slope turbidity current channel reservoir and summarized the architecture pattern. McHargue T. concluded that at least three scales of channel growth and shrinkage cycles could be predicted

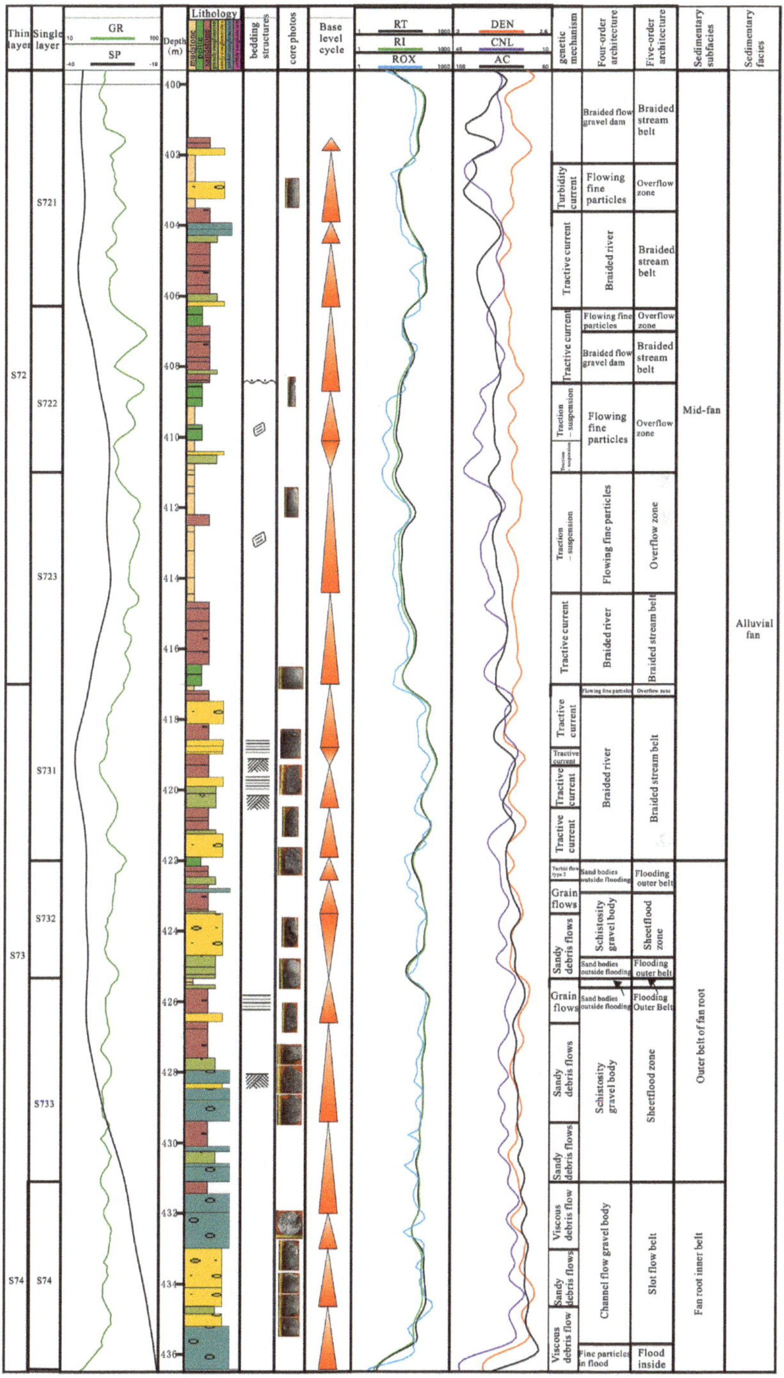


**Figure 3.22** Architectures of alluvial fan reservoir in Well J7 in the Lower Karamay Formation in an area at the northwestern margin of the Junggar Basin.

for most turbidity sediments, namely, architectural element, complex set, and sequence.

4. The present reservoir architecture characterization is too simple and less inclusive. Initially, reservoir architecture characterization was achieved through sedimentology. This does not mean that such characterization just involves sedimentologic matters. The reservoir development is a comprehensive geological process, and the reservoir architecture development also involves many aspects, including sedimentation, diagenesis, and tectonism, which are all covered in the range of reservoir architecture characterization. Therefore, to accurately characterize the reservoir architecture, comprehensive research must be conducted, namely, that in addition to sedimentology, the elements such as diagenesis and structural analysis should be considered. If only sedimentation is involved, the results will unlikely be consistent with the actual geology underground. Hesselinka et al. (2003) took the Dutch section of Rhine as an example to analyze the effects of human activities on the alluvial architectures. Backert et al. (2010) exemplified the effects of fault growth on reservoir architecture by the Gilbert fan delta sediment in the Kerinitis fault, Greece.
5. There is not any definite correlation between reservoir architecture and production performance, which seriously limits the application of reservoir architecture in production practice. This also greatly restricts the progress of reservoir architecture characterization. In essence, the primary goal of reservoir architecture characterization in the development stage is to clarify the internal reservoir architecture so as to guide the prediction of the remaining oil. A reservoir architecture characterization without considering the production practice will undoubtedly be meaningless.
6. The present reservoir architecture characterization mainly targets clastic and carbonate reservoirs but rarely deals with special lithologies such as volcanic and mudstone reservoirs.

#### 3.3.1.3.2 Trend

Further reservoir understanding can be obtained as the field development enters into the middle–late stage. Various problems in production practice also require solutions by fine reservoir architecture characterization. Overall, the trend of reservoir architecture characterization can be concluded to have the following aspects:

1. The advancement of production technology and availability of new data enable the reservoir architecture characterization to be finer and more accurate. Current reservoir architecture characterizations mostly rely on field outcrop, modern sediment, and log data. As the promotion of horizontal well technology in China's major oil and gas fields takes off, the application of horizontal well data for understanding the reservoir architecture and establishing a fine architecture pattern will be a new option for reservoir architecture characterization. Horizontal well data can directly reveal the combination of various architectural elements in the underground reservoir and also show a particular advantage in the identification of different orders of architecture boundaries. Moreover, the infilling well data are increasingly applied in reservoir architecture characterization. The well spacing is up to 200–300 m, and even 100 m locally. These abundant data can play a vital role in characterizing the reservoir architectural elements at a hundred-meter scale.
2. Analogy and applicability studies are conducted for modern sediment and field outcrop data and underground reservoir architecture patterns. In this way, the understandings obtained from both aspects of the study can be adequately applied to underground reservoir architecture characterization, thereby guiding the production practice. As mentioned above, the analogy between the underground reservoir architecture pattern and outcrop data is not sufficient now. For example, outcrop and underground reservoir are both braided-river deposits, but they are certainly different in channel scale, channel bar type, and in other aspects. Thus the researchers should make a comprehensive analysis and comparison of sedimentary setting, structural development, geological time, material supply, climate, and others, to ascertain what outcrop data can be used as a reference for the characterization of underground reservoir architecture, rather than using such data directly. In this way, any errors or even mistakes can be minimized.
3. Reservoir architecture characterization evolves from the pure analysis of sedimentology to the comprehensive analysis of sedimentology, diagenesis, and tectonics. Rusciadelli and Simone (2007) investigated how differential compaction controlled the architecture of the Maiella carbonate platform margin deposits in the central Apennines, Italy, through comprehensive field outcrop observation, pore evaluation, and modeling. Ercilla et al. (2008) studied the Galician shoal area in northwestern Spain, at the northwestern

continental margin of Iberia, and the adjacent abyssal plain high-resolution seismic sequence, indicating that the sedimentary architecture is controlled by the fault system. Reservoir architecture is a result of multiple geological processes. Therefore more and more scholars are aware of the importance of comprehensive analysis to reservoir architecture characterization. Such comprehensive analysis will certainly become a prospecting trend.

4. The reservoir architecture characterization will extend to other lithologies beyond clastic and carbonate rocks. Under the background that the hydrocarbon exploration and development of clastic and carbonate reservoirs are increasingly severe and the rapid economic development has triggered a dramatic increase in hydrocarbon consumption, the volcanic rocks, shale, and metamorphic rocks stand out before the scholars (Huaiyou et al., 2011; Xiaofeng et al., 2012). So far, these reservoirs have been rarely investigated. For volcanic rocks, the crater and volcanic conduit control the development of volcanic rock bodies and the lithologies and lithofacies of volcanic reservoirs and thus control the development of volcanic reservoirs. The characterization of volcanic reservoir architecture is essentially the dissection of the volcanic rock body. Only when the latter is completed can the architecture characterization be ensured for the volcanic reservoir, which is characterized by multiple controlling geological factors, variable lithofacies in space, well-developed fractures, and difficulty in lateral reservoir prediction. This will be attractive for more and more scholars.
5. Reservoir architecture characterization transits from qualitative and semiquantitative to quantitative. One of the essential goals for reservoir architecture characterization is to determine the reservoir architecture pattern with available data so that the distribution of remaining oil can be predicted to provide a reference for practical production. The reservoir architecture characterization in the middle–late development stage, which is mainly designed to tap the potential of the remaining oil, is beneficial to many measures, such as quantitative characterization of remaining oil, infilling deployment, refined separated-layer water injection, and conversion of a producer to an injector. Only when the reservoir architecture is quantitatively characterized can the lateral and vertical development and superimposition patterns of architectural elements be accurately identified, thereby facilitating the quantitative characterization of the remaining oil. Clearly, a quantitative reservoir architecture characterization will be an inevitable option.

6. Reservoir architecture characterization may be extended to other research on reservoirs. By definition, reservoir architecture characterization is to analyze the architecture of the reservoir. Therefore reservoir architecture characterization itself is one of the useful means to study reservoir distribution and superimposition. It can help address many problems in other research efforts on the reservoir. Weiguo et al. used the 3D facies architecture and ichnologic analysis to investigate the asymmetry of the Capital Notom delta in Utah, USA. In this book, the results of reservoir architecture characterization are used to analyze the connectivity of sand bodies in the development well cluster. It is found that reservoir architecture boundaries greatly influence and control the interwell sand body connectivity. The detail will be described below. The fine and accurate reservoir architecture characterization will be extended in extent and coverage.

### 3.3.2 Architectures of the Lower Karamay Formation alluvial fan reservoir in an area at the northwestern margin of the Junggar Basin

In this study, based on the classification of sedimentary facies, the architectures of the Lower Karamay Formation alluvial fan reservoir in an area at the northwestern margin of the Junggar Basin were divided after a comprehensive analysis of geological and geophysical data, and the lithoelectric properties and markers of four orders of architectures were identified.

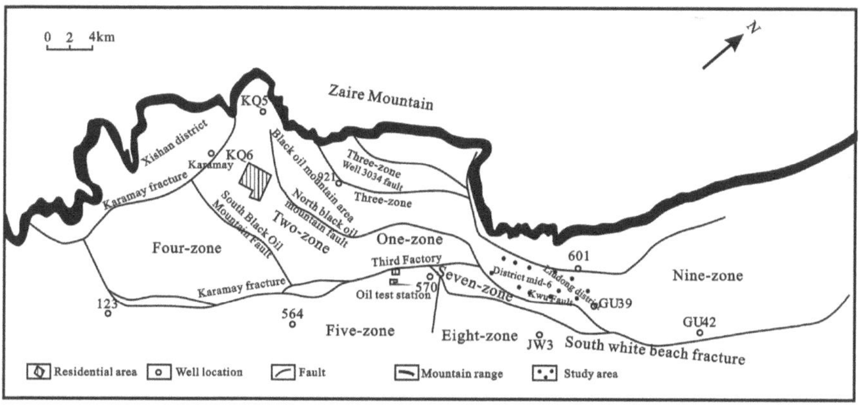


**Figure 3.23** Structural location of Karamay Oilfield, the Junggar Basin. *From Zhan, Z., Shenghe, W., Changfu, X., et al., 2010b. Lithofacies and reservoirs of allluvial fan in the Lower Keramay Formation in the Block6 of Karamay Oilfield, the Junggar Basin. Oil & Gas Geology 31 (4), 463–471.*

Moreover, for the first time, the control of architecture on the reservoir injection-production adjustment was analyzed.

#### 3.3.2.1 Geology

The Karamay Oilfield at the northwestern margin of the Junggar Basin is adjacent to the Zaire Mountain in the north and presents as a NE-SW monoclinic structure with a banded shape, about 50 km long and 10 km wide, descending stepwise from the northwest to the southeast (Fig. 3.23). The oilfield is cut by faults into nine blocks and several fault blocks. The study area is bounded by Block 7 and the Karamay-Wuerhe Fault in the southwest and adjacent to Block 9 in the east. It is divided by the Baijiantan Fault into two blocks, that is, the central block and the east block. From bottom to top, there are Carboniferous, Triassic, Jurassic, and Cretaceous strata. The Triassic includes the Baikouquan Formation, Lower Karamay Formation, Upper Karamay Formation, and Baijiantan Formation. The Lower Karamay Formation and Upper Karamay Formation are the major oil-bearing layer series. The Lower Karamay Formation, the target interval in this study, is divided into two sand groups: S6 and S7, which are subdivided into S61, S62, S63, S71, S721, S722, S723, S731, S732, S733, and S74. A total of 9 layers (S63, S71, S721, S722, S723, S731, S732, S733, and S74) were selected for analysis. The Lower Karamay Formation is 350–850 m deep and 50–70 m thick. There are 1085 production and water injection wells in the study area, with an average well spacing of about 110 m (Zhan et al., 2010b). The Triassic is a set of very thick greyish-green–brown-red conglomerates with a thickness of 300–2500 m. The Lower Triassic is only found in the eastern part of the oilfield and is almost overwhelmed by a conglomerate and psephitic sandstone, being 130–200 m thick. The Middle Triassic shows a wide distribution, with positive depositional cycles of thick conglomerate–sandstone, conglomerate and mudstone interbed, and fine siltstone–mudstone in the lower part, and a set of alternative glutenite and mudstone deposits in the upper part, totally 50–450 m thick (Zaixing, 2003). According to the lithology identification of sealed-coring samples, the target interval is composed of cobblestone, pebblestone, glutenite, sandy conglomerate, microconglomerate, coarse sandstone, medium sandstone, pebbly sandstone, fine sandstone, siltstone, conglomeratic claystone, silty mudstone, and mudstone (Fig. 3.24). For the convenience of analysis, these lithologies are integrated as seven lithologies, namely, cobblestone, pebblestone, microconglomerate, glutenite (including sandy conglomerate), sandstone (coarse sandstone, medium sandstone,

Figure 3.24 Lithologies observed in cores taken from the Lower Karamay Formation in an area at the northwestern margin of the Junggar Basin. (A) Well J1, light greyish brown pebblestone, 411.80–411.90 m; (B) Well J1, offwhite microconglomerate with coarse gravels, 416.26–416.38 m; (C) Well J1, dark brown pebblestone, 398.61–398.80 m; (D) Well J7, a light brown sandy conglomerate with medium gravels, 421.44–421.60 m; (E) Well J7, light greyish brown sandy conglomerate, 426.47–426.57 m; (F) Well J1, celandine green silty mudstone, 397.66–398.80 m; (G) Well J50, glutenite, × 50, 522.09 m; (H) Well J9, pebblestone, × 50, 576.30 m; (I) Well J6, fine sandstone, × 25, 386.62 m.

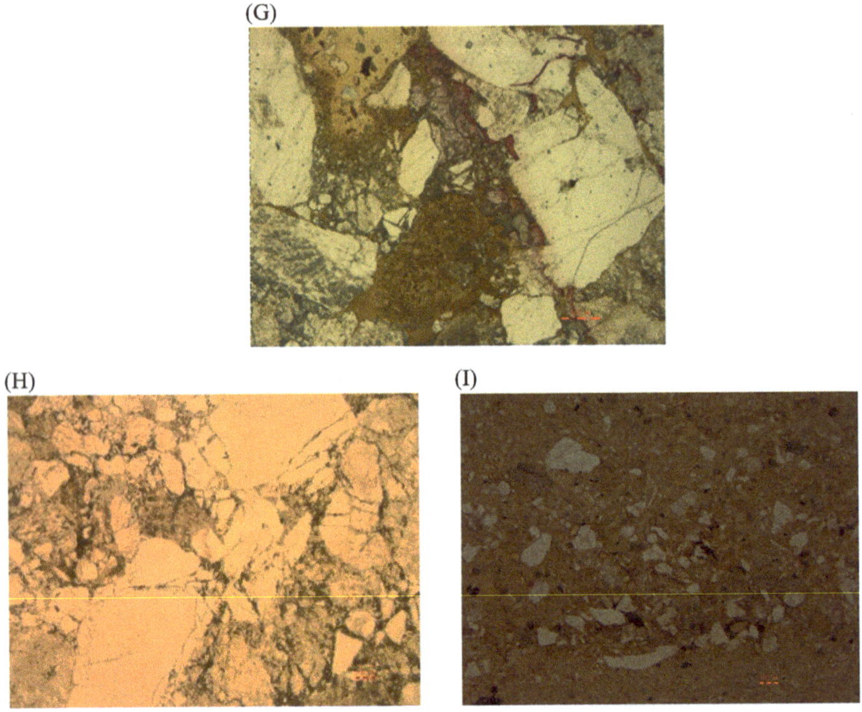


**Figure 3.24** (*Continued*)

pebbly sandstone, fine sandstone), pelite (siltstone, pebbly mudstone, and silty mudstone) and mudstone.

According to the lithologic profile (Fig. 3.25), the conglomerate reservoir exhibits strong heterogeneity and variable lithology in a vertical direction. Even in the relatively large intervals of microconglomerate and sandstone, the reservoir physical properties may change greatly with the grain size, shale content, and other parameters; such changes will reflect evident logging responses. Generally, glutenite and sandstone are wide, stable, and continuous, while other lithologies show relatively poor behaviors. In S74, S733, S723, S722, and S721, there are thin sandstone, glutenite, microconglomerate, pebblestone, and other lithologic stripes. The stripes with long extension (more than two well spacings) can: (1) serve as dominant flow paths if they have good physical properties, or (2) act as physical barriers if they have poor physical properties. The two cases have positive and negative impacts, respectively, on waterflooding. Cobblestone only appears in S74. Pelite and mudstone intercalations, which are small in thickness and distribution, are less effective in S74,

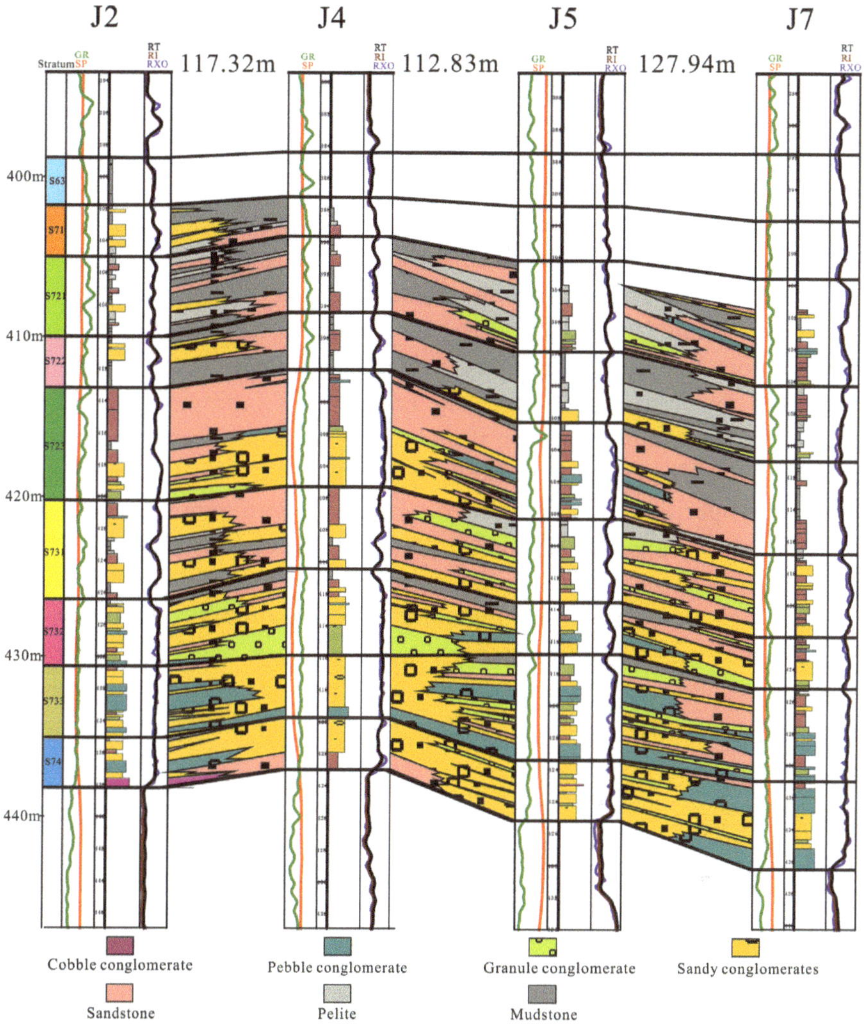


**Figure 3.25** Lithological profile of conglomerate reservoir in the Lower Karamay Formation in an area at the northwestern margin of the Junggar Basin.

S733, and S732 but significantly effective in S731, S723, S722, and S721. On the lithologic profile, the conglomerate reservoir can be divided into three structures (from the bottom to top): (1) S74–S732, dominated by glutenite, and sandwiched with thin sandstone, microconglomerate, pebblestone, and cobblestone; (2) S731–S723, dominated by sandstone, and sandwiched with thin microconglomerate, pebblestone, and glutenite; and (3) ****, dominated by a large set of mudstone, and sandwiched with thin sandstone, glutenite, microconglomerate, and pebblestone. Upwardly, the lateral facies control level increases with the decrease of mud content and the increase of sand content.

### *3.3.2.2 Research approach*

The conglomerate reservoir in the northwestern margin of the Junggar Basin is featured by great facies variation and strong heterogeneity, which impose challenges to the reservoir characterization. The Karamay Oilfield was discovered in 1955 and put into development in 1958. So far, abundant outcrop, drilling, logging, seismic, and production performance data have been accumulated (Qingchang et al., 1997), providing a reliable database for this study. Based on the objective of architecture characterization and available data, the research approach in this study was designed (Fig. 3.26). Specifically, the data used include modern sediment, field outcrop (Fig. 3.27), logging, seismic, drilling core, test, and production performance data. The research contents mainly involve four parts. The first part is to complete the fine stratigraphic classification and correlation and establish the fine bed classification and correlation database using the abovementioned data. The second part is to establish the reservoir

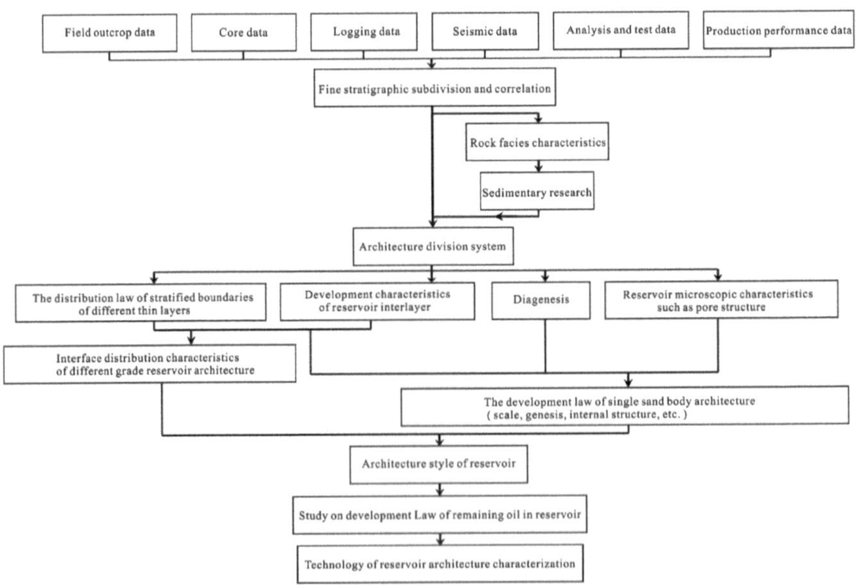


**Figure 3.26** Process of reservoir architecture characterization for the Lower Karamay Formation in an area at the northwestern margin of the Junggar Basin.

(A)

(B)

**Figure 3.27** Modern sediment and field outcrop data for architecture characterization of the conglomerate reservoir in the Lower Karamay Formation in an area at the northwestern margin of the Junggar Basin. (A) Modern sedimentary characteristics of alluvial fan; (B) Field outcrop profile characteristics of alluvial fan reservoir configuration.

architecture classification system through lithofacies and sedimentology analysis within the fine isochronous stratigraphic framework. It is critical to determine whether the reservoir architectures are classified to the fifth-order or fourth-order. Only when a reasonable architecture hierarchy is defined can the subsequent activities move on. The third part is to establish the architecture pattern/style. On the one hand, the reservoir architecture boundaries are characterized through layer boundary analysis and intercalation characterization. On the other hand, the reservoir architectural elements with different geneses, that is, single sand body, including size, genesis, internal structure, etc., are identified depending on the characteristics of reservoir architecture boundary and the microscopic reservoir properties such as diagenesis and pore structure. The fourth part is to summarize the reservoir architecture patterns and analyze the remaining oil distribution to develop a set of reservoir architecture characterization technology for the efficient development and the remaining oil recovery. Various data play different roles in reservoir architecture characterization. The modern sediment and field outcrop data are used to establish the architecture pattern, mainly through the quantitative analysis of the spatial superimposition pattern and development scale of architectural elements. The core and log data are combined to analyze the geological geneses of different reservoir architectures. The logging and seismic data are combined to quantitatively characterize the scales of different architectural elements—log data for vertical scale and seismic data for lateral scale, whereas the test and production performance data are combined to characterize the reservoir architecture connectivity.

### *3.3.2.3 Scheme and result of architecture classification*

#### 3.3.2.3.1 Architecture classification scheme

A reasonable reservoir architecture classification scheme is considered as the premise and guarantee for successful architecture characterization. Firstly, based on the sedimentary analysis, the conglomerate reservoir is classified into four subfacies, that is, upper fan inner zone, upper fan outer zone, middle fan, and lower fan. Then, according to lithology, log pattern, and sedimentary model, the target interval is divided into 13 types of architectural elements, that is, trough flow gravel, trough beach sand-gravel, sheetflood inner sand, sheetflood inner fine grain, laminar flow gravel, sheetflood outer sand, sheetflood outer fine grain, braided flow channel, braided flow sand-gravel bar, cross flow sand, cross flow fine grain, runoff channel, and interchannel fine grain (Table 3.3). Trough flow gravel, laminar flow gravel, braided flow channel, and braided flow sand-gravel bar are dominant.

According to the single-well reservoir architecture classification, the lithoelectric properties of architectural elements in the alluvial fan conglomerate reservoir were summarized (Table 3.3), which are also the most fundamental basis for the architecture classification in this study.

1. The trough flow gravel architectural element consists of cobblestone, pebblestone, and pebbly glutenite. It is stripped laterally, 2–8 m

**Table 3.3** Classification of the architecture of the Lower Karamay Formation alluvial fan conglomerate reservoir in an area at the northwestern margin of the Junggar Basin.

| Subfacies | Fourth-order architectural element | Lithology | SP | RT | Bed thickness |
|---|---|---|---|---|---|
| Upper fan inner zone | Trough flow gravel | Cobblestone, pebblestone, pebbly glutenite | High, funneled, or inverted trapezoid | High, funneled, or inverted trapezoid | 2–8 m |
| | Trough beach sand-gravel | Pebblestone, glutenite | Medium to high, funneled, or inverted trapezoid | Medium to high, funneled, or inverted trapezoid | <2 m |
| | Sheetflood inner sand | Pebbly sandstone, coarse sandstone | Low to medium, funneled, or inverted trapezoid | Low to medium, funneled, or inverted trapezoid | <3 m |
| | Sheetflood inner fine grain | Siltstone, argillaceous siltstone, pebbly mudstone | Low, flat | Low, flat | <2 m |
| Upper fan outer zone | Laminar flow gravel | Pebblestone, medium-gravel-bearing microconglomerate, mud-bearing glutenite | High, funneled, or inverted trapezoid | High, funneled, or inverted trapezoid | 2–7 m |
| | Sheetflood outer sand | Pebbly sandstone, medium-fine sandstone | Low to medium, funneled, or inverted trapezoid | Low to medium, funneled, or inverted trapezoid | <3 m |
| | Sheetflood outer fine grain | Siltstone, argillaceous siltstone, pebbly mudstone | Low, flat | Low, flat | <2 m |

(*Continued*)

**Table 3.3** (Continued)

| Subfacies | Fourth-order architectural element | Lithology | SP | RT | Bed thickness |
|---|---|---|---|---|---|
| Middle fan | Braided flow channel | Sandy conglomerate, pebbly sandstone, coarse sandstone, medium sandstone | Medium to high, bell- or box-shaped | Medium to high, bell- or box-shaped | 2–7 m |
| | Braided flow sand-gravel bar | Microconglomerate, sandy conglomerate, pebbly sandstone | Medium to high, funneled, or inverted trapezoid | Medium to high, funneled, or inverted trapezoid | 2–7 m |
| | Cross flow sand | Pebbly sandstone, fine sandstone | Low to medium, bell- or box-shaped | Medium to high, finger-shaped | <2 m |
| | Cross flow fine grain | Pebbly siltstone, pebbly mudstone | Low, flat | Low, flat, or slightly jagged | 2–5 m |
| Lower fan | Runoff channel | Fine sandstone, siltstone | Low to medium, bell- or box-shaped | Medium to high, finger-shaped | <2 m |
| | Interchannel fine grain | Silty mudstone, mudstone | Low, flat | Low, flat, or slightly jagged | 1–4 m |

thick, and poorly sorted. The volume fraction of conglomerate (glutenite) is high (>90%). The coarsening-upward electric properties are derived, with funneled or inverted trapezoidal SP (spontaneous potential log) and RT (resistivity log) logs, and RT >70 Ω•m.

2. The trough beach sand-gravel architectural element consists of pebblestone and glutenite, which is the transition zone between the fan-head trench and the sheetflood inner zone. It is elongated, striped, thin (<2 m), and poorly sorted. Massive and graded beddings are observed. The volume fraction of glutenite is 70%–90%. The RT log is finger-shaped with low amplitude and a value greater than 60 Ω•m.
3. The sheetflood inner sand architectural element consists of pebbly sandstone and coarse sandstone. It is thin and poorly sorted, with carbon dust occasionally. The RT log is funneled or inverted trapezoidal, with a value greater than 80 Ω•m.
4. The sheetflood inner fine grain architectural element consists of siltstone, argillaceous siltstone, and pebbly mudstone. It is thin (<2 m). The RT log is flat, with a value less than 20 Ω•m.
5. The laminar flow gravel architectural element consists of pebblestone, medium-gravel-bearing microconglomerate, and mud-bearing glutenite. It is thick (2–7 m). Graded bedding and quasi-parallel bedding are observed, being reverse graded. The RT log is generally funneled and inverted trapezoidal, with a value greater than 100 Ω•m.
6. The sheetflood outer sand architectural element consists of pebbly sandstone and medium-fine sandstone. It is generally thin (<3 m). Graded bedding and cross beddings can be observed. The RT log is funneled or inverted trapezoidal.
7. The sheetflood outer fine grain architectural element consists of siltstone, argillaceous siltstone, and pebbly mudstone. It is thin (<2 m), with carbon dust occasionally. The RT log is funneled or inverted trapezoidal.
8. The braided flow channel architectural element consists of sandy conglomerate, pebbly sandstone, coarse sandstone, and medium sandstone. It is striped in the lateral direction with a width of 80–400 m and lenticular on section with a thickness of 2–7 m. It is moderately sorted and normal graded, with cross bedding and bottom scour surface. The RT log is bell- or box-shaped, with a value greater than 100 Ω•m.

9. The braided flow sand-gravel bar architectural element consists of microconglomerate, sandy conglomerate, and pebbly sandstone. It is relatively thick (2–7 m), and reverse graded, with cross bedding. The RT log is funneled or inverted trapezoidal, with a value greater than 60 Ω•m.
10. The cross flow sand architectural element consists of pebbly sandstone and fine sandstone. It is relatively thin (<2 m), with massive bedding. The RT log is finger-shaped with low amplitude and moderate value.
11. The cross flow fine grain architectural element consists of pebbly silty mudstone and pebbly mudstone. It is thick (2–5 m), with horizontal bedding. The RT log is flat, with a low value.
12. The runoff channel architectural element consists of fine sandstone and siltstone. It is generally thin (<2 m), with ripple bedding and massive bedding. The RT log is finger-shaped, with a value of about 300 Ω•m.
13. The interchannel fine grain architectural element consists of silty mudstone and mudstone. It is greatly variable in sedimentary thickness (1–4 m), with horizontal bedding and carbon dust. The RT log is flat, with a low value.

The sheetflood inner sand and sheetflood outer sand are classified as the four-order classification of architectures of the upper alluvial fan. Essentially, they are almost identical, except for their positions in the fan. This study indicates that the upper fan can be distinctly divided into two architectural sub-elements, that is, inner zone and outer zone. In the inner zone, closer to the provenance, the major trough and the edge trough are evident, and the main channel position can be clearly identified. In the outer zone, farther to the provenance, the laminar flow zone is continuous in a large area, the main channel position is not obvious, and the terrain is gentler than that in the trough flow zone. This division of two architectural sub-elements is reasonable and scientific. It represents a fine dissection of architecture and also agrees with the genesis of sedimentary units in the alluvial fan. The markers for identifying the inner and outer zones and their boundary have been described above and will not be repeated here.

#### 3.3.2.3.2 Architecture classification result

According to the architecture classification scheme mentioned above, the architectural elements of the target interval were classified at the levels of

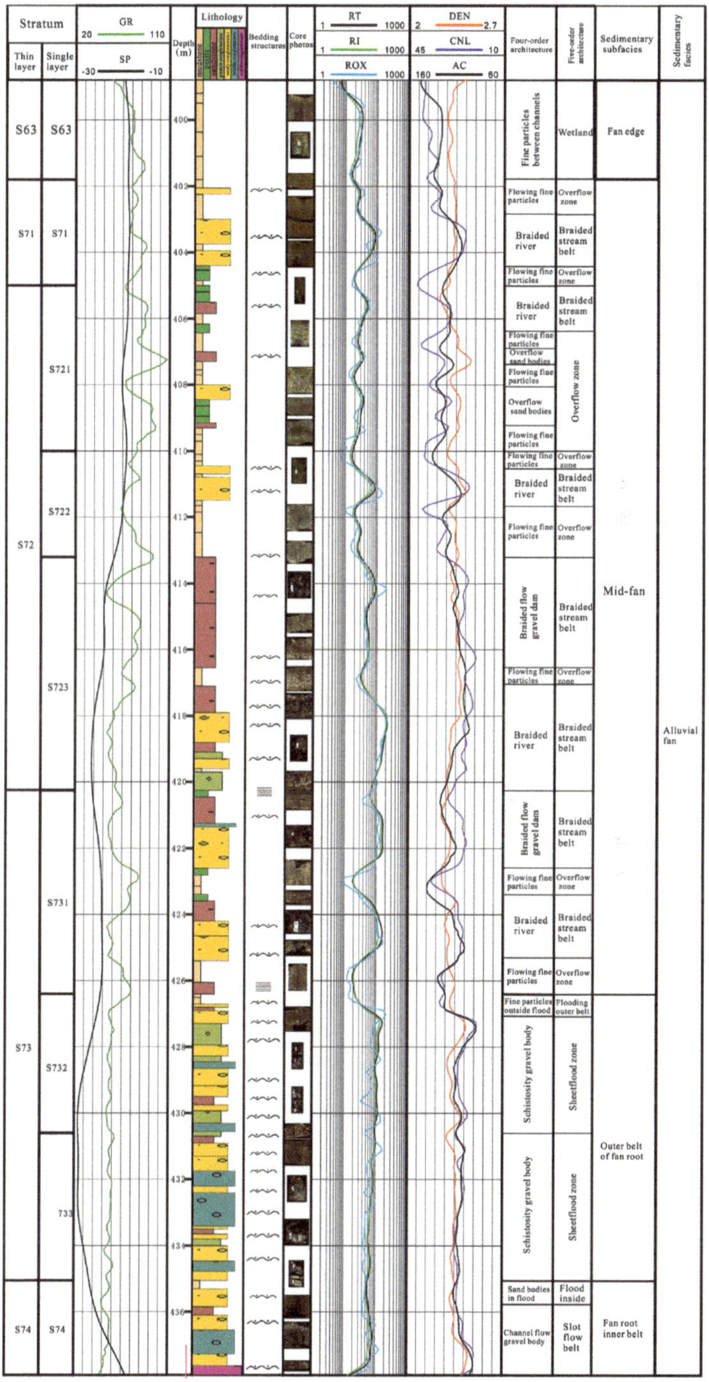


**Figure 3.28** Architecture classification of Well J2 in the Lower Karamay Formation alluvial fan reservoir in an area at the northwestern margin of the Junggar Basin.

a single well, section, and plane and in using the data of coring wells (Fig. 3.28). S74 corresponds to the upper fan inner zone, which is classified into four fourth-order architectural elements, that is, trough flow gravel, trough beach sand-gravel, sheetflood inner sand, and sheetflood inner fine grain, of which the trough flow gravel is dominant and contains coarse grains generally. S733 and S732 correspond to the upper fan outer zone, which is classified into laminar flow gravel, sheetflood outer sand, sheetflood outer fine grain, and other fourth-order architectural elements, of which the laminar flow gravel is dominant and contains coarse grains generally, with a continuous distribution of conglomerate or glutenite. S731, S723, S722, S721, and S71 correspond to the middle fan classified into braided flow channel, braided flow sand-gravel bar,

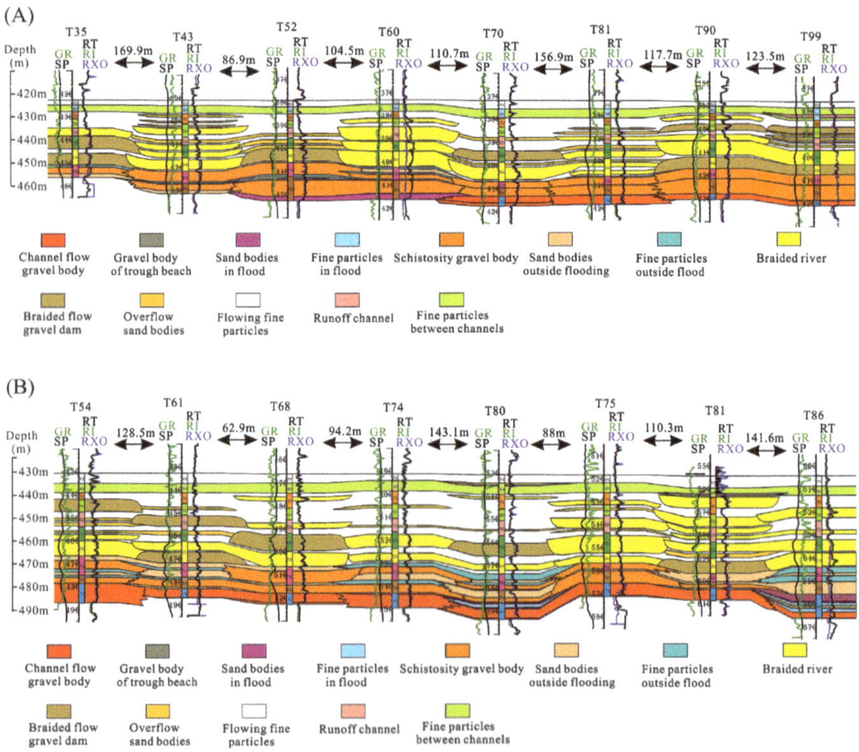


**Figure 3.29** Architecture development characteristics of alluvial fan conglomerate reservoir in section. (A) Well section with the same orientation as sedimentary source direction; (B) Well section with the vertical orientation as sedimentary source direction.

cross flow sand, cross flow fine grain, and other fourth-order architectural elements, of which the braided flow channel and braided flow sand-gravel bar are dominant and contain coarse grains generally. As the fan shrinks upwards, fine-grained deposits increase, and coarse-grained deposits decrease, showing a feature of retrogradation. S63 corresponds to the lower fan, which is classified into runoff channel and interchannel fine grain, of which the interchannel fine grain is dominant while the runoff channel shows a small scale.

#### *3.3.2.4 Reservoir architecture characteristics*

##### 3.3.2.4.1 Vertical development

Vertically, as shown in Fig. 3.29, the trough flow deposits, including trough flow gravel, trough beach sand-gravel, sheetflood inner sand, and sheetflood inner fine grain, are poorly connected in local areas due to erosion perpendicular to the provenance. The trough flow gravel is dominant and continuous but partially separated by an interface. The braided flow deposits are dominated by braided flow channel and braided flow sand-gravel bar, which are cross-cut and superimposed; from bottom to top, the sand bodies become less connected and superimposed with the shrinking of fluvial scale. The cross flow deposits exist, with small thickness, between braided flow channel and braided flow sand-gravel bar. The runoff deposits in the lower fan are dominated by interchannel fine grain but contain less runoff channel.

Perpendicularly to the provenance, the sand bodies exhibit better connectivity and longer extension under the control of sedimentation. The trough flow zone, braided flow zone, cross flow zone, and runoff zone are found with the development patterns basically consistent with that illustrated on the section parallel to the provenance. Close to the central fan, the sand bodies are large and connected, with less fine-grained deposits. At the position near the fan edge, the trough flow zone and laminar flow zone in the lower part of the target interval, as well as the braided flow zone, show obviously a higher content of fine-grained deposits and, also, the presence of intercalation. According to the architecture section of the coring well, the contents of the conglomerate, glutenite, and coarse sandstone of the upper fan, the middle fan, and the lower fan facies at the positions near the central fan are apparently higher than that at the fan edge. As closer to the fan edge, the contents of fine sandstone and mudstone increase significantly.

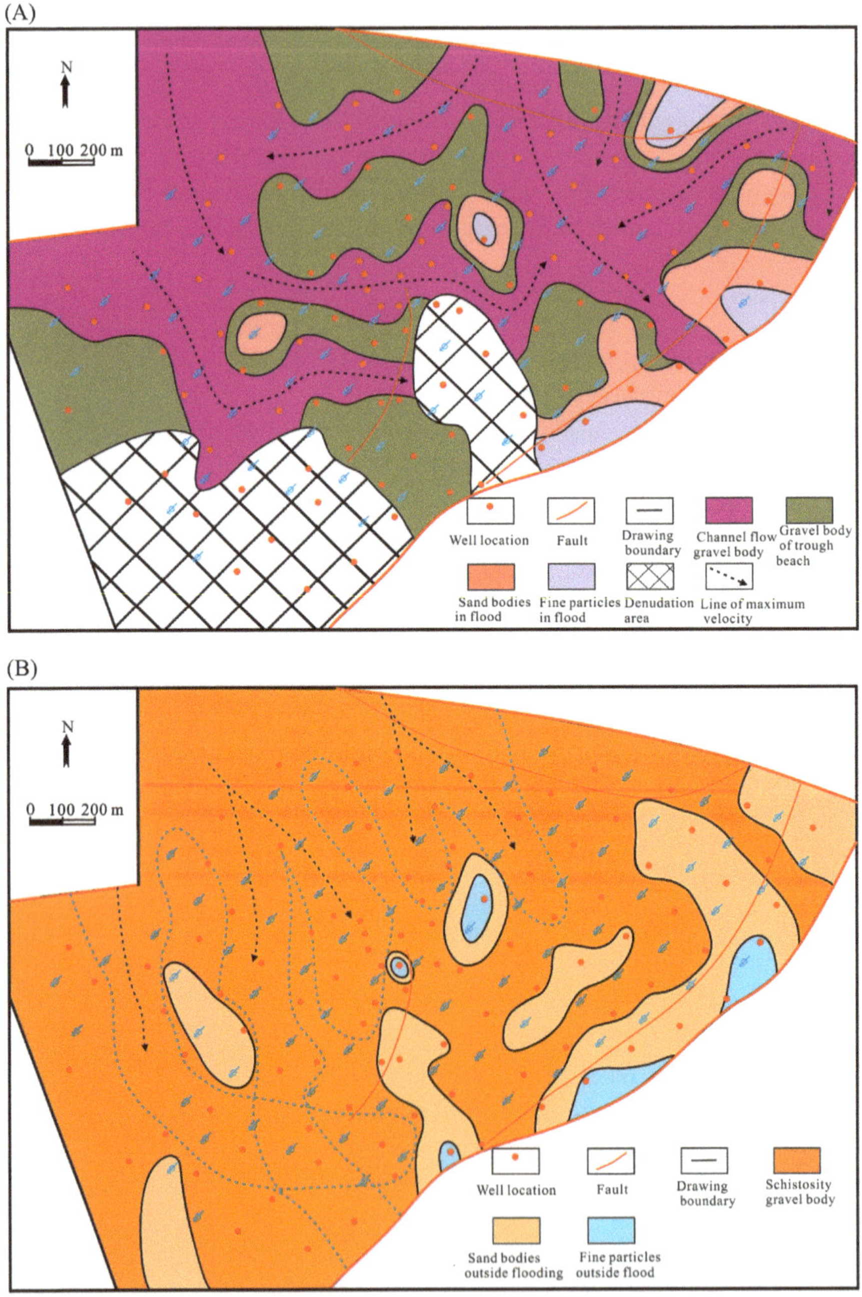


**Figure 3.30** Lateral development of architectural elements of the Lower Karamay Formation alluvial fan conglomerate reservoir in an area at the northwestern margin of the Junggar Basin. (A) Fan root inner belt; (B) outer belt of fan root; (C) mid-fan; (D) fan edge.

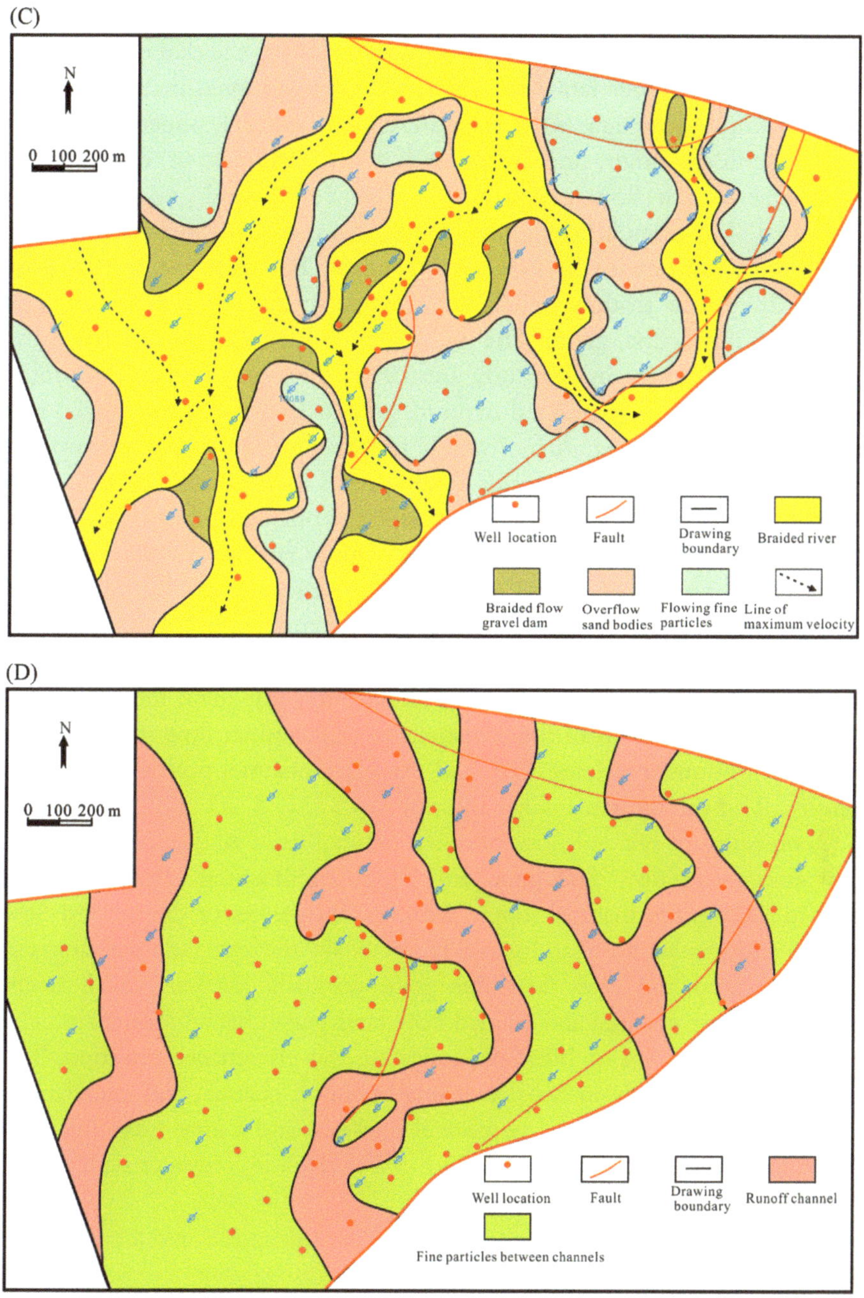


**Figure 3.30** (*Continued*)

#### 3.3.2.4.2 Lateral development

The reservoir architectures develop differently in the lateral direction (Fig. 3.30). The upper fan inner zone is mainly observed in S74, consisting of trough flow gravel, trough beach sand-gravel, sheetflood inner sand, sheetflood inner fine grain, and other architectural elements, of which the trough flow gravel and trough beach sand-gravel are most developed. The trough flow gravel deposits in the study area are generally distributed as stripes in the NE-SW direction. Between the trough flow gravels, there are trough beach sand-gravel, sheetflood inner sand, and sheetflood inner fine grain. The materials are mainly sourced from the northwest and north. In the study area, from the northwest to the southeast, the ranges of fine-grained deposits such as sheetflood inner sand and sheetflood inner fine grain increase as farther they move away from the provenance. The upper fan outer zone appears mainly in S733 and S732. Despite the similar feature of coarse-grained deposits in dominance, it is different from the upper fan inner zone. The upper fan outer zone consists of laminar flow gravel, sheetflood outer sand, sheetflood outer fine grain, and other architectural elements, with the laminar flow gravel in dominance. From the northwest to the southeast, the ranges of fine-grained deposits such as sheetflood outer sand and sheetflood outer fine grain increase as farther they move away from the provenance. The laminar flow gravel deposits are continuous in a large area, and the main channel position is hard to identify. The middle fan mainly develops in S731, S723, S722, S721, and S71, consisting of braided flow channel, braided flow sand-gravel bar, cross flow sand, cross flow fine grain and other architectural elements. From S731 to S71, as the material supply weakens, the ranges of the braided flow channel and braided flow sand-gravel bar gradually decrease, and the ranges of cross flow sand and cross flow fine grain gradually increase. In all layers, the braided flow channel and braided flow sand-gravel bar are dominant. The overall trend is in SN or NE-SW, indicating the provenance direction. The lower fan exists mainly in S63, consisting of runoff channel and interchannel fine grain, with small channel width and thickness and a long extension. The runoff channel is generally in SN or NE-SW, which is consistent with that of the upper fan and the middle fan.

#### 3.3.2.4.3 Scales of reservoir architectures

In this study, the geometry, length, width, and other quantitative information of reservoir architectures were preliminarily analyzed according to the vertical and lateral development of architectural elements (Table 3.4).

**Table 3.4** Scales of architectural elements of the Lower Karamay Formation alluvial fan conglomerate reservoir in an area at the northwestern margin of the Junggar Basin (fifth-order vs fourth-order).

| Subfacies | Fifth-order architectural element | Fourth-order architectural element | Length/m Range | Average | Width/m Range | Average | Width/ thickness |
|---|---|---|---|---|---|---|---|
| Upper fan inner zone | Trough flow zone | Trough flow gravel | >1300 | / | 70–700 | 260 | / |
| | | Trough beach sand-gravel | 30–750 | 500 | 120–550 | 230 | / |
| | Sheetflood inner zone | Sheetflood inner sand | 70–400 | 250 | 30–400 | 100 | / |
| Upper fan outer zone | Laminar flow zone | Laminar flow gravel | >1900 | / | >2500 | / | / |
| | Sheetflood outer zone | Sheetflood outer sand | 140–650 | 430 | 30–800 | 100 | / |
| Middle fan | Braided flow zone | Braided flow channel | >1800 | / | 80–1200 | 240 | 61 |
| | | Braided flow sand-gravel bar | 150–730 | 280 | 80–700 | 210 | / |
| | Cross flow zone | Cross flow sand | 120–300 | 270 | 20–500 | 160 | / |
| Lower fan | Runoff zone | Runoff channel | >2300 | / | 90–260 | 200 | 47.5 |

For the upper fan inner zone, the trough flow gravel extends with the length of greater than 1300 m and the width of 0–700 m; the trough beach sand-gravel extends with the length of 30–750 m and the width of 120–550 m; and the sheetflood inner sand extends with the length of 70–400 m and the width of 30–400 m. The laminar flow zone extends with a length greater than 1900 m and a width exceeding 2500 m. The sheetflood outer sand, as a thick wedge shape, extends with the length of 140–650 m and the width of 30–800 m. For the braided flow zone, the braided flow channel extends with a length greater than 1800 m and the width of 80–1200 m; the braided flow sand-gravel bar extends with the length of 150–730 m, and the width of 80–700 m. The cross flow sand extends with a length of 120–300 m and a width of 20–500 m. The runoff channel extends with a length greater than 2300 m and a width of 90–260 m. These quantitative information provide a reliable geological basis for understanding the architectures of the alluvial fan conglomerate reservoir and also present a reference for the efficient reservoir development and adjustment of reservoir exploration efforts.

#### *3.3.2.5 Correspondence between architecture classification and lithology classification*

Controlled by the sedimentary genesis, lithology is one of the most basic reservoir properties, and it is significant for the value and distribution of porosity, permeability, and other physical properties. Many scholars have predicted and evaluated the reservoirs through lithology analysis (Weiwei et al., 2010; Weilin et al., 2012). Some scholars have also probed into the relationship between lithology and sedimentary facies or reservoir physical properties. Ye et al. (2011) used the lithology statistical method to identify the sedimentary facies by taking the delta deposits in the Pearl Mouth Basin as an example. His results show that the traditional sedimentary facies analysis is greatly influenced by human factors, which may lead to different understandings on the same sedimentary structure, log pattern, and seismic reflections; in contrast, lithology is relatively easy to identify and can accurately be obtained using a precise device, so that the lithology can be used to identify sedimentary facies to avoid any differences in human identification of facies marker. In addition, lithology is closely related to reservoir architecture and physical properties.

Lithology is considered as the material basis for the reservoir architecture and can be gained from the core test of the coring well or electric log of the noncoring well. Thus lithology analysis is optional for reservoir architecture characterization. The continental conglomerate reservoirs are mostly

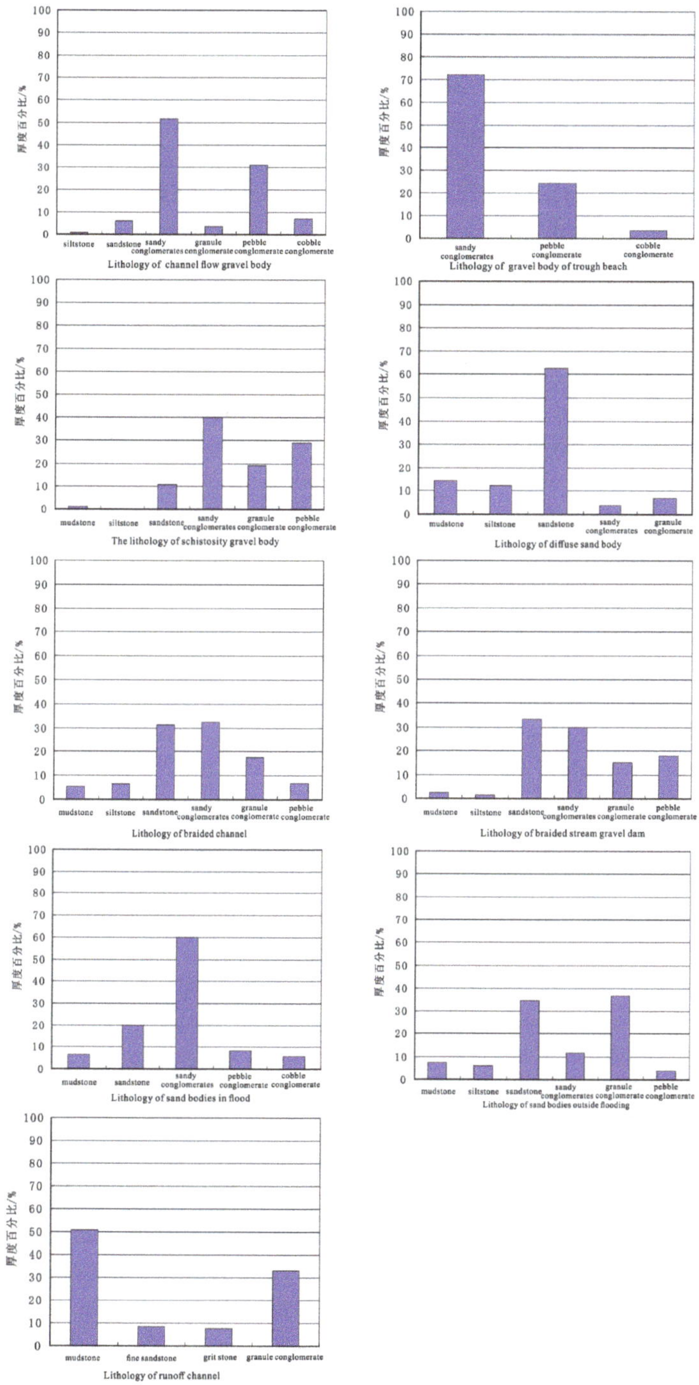


**Figure 3.31** Correspondence between lithology and architecture in the Lower Karamay Formation in an area at the northwestern margin of the Junggar Basin.

formed in rapid, unstable, frequent, and strong water flow alluvial–deluvial sedimentary environment, resulting in theirs having a more heterogeneous lithology than other reservoirs. This heterogeneity is mainly reflected by significant lateral variation and complex vertical association (Qingchang et al., 1997). Accordingly, in this study, through fine lithology dissection of cores taken from sealed-coring wells, we mapped the single well and vertical lithology sections to analyze the spatial development of lithology. On this basis, we investigated the correspondence between lithology and reservoir architecture. Finally, the reservoir architecture (including architecture distribution) of the whole study area including noncoring wells, was characterized, and the reservoir was predicted.

The correspondence between the fourth-order architecture and lithology was statistically gained using the data of 8 sealed-coring wells (Fig. 3.31). The trough flow gravel and trough beach sand-gravel are mainly composed of coarse-grained deposits such as glutenite and pebblestone, with one or two rock types in absolute dominance. In view of thickness, the dominant rocks usually exceed 80% or even approach 90%. Therefore the trough flow gravel and trough beach sand-gravel generally form high-permeability stripes. The laminar flow gravel, in large-area continuous distribution, shows a significant variation in lithology, which mainly consists of glutenite, microconglomerate, and pebblestone. There are no dominant rock types for the laminar flow gravel. In addition to the coarse-grained deposits, it also includes fine-grained deposits such as a small amount of mudstone. The fine-grained deposits with a small proportion show a great influence on reservoir properties. They usually exist as the architecture boundaries, which aggravate the spatial reservoir heterogeneity and increase the challenges in efficient reservoir development. Overall, the coarse-grained deposits are dominant, including glutenite, microconglomerate, and pebblestone.

The braided flow channel and braided flow sand-gravel bar architectural elements usually consist of glutenite, sandstone, and others. These rock types present wide distribution, but none of them are dominant. This also reflects the strong reservoir heterogeneity and variable rock types in the study area. Although siltstone, mudstone, and pebblestone deposits take less content—with the thickness proportion $<10\%$—their influences on reservoir properties are worthy of sufficient attention. The siltstone and mudstone (fine-grained deposits) usually stay as the architecture boundaries at different orders, thus involving the reservoir connectivity. A comparison shows that the braided flow channel and braided flow sand-gravel bar contain both sandstone and conglomerate, while the laminar flow gravel and trough flow gravel are dominated by glutenite and pebblestone with much larger grain sizes.

For cross flow sand, sandstone is dominant, with the thickness proportion greater than 60%. A variety of other rock types, from siltstone to microconglomerate, can be observed. The multiple rock types with wide distribution and significant variation also influence the physical properties of this architectural element. The grain sizes of sheetflood outer sand and sheetflood inner sand are apparently coarser than that of cross flow sand and mainly consist of glutenite, microconglomerate, sandstone, and others. The fine-grained deposits such as siltstone and mudstone are less, with the thickness proportion lower than 10%. The presence of siltstone and mudstone cause significant impacts to bear on the physical properties of these architectural elements. For the fine-grained deposits such as interchannel fine grain, cross flow fine grain, sheetflood outer fine grain, and sheetflood inner fine grain, mudstone is dominant, with the lowest thickness proportion greater than 65%. There are fewer other rock types, such as siltstone, with the thickness proportion almost less than 10%.

The analysis of correspondence between reservoir lithology and architecture can provide a further understanding of the spatial development of reservoir architecture from the genetic perspective and also help the scholars to clarify and predict reservoir architectural elements by lithology identification and analysis.

#### *3.3.2.6 Control and influence of reservoir architecture on oilfield development*

Well J3 was taken as an example to analyze the correspondence between physical properties (mainly the porosity and permeability obtained from the core test of sealed-coring well) and architecture/lithology (Fig. 3.32). Generally, high porosity and high permeability correspond to the architectural elements of braided flow channel, braided flow sand-gravel bar, laminar flow gravel, and trough flow gravel. Porosity is well correlated to permeability; namely, the position with high porosity usually presents a high permeability. The thick trough flow gravel, laminar flow gravel, braided flow sand-gravel bar and braided flow channel show multiple stages, and high permeability and high porosity are generally concentrated in the stage with strong hydrodynamics. The laminar flow gravel in the upper fan outer zone usually has higher porosity and permeability than the trough beach sand-gravel in the upper fan inner zone. The trough

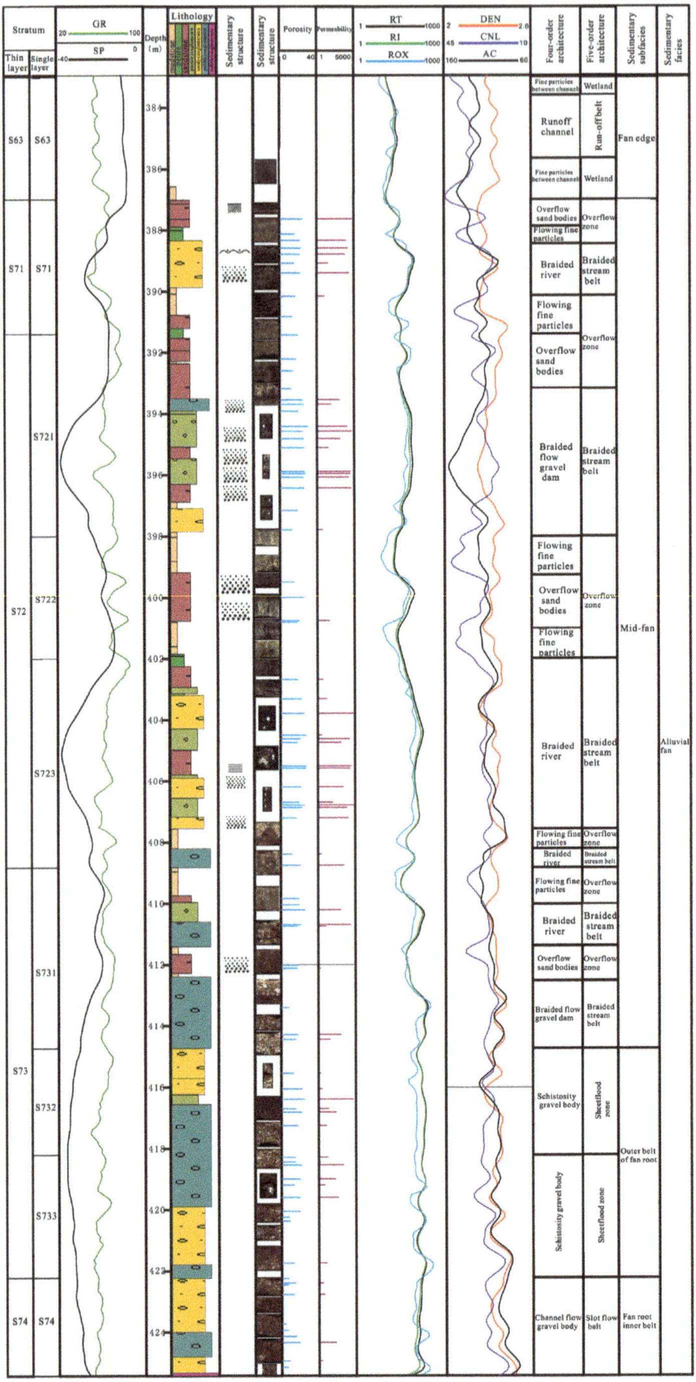

**Figure 3.32** Lithology-architecture correspondence for Well J3 in the Lower Karamay Formation in an area at the northwestern of the Junggar Basin.

flow gravel in the upper fan inner zone and trough beach sand-gravel in the upper fan outer zone have lower porosity and permeability than the braided flow channel and braided flow sand-gravel bar in the middle fan. The braided flow channel and braided flow sand-gravel bar in the middle fan share similar porosity and permeability. The cross flow sand architectural element includes high-permeability stripes locally, where the physical properties are lower than that of braided flow channel and braided flow sand-gravel bar. The microconglomerate shows the best physical properties according to the comparison. The physical properties of glutenite and pebblestone are inferior to the microconglomerate. The cobblestone only appears in a few horizons, and the limited samples were not compared. There are also high-permeability stripes in sandstone. As described above, the target interval in the study area can be divided into three sets: (1) the lowermost glutenite-dominated strata with microconglomerate and pebblestone high-permeability stripes, containing water layer and oil—water layer in the lower part and oil layer in the upper part, which is the key set of strata for present development; (2) the sandstone-dominated strata with glutenite high-permeability stripes, containing oil layer and locally dry layer, and presenting poor physical properties under the control of lithology, part of which is taken as key strata for development; and (3) mudstone-dominated strata with sandstone high-permeability stripes, containing dominantly dry layer and locally oil layer with poor properties, which is considered as a replacement for future recovery. All these high-permeability stripes should be stressed in the waterflooding development, which can be improved by way of separate-layer water injection and managed-pressure water injection.

In addition, a well cluster was taken to illustrate how the architecture affects the reservoir properties (e.g., connectivity). Totally, one producer (T6070) and four water injection wells (T6060, T6080, T6059, and T6082) were selected (Fig. 3.33). According to the production performance, the producer shows poor waterflooding response, indicating poor reservoir connectivity between the producer and the water injection wells. According to the geological analysis, all five wells in S732 and S733 are dominated by widely and continuously distributed laminar flow gravel deposits. This cannot explain this poor reservoir connectivity. The single-well and interwell architecture analysis demonstrate that there are fourth-order architecture boundaries in laminar flow gravels between Well T6060 and Well 6081 and between Well T6059 and Well T6070, but another type of architecture boundary between Well T6082 and Well

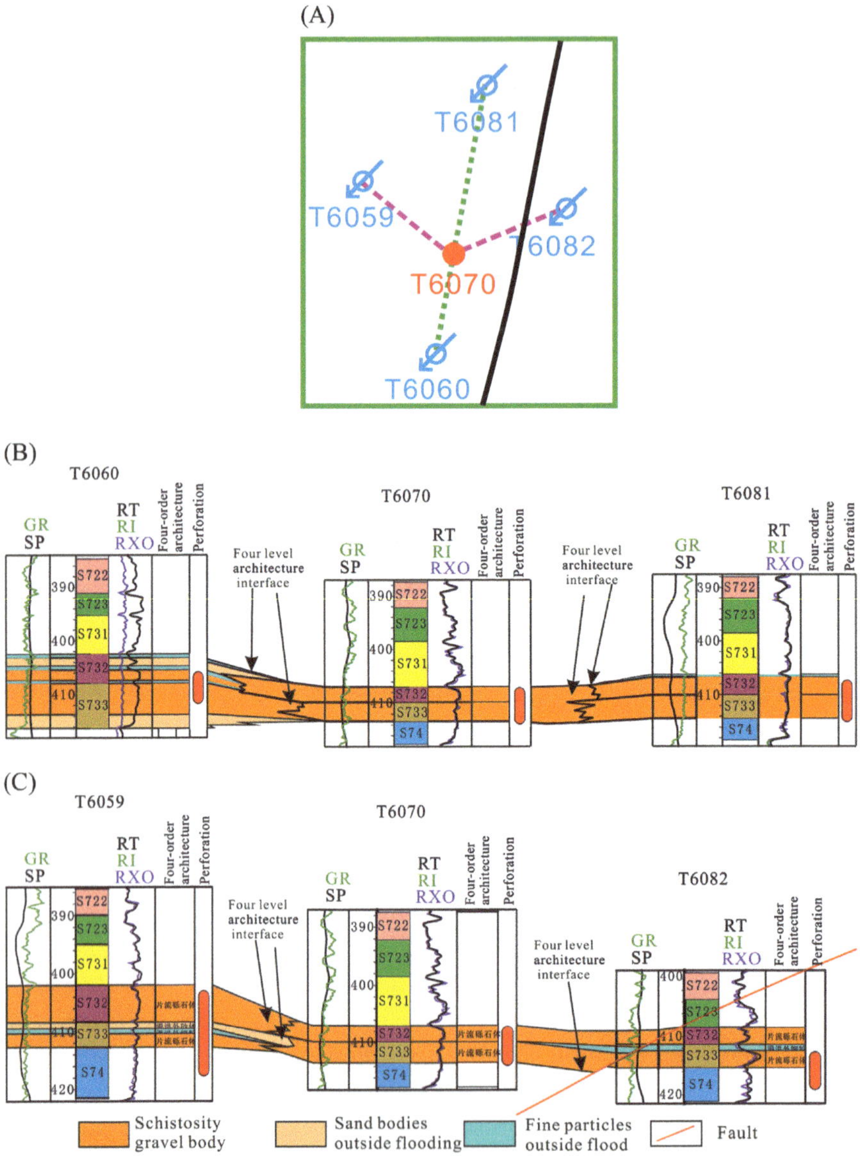


**Figure 3.33** Influences of reservoir architecture on reservoir properties and reservoir development in the Lower Karamay Formation in an area at the northwestern margin of the Junggar Basin. (A) Well position of a well group; (B) North-south profile; (C) East-West section.

T6070. This reveals that the reservoir connectivity in this wellblock is mainly dependent on the reservoir architecture and also further proves the significance of reservoir architecture characterization to efficient reservoir development. It can be concluded that the reservoir architecture plays a crucial role in the reservoir properties (connectivity, homogeneity, etc.). In the reservoir development, it is necessary to identify the position and characteristics of reservoir architecture boundaries and figure out reasonable measures to realize efficient development and enhance oil recovery.

## 3.4 Genetic classification and quantitative characterization of reservoir pore structure

The pore structure of reservoir rocks refers to the geometry, size, distribution, and connectivity of pores and throats in the rocks. It is of immense significance for clarifying the reservoir's internal structure and thereby facilitating oil and gas exploration and development. As an important part of reservoir geology study, pore structure has been one of the research foci (Shaoxian, 1998; Shenghe and Qihua, 1998; Ehrenberg and Nadeau, 2005; Sagan and Hart, 2006; Aguilera, 2006; Ehrenberg et al., 2009; Yuqing et al., 2010). Zhetan and Yuncheng (1986) pointed out that pore structure is the core of microscopic physical properties under the lens in the study of oil and gas reservoirs. Corbett and Jensen (2000) analyzed the zonal distribution of lithology, porosity, and permeability of Arab D reservoir in Uthmaniyah, Saudi Arabia. Ezat Heydari (2000) took the limestone reservoir of Smackover Formation in Upper Jurassic in the Mississippi area as an example to analyze the decrease of pore size and fluid flow and concluded that the decrease of pore structure is mainly caused by mechanical compaction and cementation of clay minerals. Chaika and Dvorkin (2000) took the samples from the Monterey Formation of California as an example to analyze the characteristics of porosity reduction during diagenesis of diatomitem and concluded that the increase of silica was the main reason for porosity reduction. David W. Morrow (2001) took the Permian dolomite platform in western Texas, USA, as an example to study the distribution law of reservoir porosity and permeability. Bloch et al. (2002) analyzed the cause of high porosity in deeply buried sandstone reservoirs and predicted the reservoir development characteristics. The reservoir quality model analysis showed that this kind of reservoir was subjected to high fluid pressure since the beginning of shallow burial and, thus, high porosity can be preserved; besides, quartz cementation is the main reason for the decrease of reservoir porosity. Jones and Xiao (2005) analyzed and evaluated the formation process of the pore structure of dolomite based on a 2D geological model. He pointed out that surface temperature plays an important role in controlling dolomitization and influences the correlation between porosity and permeability to a certain extent. He further indicated that the static and dynamic characteristics of porosity and anhydrite preservation in main dolomite reservoirs could be predicted by using the model. Ehrenberg et al. (2006) analyzed the relationship between porosity and permeability of interbedded limestone and dolomite reservoirs, revealing that sedimentation and diagenesis control the pore structure of carbonate reservoirs. Sruoga and Rubinstein (2007) took the Austral and Neuque'n basins in Argentina as examples to analyze the control of pore formation process on porosity and permeability of volcanic reservoirs and showed that a clear understanding of the formation process of primary and secondary pores could guide the prediction of volcanic reservoir quality, thereby benefiting the oil and gas exploration and development in many regions of the world. Fabricius et al. (2008) took 70 chalk samples from 4 sets of overpressure formations in the North Sea area of central Denmark to analyze the effects of mineral changes, fluid pressure, and early oil filling on the correlation between the reservoir porosity, on the one hand, and interval transit time and the burial depth, on the other, and hit study results showed a high correlation between pore volume and interval transit time when fluid pressure and pore volume compressibility are taken into account. Nabawy et al. (2009) characterized the pore and throat characteristics of the Paleozoic–Cretaceous Nubian high-porosity and high-permeability water-bearing sandstone reservoirs in southern Egypt through laboratory experiments. Camp (2011) made a comparative analysis on pore-throat size in sandstone, tight sandstone, and shale. Ehrenberg et al. (2012) indicated that the viewpoint that the pores of carbonate rocks were formed in the middle stage of diagenesis is incorrect, and the pore development degree of carbonate rocks could not be predicted by the knowledge that secondary pores increased significantly in the middle stage of diagenesis. Huanqing et al. (2013a,b,c,d) classified and evaluated the pore structure of volcanic reservoir in Ying 1 Member in Xudong area, the Songliao Basin, by cluster analysis. Huanqing et al. (2013a,b,c,d) summarized the research progress of reservoir pore structures and the contents and methods of pore structure research. Nguyen et al. (2013) took the Triassic fluvial sedimentary reservoir in the central graben of the North Sea as an example to analyze the influence of fluid pressure and diagenetic cementation on the preservation of reservoir pore structures. Mastalerz et al. (2013) analyzed the pore structure of New Albany shale in Devonian and Mississippi systems by maturity and studies on petrology, natural gas adsorption, and mercury injection, and pointed out that organic matter transformation and hydrocarbon migration are the key factors for porosity differences. Jin et al. (2014), taking the Chang 8 pay zone in the Jiyuan area of the Ordos Basin as an example, analyzed and quantitatively evaluated the genesis mechanism of the pore structure of tight sandstone reservoir. At present, the reservoir pore structures are mainly identified and classified by way of microscopic thin-section observation, tests, and laboratory experiments. The tectonism, sedimentation, and diagenesis are analyzed to clarify the genesis of reservoir pore structures, the geological model is used to simulate and predict the reservoir pore development, and the mathematical methods such as cluster analysis are adopted to quantitatively classify and evaluate the reservoir pore structures. In this way, the reservoir microcharacteristics such as physical properties are recognized. The study objects include clastic rock, carbonate rock, and volcanic rock. Prediction by geological modeling is commonly used abroad, while tests and experiments are popular in China. Globally, the pore structure analysis of tight oil and gas reservoir or shale oil and gas reservoir is the hotspot of reservoir pore structure studies. In the study area, influenced by the sedimentary environment of the fan delta front, underwater distributary channel frequently redirects, and the reservoir heterogeneity is strong. To realize the conversion of reservoir development mode from steam stimulation to steam flooding, it is necessary to deeply understand the spatial variation of reservoir physical properties. Thus the pore structure analysis is particularly important. Through investigation, it is found that there are rare studies on heavy oil thermal recovery reservoirs through quantitative characterization of pore structure on the basis of geological genesis. In this study, we classified and evaluated the pore structures based on the existing studies around the world. This classification and evaluation enable the researchers to deeply understand the spatial variation of reservoir physical properties and the spatial connectivity of reservoirs, thereby providing a scientific basis for the conversion of development mode and the enhancement of oil recovery by way of interwell infilling, the transformation of producers to injectors, and fine separate-layer water injection; and also by theirs, that is, classification and evaluation, serving as a reference for analyzing the pore structure of reservoirs with similar geological setting and development mode.

## 3.4.1 Studies on pore structure

### *3.4.1.1 Current status*

#### 3.4.1.1.1 Study contents

The studies on reservoir pore structures are different in aspects and priorities depending on their objectives. The study contents in the following seven aspects are summarized in this book.

1. Analysis of factors influencing the reservoir pore structures

   The factors influencing different types of reservoir pore structures are analyzed from the perspective of pore structure evolution so as to thoroughly understand the reservoir pore structures and reservoir properties. Min and Qintian (2003) analyzed the relationship between reservoir pore structure and oil displacement efficiency in the Panhe faulted block area, and they divided the reservoir pore structure into four types: coarse throat with ultra-high permeability and ultra-high porosity, coarse throat with high permeability and high porosity, medium throat with medium permeability and high porosity, and fine throat with low permeability and medium porosity. Yueming et al. (2007) evaluated the diagenesis of Upper Paleozoic reservoirs in Daniudi Gas Field in the Ordos Basin, and they divided the reservoir pores into four types: (residual) primary intergranular pore, intergranular (intercrystalline) micropore, intergranular (intragranular) dissolved pore (including moldic pore), and micro-fissure (including those cutting through grains and those at grain margins). Taking the Cretaceous sandstone in the Sawan Basin, Pakistan, as an example, Berger et al. (2009) studied the neritic volcanic clastic sandstone, and they concluded that the well-developed quartz-cemented chlorite rims improve the pore preservation by more than 20%, contributing a good reservoir permeability, and the role of chlorite in the pore preservation process was limited by the sedimentary environment of neritic facies. Chengbo et al. (2010) studied the physical properties and micro pore structure characteristics of the low-porosity and low-permeability sandstone reservoir in Quan 4 Member in the Longxi area, the Songliao Basin, through core observation, microscopic thin-section, scanning electron microscopy, and X-ray diffraction, as well as the use of analysis of physical properties and mercury intrusion, and they concluded that the reservoir is apparently characterized by low porosity and low permeability, and composed of residual intergranular pore and dissolved pore, with small pore-throat radius, relatively large proportion of small pore throats which are poorly sorted and connected, and stronger heterogeneity. Fan Aiping et al. (2011) analyzed the pore structures of Shan 1 Member and He 8 Member sand reservoirs in the East 2 zone of Sulige Gas Field, the Ordos Basin, and considered that sedimentation and diagenesis are the main factors controlling the low porosity and low permeability in the study area. Genesis analysis is the most basic content of pore structure research. It can help the researchers master the characteristics of reservoir pore structure from a deep level accurately and has been highly valued by researchers.

2. Quantitative characterization and classification and evaluation of reservoir pore structures

Quantitative characterization and classification and evaluation of reservoir pore structures mainly rely on mathematical methods such as geostatistics and cluster analysis to characterize and predict the reservoir pore structures, thereby providing the basis for reservoir prediction and oil and gas development. Xijin et al. (2002) divided the pore structures of the Jurassic reservoirs at the northwestern margin of the Junggar Basin through cluster analysis of mercury injection samples, and they quantitatively and qualitatively evaluated the reservoir pore structure types by the statistics of various parameters and the comparison of log patterns. Nelson (2009) analyzed the sizes of pore throats in sandstone, tight sandstone, and mudstone, and he pointed out that the pore-throat diameters are generally larger than 2 μm in conventional sandstone, 0.03–2 μm in tight gas sandstone, and 0.005–0.1 μm in mudstone. Weger et al. (2009) quantitatively studied the pore structure of carbonate reservoir, and they analyzed its influence on acoustic velocity and reservoir permeability; their study results showed that the pores of internal debris or internal crystal debris and separated crystal clusters cannot always be separated by sonic log data, the P-wave velocity is not the only proportion controlled by spherical porosity, and the geometric characteristics of a large number of micropores can be estimated by acoustic velocity data and used to improve the permeability evaluation effect. Wenquan et al. (2010) applied the quantitative reverse inversion method to quantitatively study the pore evolution of deep low-permeability gas reservoirs in rifted basins, taking the deep glutenite gas reservoirs in the northern steep slope zone of the Dongying Sag as an example. Wuren et al. (2010) classified and evaluated the pore structures of the Upper Triassic Xujiahe Formation sandstone reservoir in central Sichuan Basin, and they divided the reservoir pore structures into four types: coarse-skewed bimodal type, medium-skewed bimodal type, fine-skewed bimodal type, and single-peak type. The quantitative characterization and classification and evaluation is a difficult and also a hot spot in reservoir pore structure characterization. It can facilitate accurate reservoir prediction and effective reservoir development.

3. Reservoir evaluation based on pore structure analysis

   The pore structure is an important reservoir property. Through the pore structure analysis, the reservoir can be deeply recognized and evaluated. Haifa et al. (2006) analyzed the micro pore structure of the tight sandstone reservoirs in the He 2 + 3 Members in Daniudi Gas Field, and then they classified and evaluated the reservoirs. Xiaoli and Zhengwen (2006) analyzed the characteristics of Chang 8 reservoir in the Triassic Yanchang Formation in the Longdong area, the Ordos Basin, indicating that the reservoir physical properties are generally poor, belonging to low-porosity and low-permeability reservoir, which mainly contains intergranular pore, dissolved pore, and intergranular pore, and medium-fine to fine throats, and then they made a classification and evaluation.

4. Influence of reservoir pore structure on fluid behavior

   The properties of pore structure significantly control and influence the behaviors of fluids in the reservoir. Heydari (2000) studied the pore loss, fluid flow, and material transfer of the Upper Jurassic Smackover Formation limestone reservoir in Mississippi, USA. Hollis et al. (2010) analyzed the characteristics of the pore system in the giant heterogeneous carbonate reservoir in northern Oman, and it revealed that pore heterogeneity in the reservoir decides the fluid flow behaviors, while conventional petrophysical parameters cannot effectively indicate the oil displacement efficiency. Wenxiang et al. (2011) analyzed how the micropore structure parameters of ultra-low-permeability reservoir affect the flow behaviors, with the Chang 6 reservoir in the Ordos Basin as an example. Jamaloei et al. (2011) analyzed the effect of pore wettability on the microstructure of remaining oil, that is, the pore-scale fluid flow patterns, when surfactants are used to improve the water flooding performance in heavy oil reservoirs, and their results

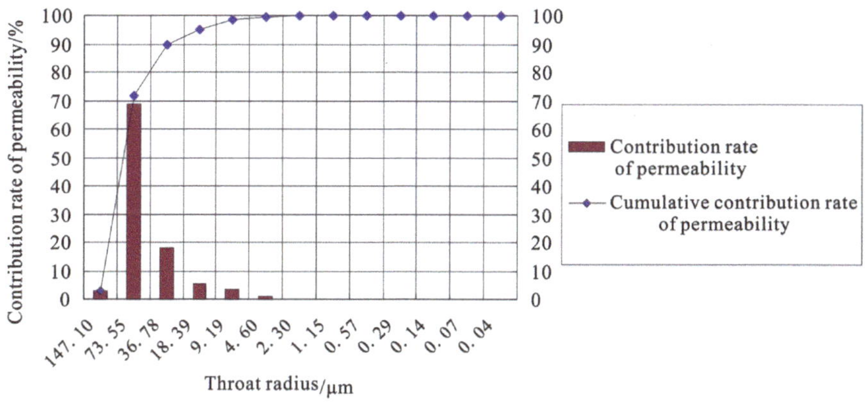


**Figure 3.34** Contribution of pore-throat radius to reservoir permeability of the Lower Karamay Formation glutenite reservoir in an area at the northwest margin of the Junggar Basin. *Note*: Passel formula method; Well J3, 410.7 m.

showed that counting the length and diameter of oil droplets allows us to correctly recognize the viscosity, capillary force, and inertial force at the pore level. Benyamin Yadali Jamaloei's study validated the previous view that the size and shape of oil droplets are related to oil capture and movement at the pore level. This study can help rebuild the remaining oil distribution model under different wettability conditions.

**5.** Influence of reservoir pore structure on development

As mentioned before, the pore structure is one of the most important reservoir properties, so it is closely related to reservoir evaluation and effective development (Fig. 3.34). Zhong (2000) investigated the relationship between reservoir pore structure and oil displacement efficiency by taking the middle Sha 3 Member in Xia 52 block and Sha 2 Member in Xia 32 block in Linnan Oilfield as examples. Mingguo and Xiaolan (2007) analyzed and revealed the intrinsic relationship between pore structure and oil production potential of the Xingouzui Formation sandstone reservoir in the Jianghan Basin, and they pointed out that throat size is the most important factor to control oil production. Jie et al. (2009) applied the 3D pore network model to study the effect of reservoir pore structure parameters on polymer flooding recovery, and their results showed that the water flooding recovery increases with the increase of throat radius, coordination number, and shape factor, and it decreases with the increase of pore/throat ratio. Hu et al. (2011) discussed the micropore structure of Upper Paleozoic reservoirs in northern Jingbian Gas Field and its influence on production performance, and their results revealed that the microscopic pore structure of reservoir has a significant effect on the production performance of gas wells. Chuang and Wei (2011) analyzed the relationship between microcharacteristics of the reservoir in the Fu 3 Member in Shawei Oilfield, the Subei Basin, and oil displacement efficiency, and obtained the oil displacement efficiency by water displacement experiment with real sandstone micromodel, and they also discussed the effects of physical properties and reservoir pore structure on oil displacement efficiency.

**6.** Effect of oilfield development on reservoir pore structure

The reservoir pore structure is significant for oilfield development. Meanwhile, with the implementation of development measures such as water flooding and fracturing, the reservoir pore structure changes accordingly. Analysis of these changes can allow us to more accurately understand the current reservoir properties and, thereby, contribute to the oilfield development. Sijing et al. (2000) analyzed the effect of water flooding on the pore structure of sandstone reservoirs, and their results showed that the pore structure and heterogeneity of sandstone reservoirs become very different after water flooding. Hongming et al. (2000) studied the effect of steam flooding on pore structure and mineral composition of Lianhua reservoir in Gaosheng Oilfield, Liaohe Basin, and they concluded that steam flooding increases the reservoir porosity and pore diameter, decreases the throat radius and permeability and enhances the heterogeneity of pore-throat distribution. Jihong et al. (2001) analyzed how water flooding affects the reservoir microstructure in Gudao Oilfield, and they concluded that long-term water flooding had improved the reservoir physical properties and pore structure, decreased the shale content and irreducible water saturation and enhanced the reservoir microstructure to a certain extent. Meina et al. (2004) discussed the effect of water flooding on the properties of the reservoir in the second member of Shahejie Formation in Tuo 30 fault block of Shengtuo Oilfield, and they found that water flooding can improve the reservoir pore structure to a certain extent. Jingyao and Jiaming (2011) took the Fuyu reservoir as an example to analyze the rock properties after $CO_2$ injection through laboratory experiment, which indicated that with over the rock−$CO_2$ contact time, the permeability and porosity of rock increase, and the average pore radius decreases, but the volumes of small pores and large pores increase (the former reflects a more of an addition). This study makes the results of reservoir pore structure analysis more time-effective; especially, the addition of some dynamic data enables the study results to successfully support the effective development and adjustment of oilfields.

7. Improvement of reservoir pore structure analysis methods

   Some traditional reservoir pore structure analysis methods/techniques have gradually exposed some problems in practice. The researchers have improved these methods/techniques. Yudan et al. (2005) improved the method of evaluating the pore diameter distribution of rocks based on the NMR T2 spectra. Shedid (2007) proposed a new method to quantitatively identify the pore distribution of heterogeneous reservoirs by treating crude oil with sulfur and injecting it into real core samples, and coupled with SEM and computer software, the results showed that the new method could explain and provide the required pore distribution characteristics, improve the level of microscale reservoir research, and also reproduce the behavior when the crude oil treated with sulfur flows through the core. Weizhi et al. (2009) investigated the application of NMR logging in reservoir pore structure evaluation. Through comparative analysis of laboratory NMR spectra and mercury intrusion curves, Shao Weizhi et al. put forward a method for calculating scale conversion coefficient and lateral scale coefficient between T2 spectra and mercury intrusion curves, as well as longitudinal scale coefficient of large and small pore sizes, by using the 2D piecewise equal-area method, showing the consistency between the calculated pseudo-capillary pressure curve and the capillary pressure curves measured by mercury intrusion in the laboratory to have been greatly improved. Yubao et al. (2008) used constant-rate mercury intrusion and constant-pressure mercury injection methods to study the pore structure of reservoirs in Lamadian Oilfield at the high water-cut stage, both suggesting that the pore throats of rock samples increase after long-term water injection, and large pore throats are the main flow paths. Yichao et al. (2011) took the flow unit of Kongnan reservoir in Dagang Oilfield as an example to analyze the micropore characteristics of the reservoir by constant-rate mercury intrusion. The improvement of research methods/techniques helps solve the problems therein, improves the level of research, and drives the research progress to new heights.

#### 3.4.1.1.2 Study methods

Reservoir pore structure can be analyzed by multiple methods, such as thin-section observation, 2D and 3D imaging, core experiment, and computer simulation, which are presented as follows.

1. Method of statistical analysis of physical property data (e.g., cast thin-section observation and mercury intrusion)

   This method is the most basic and commonly used one in the study of reservoir pore structures. Through microscopic observation of cast thin-sections, the genetic type and development status of reservoir pore structures can be determined intuitively. Mercury intrusion method is also called mercury injection method. Mercury is a non-wetting phase fluid for rocks. Pressure is applied to force the mercury to enter into the throats after overcoming the capillary resistance at the throats, and then the capillary force is measured to indirectly determine the size distribution of the pore throats (Shenghe and Qihua, 1998). Based on the analysis of rock slice, cast thin-section, scanning electron microscope, and mercury injection data, Jihong et al. (2001) studied the reservoir microstructure of Gudao Oilfield comprehensively and divided the pore structures of the upper Guantao Formation reservoir into four types; the study results showed that the upper Guantao Formation reservoir is dominated by type II and III pore structures, with serious microscopic heterogeneity. Min and Yuehua (2002) discussed the special geological characteristics of reservoirs in the Baolang Sumu structural belt; the cast thin-section identification and reservoir physical property analysis revealed that the target glutenite interval contains non-uniform fine throats with ultra-low to low permeability and medium porosity, and it is mainly composed of intergranular pores, with the maximum connected pore-throat radius of 5.16 μm, the main flowing pore-throat radius of 1.92 μm, and the median radius of 0.52 μm, suggesting as the category of micropore-throats that are poorly sorted. Tiantai et al. (2005) analyzed the reservoir characteristics and damage factors of the gas reservoir in Karamay structural belt based on the capillary pressure curve, and the results showed that the gas reservoir is characterized by medium to small pores and micropores, medium to low permeability, medium to fine short throats, and strong intralayer heterogeneity and plane heterogeneity. This method is advantageous in the fact that the physical property data such as cast thin-section and mercury injection are simple and intuitive (Fig. 3.35) and can be easily acquired for sandstone and carbonate reservoirs. However, thin-section is not suitable for analyzing the pore structure of special lithologic reservoirs such as conglomerate and mudstone unless it is improved.
2. Diagenesis method

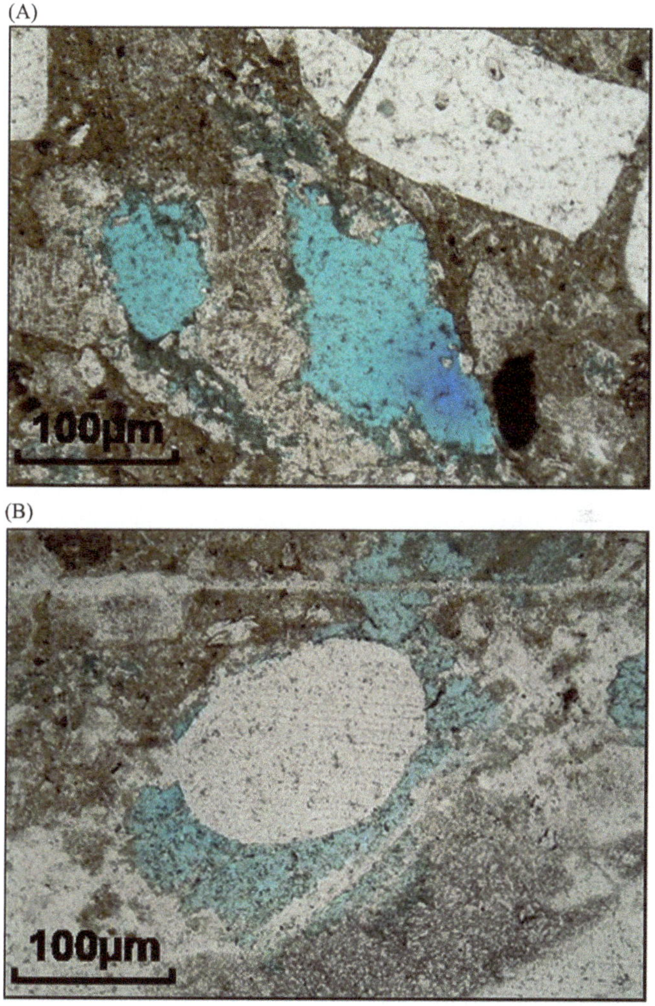


**Figure 3.35** Core and microscopic thin-section photographs illustrating the pore structures of the volcanic reservoir in Ying 1 Member in Xudong area, the Songliao Basin. (A) Moldic pore, volcanic breccia, W42, 3701.06 m, $5 \times 10$, (-); (B) Intergranular dissolved pore, volcanic breccia, W42, 3698. 08, $5 \times 10$, (-).

The diagenesis method is also widely used in the study of reservoir pore structure. The roles of diagenetic processes in the evolution of reservoir pore structure are analyzed to understand how the reservoir pore structure changes. Linyu et al. (2006) analyzed the diagenesis of Chang-81 sandstone in Baimanan area and its influence on the reservoir, and the results revealed that compaction and cementation strongly destroyed the primary pore structure of sandstone, while dissolution and cataclasis effectively improved the pore structure of sandstone. Sruoga and Rubinstein (2007) investigated how the diagenesis of the volcanic reservoir in the Neuque'n Basin, southern Argentina, had controlled the porosity and permeability, and the results showed that primary and secondary processes could be used to predict the quality of volcanic reservoirs, thereby providing guidance for oil and gas exploration and development in many parts of the world. Yefang et al. (2011) studied the effect of authigenic chlorite on pore structure in the sandstone reservoir in the Upper Triassic Yanchang Formation in the Ordos Basin. This method can deeply recognize the genesis of pore structure, but it prefers qualitative analysis rather than quantitative analysis.

**3.** Well logging analysis method

Well logging analysis method is a powerful tool for studying the reservoir pore structure, especially the quantitative evaluation of pore structure. Based on the core analysis and testing data of coring wells, the well logging interpretation model is established and applied to noncoring wells so that the reservoir porosity and permeability are quantitatively interpreted to support the study of reservoir pore structure. Yiling et al. (2004) found that the pore structure (ratio of permeability to porosity) of the reservoir has a good correlation with the pore-throat ratio, and accordingly, the porosity and permeability derived by log data are used to predict the reservoir productivity, which is in good agreement with actual production performance. Hao et al. (2005) took the Paleozoic carbonate buried-hill reservoir in Chegu area as an example to illustrate the evaluation method of the triple pore structure of complex reservoir and its application. Chaomo et al. (2007) studied the fractal pore structure of reservoir rocks based on the distribution of the NMR T2 spectrum. This method is more rapid and non-destructive than the conventional fractal pore structure research method and further expands the application scope of NMR data. Wei et al. (2009) made a comparative study on the methods of evaluating reservoir pore structure by using NMR log data, and the results showed that the method of NMR capillary pressure curve construction based on Swanson parameter is suitable for evaluating pore structure in different types of reservoirs. This method is limited by data, and not all research objects can satisfy such conditions.

**4.** The method based on the fractal dimension model

The so-called fractal is a general term for graphs, structures, and phenomena that have no characteristic length but have self-similarity

(Lihua et al., 2000). In fractal theory, fractal dimension is the most basic feature to describe a fractal. By studying the fractal dimension, the relationship between vector and fractal dimension can be established. The pore structure of reservoir rock is a fractal structure, which can be quantitatively described by fractal dimension (Qiang and Juanru, 2009). The fractal dimension of the pore structure can be calculated by capillary pressure curve to realize the quantitative analysis of the pore structure. Generally, with the increase of fractal dimension, the sorting of the throat becomes worse, the inhomogeneity of the throat becomes stronger, the connectivity of the throat becomes lower, and the displacement pressure and median pressure increase. Also, the larger the fractal dimension, the rougher the pore surface, with uneven size distribution. Xiugang et al. (2005) proposed new parameters for pore-throat fractal structure and physical property evaluation of low-permeability reservoirs. Wei et al. (2008a) analyzed the application of fractal theory in reservoir physics, and they pointed out that the pore structure of sandstone cores has not only fractal characteristics but also the electrical properties of oil and water, and the relative permeability curves of oil and water in the process of seepage inside rocks have fractal characteristics. Qiang and Juanru (2009) illustrated the application of fractal geometry in the analysis of reservoir microheterogeneity, taking the Chen-2 fault block in eastern Gaoyou sub-sag of the Subei Basin as an example, and the results show that the larger the fractal dimension, the rougher the pore surface; and more uneven the size distribution, the greater the resistance to fluid flow and the worse the reservoir performance. The method based on the fractal dimension model has obvious advantages in reservoir pore structure research. Of course, it needs to be improved depending on the specific research objects.

5. Various mathematical methods

As the pore structure research turns to be more quantitative, many mathematical methods are increasingly applied. Based on the core analysis and testing data of coring wells, different mathematical calculation methods are used to quantitatively evaluate the characteristic parameters of reservoir pore structure. Yong et al. (2007) studied the application of mathematical methods (mainly fuzzy recognition method and BP neural network method) in low-porosity and low-permeability reservoir evaluation, and they found that the fractal dimension of reservoir pore structure can quantitatively characterize the sandstone reservoir performance and can reflect the genetic characteristics of sandstone reservoir pore structure, and the pore structures of different origins have different fractal dimensions, so the fractal dimension can be used to classify and evaluate sandstone pore structure. Xupeng (2010) identified the intrinsic relationship between reservoir physical properties and micropore structure, wherein, based on the Kozeny-Carmon equation, the intrinsic relationship between reservoir macro-physical parameters and its micropore structure was deeply analyzed by further deducing and using the method of function monotony analysis. Each mathematical method has its advantages in the quantitative characterization of reservoir pore structure, but it must be based on solid geological analysis.

**6.** Physical or numerical simulation methods

The simulation methods also play important roles in the study of reservoir pore structure. Essentially, the possible changes of reservoir pore structure during the formation of reservoirs or the oilfield development are depicted through different simulation processes so as to characterize the reservoir pore structure. Wei and Ningsheng (2000) simulated the transport network of solid particles in fractal porous media, where the fractal dimension of pore size distribution was estimated by logarithmic regression between differential mercury saturation and pore radius; the law of permeability decline caused by migration and settlement of injected particles in porous media with different fractal dimensions was simulated by 3D multi-segment combined network model; and the relationship between the formation damage degree caused by migration and settlement of particles and fractal dimension of pore size distribution was discussed. Jackson et al. (2003) took the Siberian sandstone as an example to predict the wettability change and simulate its influence on fluid flow from pore scale to reservoir scale, proving that the 3D network model of real sandstone can be used as a powerful tool for predicting reservoir wettability changes and their effects on reservoir fluid flow at oilfield scale. Daley et al. (2006) analyzed the changes of elastodynamics and permeability with the change of pore pressure in fractured reservoirs, and they carried out special coupling simulation to investigate the changes of pore pressure, permeability, and elastic constants in space caused by fluid injection into fractured reservoirs; the results showed that in fractured reservoirs, with the injection of fluid, the seismic response to reservoir changes during seismic detection can be traced. Kewen and Ning (2009) numerically simulated the effects of reservoir characteristics and saturation on NMR T2 spectrum, and they established a 3D irregular network model of a complex reservoir to study the relationship between water-saturated NMR T2 spectrum and pore size distribution by micronumerical simulation, and they discussed the influence of reservoir characteristics and saturation on T2 spectrum. The simulation methods can visually reproduce the fluid behavior in the reservoir pores and its influence on the pore structure itself. Their disadvantage is that there is a certain gap between the simulation and the actual geological conditions underground. Therefore, in practice, on the one hand, we should set up the simulation conditions close to the actual geological conditions; on the other hand, we should envisage the gap between the simulation results and the actual situations.

**7.** New technologies such as 3D imaging

Many new technologies, such as computer technology and electronic display technology, are showing more and more footprints in the study of reservoir pore structure. These technologies make it possible to visually present the micropore structure of reservoirs with high definition. Fengxiang et al. (2002) used laser scanning confocal microscopy to study reservoir pore structure. Shuxin et al. (2007) comprehensively tested and analyzed the microscopic characteristics of the conglomerate reservoir in the fourth member of Yingcheng Formation in Xujiaweizi faulted-depression by using some advanced technologies such as large-diameter core physical analysis, casting acid solution SEM reservoir pore structure analysis, and laser confocal microfracture testing, together with the polarizing microscope identification and core scanning technology, and the results showed that the filling quantity and filling patterns are the main factors affecting the reservoir properties in the study area. Huaiqiang et al. (2008) studied the microcharacteristics of braided-river delta sandstone reservoir in Hongtai area of the Turpan-Hami Basin by means of Zeiss image analysis system (Axio Vision 4.0); they divided the target reservoir into five types of pore structures (A to E), and they evaluated the pore structure types to get a better understanding of pore structure of the reservoir. Zhenliang et al. (2009) introduced 3D imaging technologies for porous media micropore structure, including serial slicing technology, focused ion beam (FIB), laser confocal scanning microscopy, and Miro-CT. Guoping et al. (2010) used image morphology to simulate the diagenesis process of sandstone, established 3D digital cores, and analyzed reservoir physical parameters such as porosity, pore-throat structure, grain size distribution, and specific surface of the digital core model. These methods have greatly promoted the research of reservoir pore structure, but their applicability under different research contents and objectives needs to be further improved.

8. Geological (or geophysical) model method

   The Geological (or geophysical) model plays an increasingly important role in the study of reservoir pore structure. For this method, the geological or geophysical model is used to reproduce the development characteristics of reservoir pore structure in 3D space by means of modeling. The biggest advantage is manifested in the prediction of pore structure development in space. Liao and Cai (2000) established a model for predicting the pore-throat volume distribution of the reservoir in the Shengbei area of Turpan-Hami Oilfield, which could procure the reservoir pore structure analysis in case of inadequate mercury intrusion test samples or core data for some sections and wells in an area and thereby benefitted the correct evaluation of the heterogeneity of reservoir pore structure in the study area. Congjiao et al. (2005) established the reservoir pore structure model based on the microrandom network simulation method for Zhangtianqu Oilfield. Cancan et al. (2006) used the spherical tube model to evaluate the pore structure of rocks, and they applied the decomposition results of the spherical tube model to realize the graphical representation of the pore structure. Jinglong et al. (2009) selected seven logs (including SP, GR, and AC) reflecting the characteristics of reservoir pore structure to establish sample model and neural network model, which were used to predict the reservoir pore structure type, revealing a coincidence more than 80%. The geological (or geophysical) model method has effectively promoted quantitative research on reservoir pore structure, and it has been introduced to more and more applications.

9. Cluster analysis method

   The cluster analysis method is also widely used in the study of reservoir pore structure. Based on the genesis analysis, the parameters that can fully reflect the development characteristics of reservoir pore structure are selected for quantitative classification and evaluation of reservoir pore structure with the support of SPSS or other data processing and operation platforms. Liu Shunsheng et al. determined six parameters (porosity, permeability, median pore-throat radius, mean pore-throat radius, variation coefficient, and ejection efficiency) for classification after the correspondence analysis of 16 parameters reflecting the pore structure characteristics, and then they divided the pore structure of conglomerate

**Table 3.5** Result of pore structure evaluation based on physical properties for the volcanic reservoir in Ying 1 Member in Xudong area, the Songliao Basin.

| Parameter category | Porosity Φ/% | | | Permeability K/ $10^{-3}\mu m^2$ | | | Median Pore-throat Radius $R_{50}/\mu m$ | | |
|---|---|---|---|---|---|---|---|---|---|
| | Max. | Min. | Avg. | Max. | Min. | Avg. | Max. | Min. | Avg. |
| I | 14.4 | 10.5 | 13.0 | 1.19 | 0.04 | 0.20 | 0.299 | 0.062 | 0.161 |
| II | 9.9 | 7.7 | 8.8 | 1.12 | 0.02 | 0.10 | 0.251 | 0.021 | 0.113 |
| III | 7.5 | 4.9 | 6.3 | 1.35 | 0.01 | 0.05 | 0.128 | 0.018 | 0.06 |
| IV | 4.8 | 2 | 4.0 | 1.21 | 0.01 | 0.04 | 0.086 | 0.016 | 0.03 |

reservoirs into four categories and six subcategories by using the method of systematic cluster analysis (Qingchang et al., 1997). Hongxian et al. (2010) comprehensively analyzed the pore structure of the Lower Karamay Formation reservoir in Block 6-Mid of Karamay Oilfield using cast thin-section, X-ray diffraction, rock physical properties, SEM, and mercury intrusion data, and they classified the reservoir into four categories (I–IV) by cluster analysis. In the study of the pore structure of volcanic reservoirs in the Cretaceous Yingcheng Formation in the Xudong area of the Songliao Basin, we selected the characteristic parameters (porosity, permeability, and median pore-throat radius) that can fully reflect the pore structure to quantitatively classify and evaluate the pore structure of target reservoir by way of cluster analysis (Table 3.5). The cluster analysis method can realize quantitative evaluation of pore structure based on qualitative reservoir pore structure analysis, thereby ensuring the reservoir pore structure characterization. However, the selection of the clustering algorithm and the parameters is a challenging task and must be done with the support of rich experience.

In practice, the researcher should choose a specific method or a combination of methods depending on the purpose of research and the availability of data to finely characterize the reservoir pore structure.

### *3.4.1.2 Problems and trends in reservoir pore structure research*

#### 3.4.1.2.1 Problems

The current research on reservoir pore structure is believed to face the problems in five aspects. First, qualitative and quantitative methods are not combined adequately. Currently, the reservoir pore structure is either described through the observation of cast thin-sections of cores or quantitatively evaluated through core analysis. There are rare studies that integrate the two methods to reveal the origin of reservoir pore structure and also realize the quantitative evaluation of reservoir pore structure. Second, the traditional mercury intrusion method is partially limited in its application. Xiaoyong et al. (2009) analyzed the application of gas bubble-pressure method in testing reservoir pore structure, and they indicated that the diameter of main pore throats measured by this method is 2–4 times of that obtained by the mercury intrusion method, thus concluding that this method can test the distribution characteristics of effective throats involved in flow in reservoirs and it is more suitable for practical engineering needs. Yongli and Zhiguo (2011) quantitatively evaluated the pore-throat structure difference of low-permeability sandstone reservoirs by constant-rate mercury intrusion technology. Their study indicated that the smaller the core permeability, the more concentrated the throat distribution. With the increase of permeability, large throats increase, the throat distribution becomes wider, and the radius of the main throat also increases; however, larger throats have dual effects on the exploitation effect of low-permeability sandstone reservoir. Third, reservoirs with special pore structures, such as conglomerate and shale, are rarely studied. These reservoirs are very different from conventional clastic reservoirs in pore structure due to the influence and limitation of genetic conditions. So far, they have been less studied, without adequate technologies and knowledge. Fourth, there are no recognized parameters and feasible methods for the quantitative evaluation of reservoir pore structure. In China, the sedimentary reservoirs are dominantly continental, which are characterized by fast facies variation, strong heterogeneity, and rapid change of pore structure in space. To evaluate these reservoirs in different regions/horizons, we should select appropriate parameters and methods. However, there are no uniform criteria for such selection. Fifth, new methods/techniques have not been promoted sufficiently in reservoir pore structure characterization. Currently, most of the studies are based on the observation of cast thin-sections and statistical analysis of mercury intrusion data, while 3D digital imaging and digital core technology have not been widely applied.

#### 3.4.1.2.2 Trends

It is considered that the reservoir pore structure research moves forward in five trends. First, qualitative and quantitative methods are combined for the classification and evaluation of reservoir pore structures. Classification and evaluation is the most basic part of reservoir pore structure research;

specifically, the reservoir pore structure is comprehensively understood to deeply recognize the reservoir microcharacteristics. For this purpose, a qualitative or a quantitative method can be adopted, but their organic combination is more preferred. Second, the pore structure is studied to support the evaluation and accurate prediction of the reservoir. The pore structure is the most basic attribute of the reservoir, and it is also influential to permeability. Understanding the pore structure can enable reservoir classification and evaluation. Dou et al. (2011) discussed the pore classification and permeability heterogeneity evaluation of carbonate reservoirs based on petrophysics by taking the upper San Andres reservoir in the Permian Basin, West Texas, as an example, and the results showed that the lower the refractive index of core samples, the higher the acoustic velocity. Variations in the relationship between porosity and wave impedance or between porosity and permeability can be distinguished by clear geological explanations such as pore types and rock structure variation to verify the prediction accuracy of porosity and permeability. Third, reservoirs with special pore structures may become more attractive as a focal point of study. At present, the sandstone and carbonate reservoirs are mostly involved, but the conglomerate and shale reservoirs with special pore structures have been rarely reported. As the new reserves are found harder and harder to recover, the researchers have begun to deal with and will pay more attention to the pore structures of tight oil and gas reservoirs. Fourth, multiple methods are integrated to comprehensively study the reservoir pore structure. The strengths of all methods should be mobilized and combined organically to ensure the best pore structure research. Cerepi et al. (2002) combined 2D quantitative image analysis and 3D lithophysical tools to analyze the micro geological characteristics of pores in low-resistivity sandstone reservoirs, and they concluded that the micro geological characteristics of pores are greatly meaningful for the physical properties of low-resistivity sandstone reservoir. Fifth, new methods/techniques are applied to improve the accuracy of reservoir pore structure research. Many new methods and techniques are growing rapidly, and they can be considered for pore structure characterization. Jia et al. (2007) analyzed the diatom-bearing reservoir samples with high-resolution images to obtain the information about pore type, size and pattern, and have used the improved 3D two-phase flow network model to calculate the reservoir permeability and relative permeability and mapped the capillary pressure curve; the results demonstrated that the relative permeability calculated by the model agrees with that obtained from laboratory analysis. When studying the pore structure of volcanic reservoirs in the Xudong area of the Songliao Basin, we used CT scanning technology to observe the core intuitively. The CT images clearly illustrate that the heterogeneity of reservoir pore structure is very strong, which is consistent with the porosity and permeability data of core analysis.

## 3.4.2 The pore structure of Yulou reservoir in an area of Western Sag, the Liaohe Basin

### *3.4.2.1 Genetic classification*

The geology of the study area has been expounded in Section 1.1. Here, based on the core observation, thin-section, and core test data of seven coring wells and the reservoir physical properties obtained from fine logging interpretation of 400 wells, the genesis of reservoir pore structure is

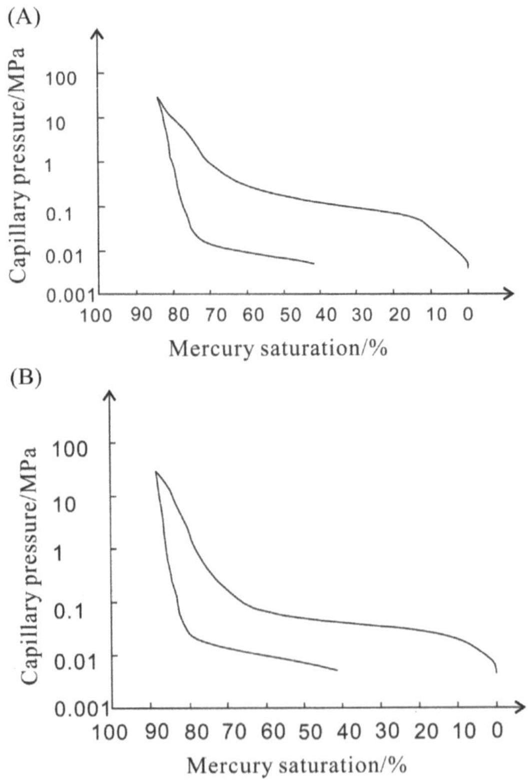


**Figure 3.36** Mercury intrusion curve of Well A1 in an area of Western Sag, the Liaohe Basin. (A) In the I reservoir group, depth:923.18 m; (B) In the I reservoir group, depth:1001.99 m.

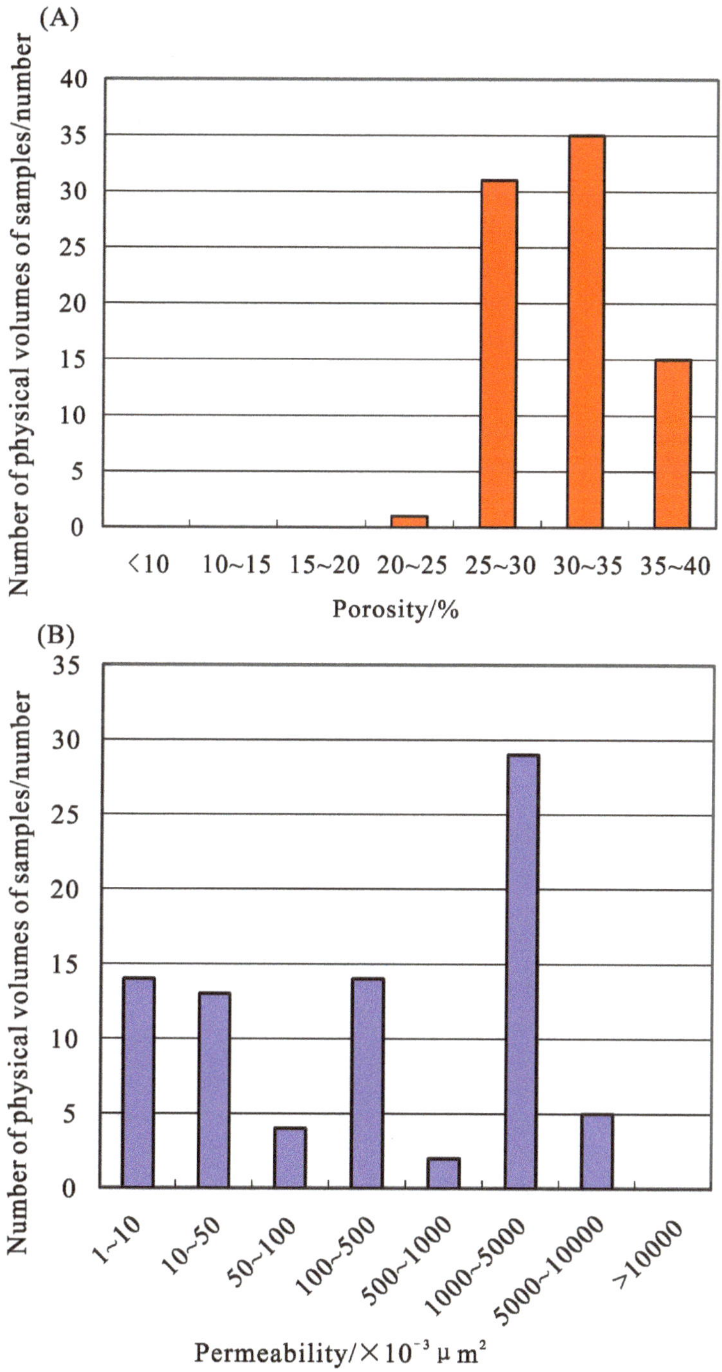


**Figure 3.37** Physical properties of Yulou reservoir in an area of Western Sag, the Liaohe Basin. (A) Porosity; (B) permeability.

analyzed and classified. Thus the development of reservoir pore structure is understood from macrolevel and microlevel perspectives, thereby providing a reference for changing the reservoir development mode. Microscopically, as shown on the mercury intrusion curve (Fig. 3.36) of a sealed-coring well in the study area, the pore throats are well-developed in the target interval, roughly skewed, and well sorted. The physical property analysis of coring wells (Fig. 3.37) shows that the Yulou reservoir in the study area exhibits the porosity of 25%–40% (31.25% on average) and the permeability of 1–5000 mD (1829.3 mD on average), indicative of a high-porosity and high-permeability reservoir. According to the core and

**Table 3.6** Genesis-based classification of pore structure of Yulou reservoir in an area of Western Sag, the Liaohe Basin.

| Pore type | | Genesis |
|---|---|---|
| Primary pore | Intergranular pore | Pores between particles formed during sedimentation |
| | Intragranular pore | Generally, intergranular micropores in debris |
| | Matrix micropore | Micropores in clay matrix |
| | Cleavage fracture | Common flake or wedge cleavage fractures in feldspar and mica minerals, with width mostly less than 0.1 μm and some up to 0.2 μm |
| Secondary pore | Intergranular dissolved pore | Pores between particles regenerated from dissolution, mainly the pores between particles formed by the dissolution of cement and matrix at the edge of or between particles |
| | Intragranular dissolved pore | Pores formed in particles by dissolution |
| | Moldic pore | Pores formed by the complete dissolution of particles, bioclastic, or substitutes. Their shape is identical to that of the original component |
| | Superlarge intergranular dissolved pore | Dissolved pores whose diameter exceeds the diameter of adjacent particles. Generally, particles, cement, and substitutes are dissolved |
| | Structural fracture | Fractures formed by tectonism |
| | Dissolved fracture | Mainly formed when the structural fractures are reworked by dissolution |

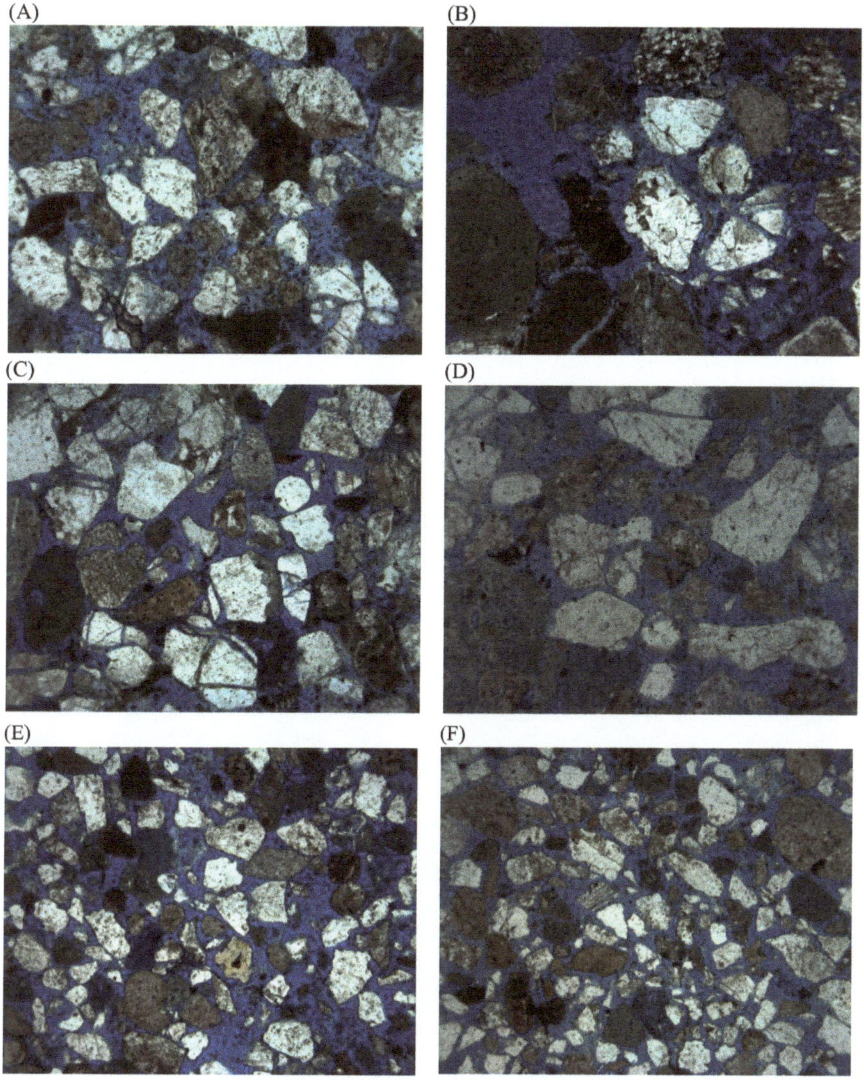


**Figure 3.38** Pore structure of Yulou reservoir in an area of Western Sag, the Liaohe Basin. (A) Well A2, intergranular pore, 945.3 m, ×50; (B) Well A2, superlarge dissolved pore, 965.72 m, ×25; (C) Well A2, structural fracture, 974.56 m, ×25; (D) Well A2, matrix micropore, 960.32 m, ×50; (E) Well A2, moldic pore, 1022.72 m, ×25; (F) Well A2, cleavage fracture, 1005.72 m, ×25.

thin-section data, we divided the pore structure of the Yulou reservoir into primary pores and secondary pores by genesis, which is subdivided into ten sub-classes: intergranular pore, intragranular pore, matrix micropore, cleavage fracture, intergranular dissolved pore, intragranular dissolved pore, moldic pore, superlarge intergranular dissolved pore, structural fracture, and dissolved fracture, corresponding to different genesis (Table 3.6). Each pore type shows unique behaviors on thin-sections (Fig. 3.38). Generally, intergranular pore and intergranular dissolved pore are dominant.

The pore structure of Yulou reservoir is also influenced by diagenesis, including cementation and dissolution (Fig. 3.39). Cementation includes both carbonate rock cementation and clay mineral cementation.

**Figure 3.39** Pore structure of Yulou reservoir in an area of Western Sag, the Liaohe Basin. (A) Well A2, carbonate mineral, 965.72 m, sandstone, 6000 ×; (B) Well A2, feldspar overgrowth up to Grade I, 979.32 m, sandstone, 100038 ×; (C) Well A2, flake illite on particle surface, 984.35 m, sandstone, 6000 ×; (D) Well A2, smectite, 989.18 m, sandstone, 4729 ×; (E) Well A2, kaolinite, vermiform, 1005.72 m, sandstone, 6000 ×; (F) Well A2, particle dissolution, 1022.72 m, sandstone, 1701 ×.

**Figure 3.39** *(Continued)*

Generally speaking, reservoir diagenesis has less impact on pore structure, and sedimentation is the major driver.

### *3.4.2.2 Development features*

Macroscopically, the quantitative plane development of porosity was mapped according to the reservoir physical properties of 400 wells derived from fine logging reinterpretation (Fig. 3.40). Taking $yI_1{}^{2a}$ as an example, the porosity distributes as stripes in NW-SE direction. Compared to the distribution of sedimentary microfacies in the same horizon (Fig. 3.41), it is found that the porosity development of $yI_1{}^{2a}$ is obviously controlled by sedimentary microfacies, and high porosity is basically observed on the main stream line of underwater distributary channel, which reflects that this zone features a strong hydrodynamic force, mainly consisting of coarse-grained sediments such as gravel and fewer mud sediments.

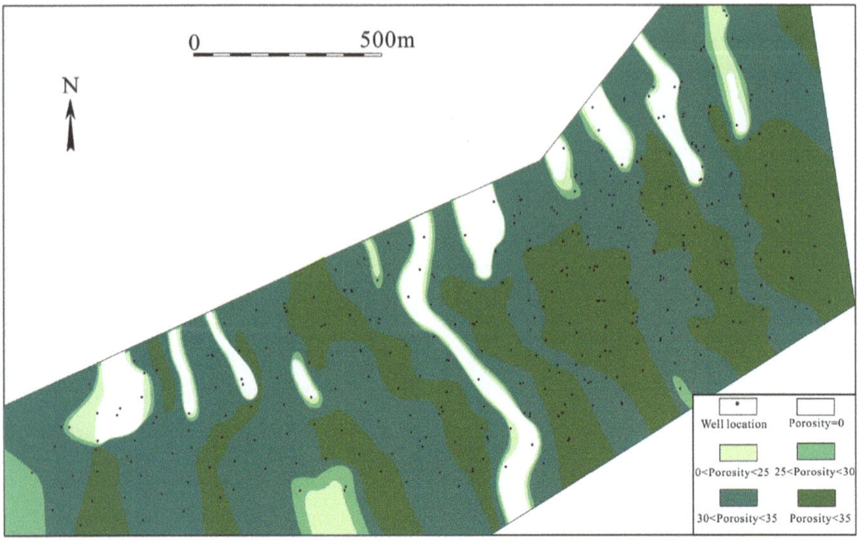


**Figure 3.40** Plane distribution of porosity in $yl_1^{2a}$ of Yulou reservoir in an area of Western Sag, the Liaohe Basin.

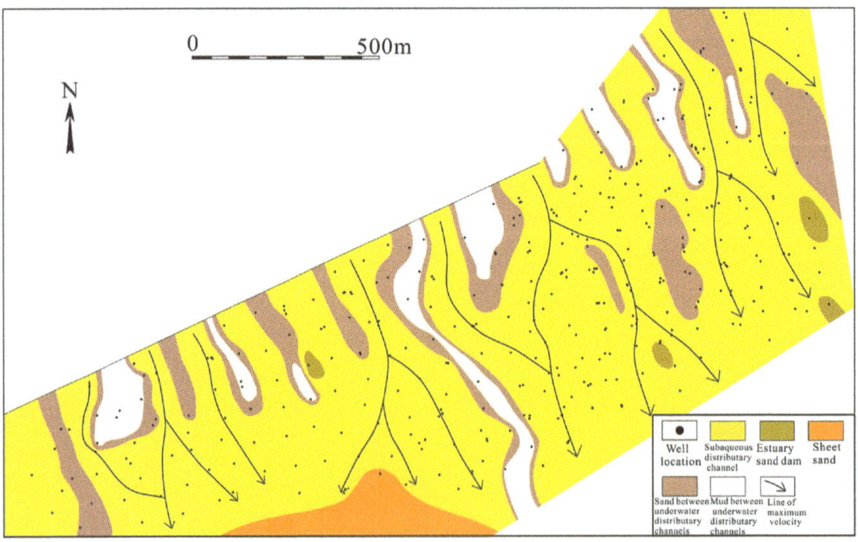


**Figure 3.41** Plane distribution of sedimentary microfacies in $yl_1^{2a}$ of Yulou reservoir in an area of Western Sag, the Liaohe Basin.

According to the horizon of thin-section sample and the plane map of porosity, the reservoir pores in space are mainly intergranular pore and intergranular dissolved pore, which is consistent with the general results of the micro genesis analysis of reservoir pore structure. Between the underwater distributary channels, the porosity is low, and the deposits mainly include thin fine sandstone, siltstone, and argillaceous siltstone. From a macro perspective, the zones with reservoir porosity greater than 35% and of 30%–35% are the most to be found in the study area, while the zones with the porosity less than 30% are very limited, which agrees with the microgenesis analysis result; namely, the target is a high-porosity and high-permeability reservoir. The remaining 28 beds present roughly similar porosity distribution to $yI_1^{2a}$.

### 3.4.2.3 Influence of reservoir pore structure on development

The pore structure is closely related to reservoir properties. Based on the core analysis data of 7 coring wells, the correlation between porosity and permeability was mapped for the target interval (Fig. 3.42), revealing a good correlation: that is, permeability increases with the increase of porosity. From the macroscopic perspective, $yI_1^{2a}$ is taken as an example to compare the plane distribution of porosity and permeability (Fig. 3.43), which indicates a good correlation between porosity and permeability. High permeability is also observed on the main stream line of the

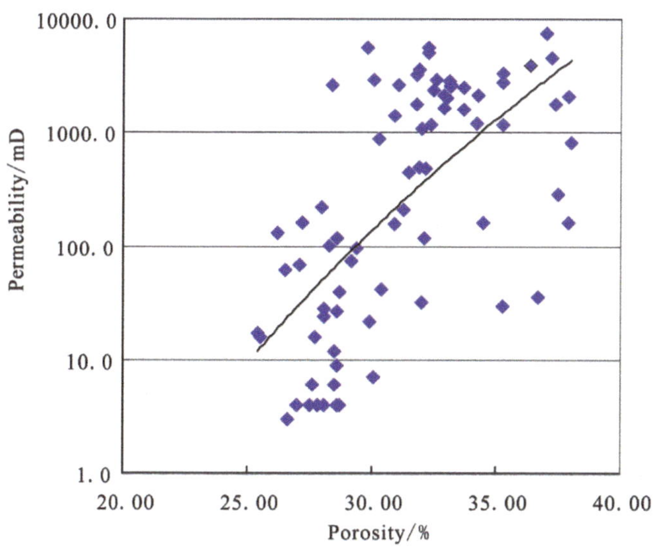


**Figure 3.42** Porosity-permeability diagram of coring well in Yulou reservoir in an area of Western Sag, the Liaohe Basin.

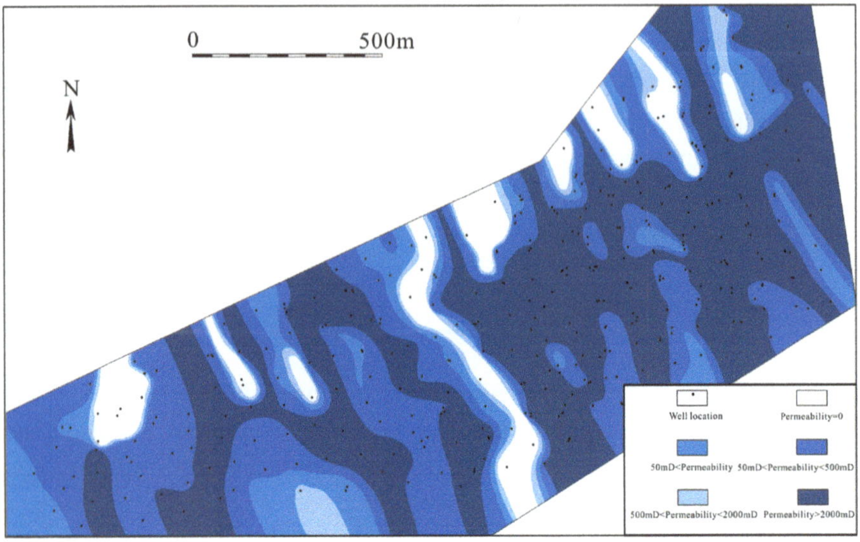


**Figure 3.43** Plane permeability distribution of Yulou reservoir in an area of Western Sag, the Liaohe Basin.

underwater distributary channel, similar to the distribution of high porosity. In addition, the zones with a permeability greater than 500 mD dominate the study area. Thus the target interval is considered to be of high porosity and high permeability (Fig. 3.43). Other beds also have this feature. The effect of porosity should be fully considered for changing the development mode. First, steam injectors and producers should be deployed in high-porosity zones, if possible. When determining such zones, we should carefully consider the microlevel and macrolevel characteristics of porosity to ensure the flooding unit is placed on the main stream line of the underwater distributary channel. Second, when designing the well pattern, we should ensure that one flooding unit is deployed in a zone with continuity of high porosity, that is, such a zone that is within the same underwater distributary channel, so as to achieve the best injection-production correspondence. Third, the proper injection pressure should be maintained during steam flooding to prevent steam from rushing along high-porosity stripes so as to expand the swept volume and improve the displacement efficiency and oil recovery. Moreover, in practice, the injector–producer straight line should be directed against the flow direction of the underwater distributary channel, if possible, in order to improve the flooding performance.

## 3.5 Reservoir heterogeneity analysis

Reservoir heterogeneity refers to the spatial non-uniformity of reservoir parameters, and it is a universal feature of reservoirs. In development reservoir evaluation, the reservoir heterogeneity is a dual heterogeneity, that is, the heterogeneity of rock containing fluids and the heterogeneity of property and occurrence of fluid in rocks (Weirong et al., 1999). Many scholars have done a lot on the definition, research contents, and research methods of reservoir heterogeneity (Yongsheng, 1993; Yinan et al., 1997a,b; Shenghe and Qihua, 1998; Longxin et al., 2000; Amro and Michel, 2001; Guangqing and Zhongxian, 2005; Shaochun et al., 2006; Mikes, 2006; Mahsanam and Diganta, 2007; Chuanqing et al., 2008; Xinghe, 2009; Nima and Ioannis, 2011; Peter et al., 2012; Benoît et al., 2013; Junwei et al.; 2014; Yuan et al., 2015; Prem et al., 2016; Ningning et al., 2016;

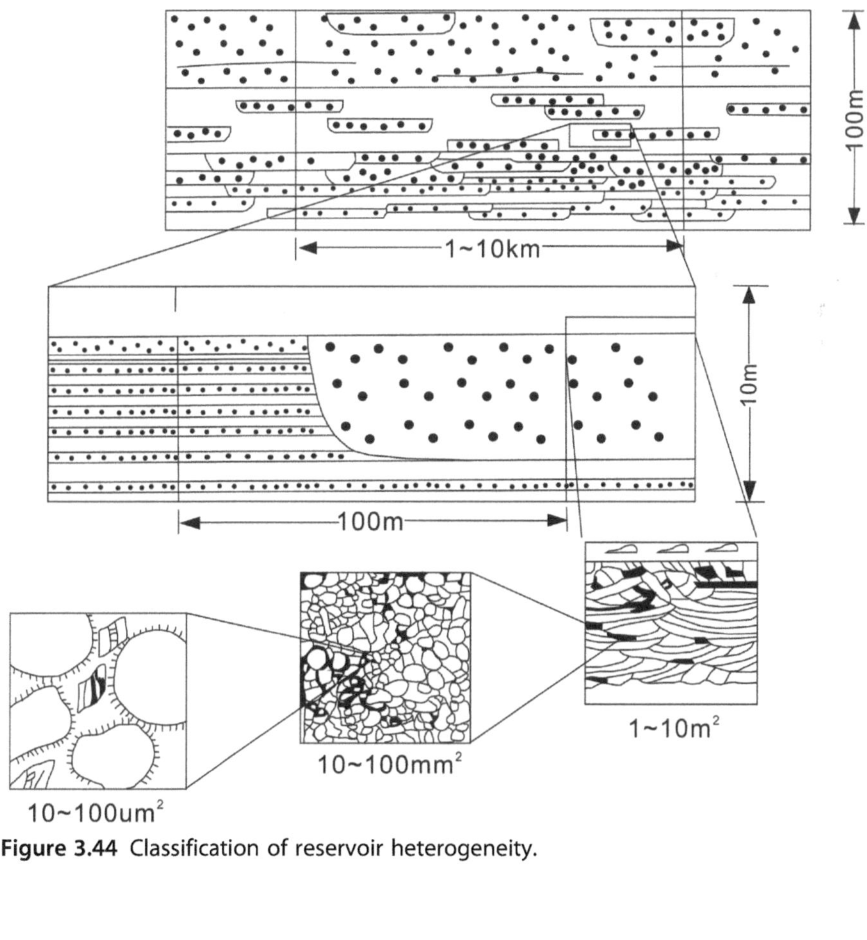


**Figure 3.44** Classification of reservoir heterogeneity.

Yue and Shiyue, 2016; Huili et al., 2016) (Fig. 3.44). Generally, reservoir heterogeneity has been emphasized by researchers abroad since it was proposed in the 1970s. They mainly discuss the classification of heterogeneity, the influence of heterogeneity on reservoir performance, and the influence of heterogeneity on oil and gas development, from the macroscopic or microscopic perspective, with the support of outcrop, laboratory experiment, geological modeling, numerical simulation, and other methods/techniques. With the deepening of studies, the accuracy and quantification level are gradually improved. In China, the most widely used classification scheme for reservoir heterogeneity is the one proposed by Yinan: intralayer heterogeneity and plane heterogeneity are mostly discussed, while microscopic heterogeneity has been rarely involved.

Reservoir heterogeneity is the main factor affecting underground oil, gas, and water movement and oil and gas recovery. Research on reservoir heterogeneity is of great significance for effective reservoir development (Shenghe and Qihua, 1998; Huanqing et al., 2011a,b,c; Huanqing et al., 2012a,b,c). M.L. Sweet et al. (1996) established a reservoir heterogeneity model of the low-permeability gas reservoir in Hyde, southern North Sea, using the geostatistical method. Madeleine Peijs-van Hilten et al. (1998) studied the reservoir heterogeneity at different scales through geological modeling, with the Countess YY incised valley reservoir in Alberta, southern Canada, as an example. Tidwell and Wilson (2000) used permeability analysis and testing data to analyze the heterogeneity of the Masilon sandstone reservoir in Ohio, USA. Pranter et al. (2005) analyzed the petrophysical characteristics according to the Madison Formation dolomite outcrop in Mississippi River, Sheep Canyon, Wyoming, USA, and pointed out that the reservoir heterogeneity caused by the lateral lithophysical property cycle in the dolomite model was 10% more important than the petrophysical difference in view of the impacts on pores and cavities. Pranter et al. (2007) modeled the mesoscale reservoir heterogeneity by simulating the fluvial point-bar outcrop of Williamson Fock Formation in the Piceance Basin, western Colorado, USA, and the results showed that the mesoscale reservoir heterogeneity is of great significance to the fluvial reservoir. Currently, outcrop, geostatistics, and geological modeling are the most popular methods abroad to study reservoir heterogeneity, and they are obviously quantitative.

In China, there are also many scholars working on the reservoir heterogeneity (Lijun and Xinjun, 2003; Xinbian and Shimin, 2003; Leyuan et al., 2005; Shaochun et al., 2006; Junwei et al., 2006; Linyu et al., 2008;

Chen et al., 2011a,b,c; Huanqing et al., 2012a,b,c). Lijun and Xinjun (2003) discussed the influence of macroscopic heterogeneity at the level of the intralayer, interlayer, and plane on the remaining oil in the process of water flooding. Xinbian and Shimin (2003) applied the variable-scale fractal technique to study the heterogeneity of fractured-vuggy carbonate reservoirs. Leyuan et al. (2005) took the Paleogene Shahejie Formation in the North Dagang structural belt of the Huanghua Depression as an example to investigate how the high-resolution sequence stratigraphic framework influences and controls the reservoir heterogeneity. Shaochun et al. (2006) took No.2 zone of Shengtuo Oilfield in the Dongying Sag of the Jiyang Depression as an example to expound the calculation, function, and significance of heterogeneity index, and they pointed out that water injection can aggravate the reservoir heterogeneity. Junwei et al. (2006) discussed the technologies for developing the reservoirs with low permeability and strong heterogeneity in Sulige Gas Field. Linyu et al. (2008) objectively analyzed the microscopic heterogeneity of Chang 8 sandstone reservoir in the Xifeng area of the Ordos Basin through real sandstone micromodel experiment. Generally, in China, reservoir heterogeneity is mainly studied by means of well logging, sequence stratigraphy, micro-experiment, and fractal technology. Particularly, the permeability variation coefficient, differential, and breakthrough factors are calculated based on the porosity, permeability, and other physical properties obtained by well logging interpretation. The studies involve the characteristics of the interlayer, intralayer, plane, and microscopic heterogeneity, and also how the reservoir heterogeneity controls the remaining oil distribution and influences the water flooding. Moreover, the studies are mainly based on drilling data and experimental analysis, but rarely apply geological modeling and numerical simulation of the outcrop. There are very few reports globally on the conversion from steam stimulation to steam flooding thermal recovery through the study on the heterogeneity of reservoirs with heavy oil thermal recovery.

## 3.5.1 Current status

### *3.5.1.1 Contents*

Reservoir heterogeneity includes both the heterogeneity of macroscopic reservoir characteristics, and the heterogeneity of reservoir microscopic pore structure. The research on reservoir heterogeneity may involve multiple contents, such as reservoir heterogeneity classification, reservoir heterogeneity evaluation, reservoir heterogeneity genesis analysis, the influence of heterogeneity on reservoir and fluid properties, and the influence of reservoir heterogeneity on oil and gas field development. Heterogeneity is the core of reservoir geology study and, also, the result of different geological processes during reservoir development. Lijun and Xinjun (2003) discussed the influence of intralayer, interlayer, and plane heterogeneities on remaining oil distribution in the process of water flooding, and they proposed that the zones with strong heterogeneity are the habitats for the remaining oil. Matthew et al. (2006) analyzed the influence of carbonate cement on permeability heterogeneity of fluvial sedimentary reservoirs through outcrop simulation, and they indicated that the spatial distribution of carbonate cement is a major factor controlling the reservoir permeability distribution. Timothy (2006) studied the influence of reservoir heterogeneity on fluid simulation and put forward that the model simulation provides a powerful tool for researchers to understand reservoir heterogeneity. Brenda et al. (2007) used reflectance spectra to study the diagenetic heterogeneity and flow path characteristics of the Jurassic Navajo sandstone, USA, and they showed that the heterogeneity of the target sandstone reservoir is mainly controlled by diagenesis and it cannot be evaluated at the outcrop scale. Lianbo et al. (2008) took the ultra-low permeability sandstone reservoir of the Upper Triassic Yanchang Formation in the Ordos Basin as an example to identify the influence of heterogeneity and multifracture development. Heath et al. (2012) analyzed the geological heterogeneity and economic uncertainty of shallow $CO_2$ buried media. Higgs et al. (2013) took Cabny Oilfield in New Zealand as an example to study the variation of reservoir heterogeneity in high-pressure gas reservoirs containing $CO_2$, and the results showed that $CO_2$ plays a very important role in the change of reservoir heterogeneity and the vertical change of reservoir heterogeneity depends on the dissolution of minerals in the sweet spot areas and on the cementation of minerals in the tight reservoir areas. To study the reservoir heterogeneity in an area at the northwestern margin of the Junggar Basin, we first analyzed the rock types (Fig. 3.45). The target interval in the study area is of alluvial fan gravel deposits, with diverse rock types that are variable in space. Thus the lithology variation can be determined to diagnose the spatial change of reservoir heterogeneity. It can be seen that reservoir lithology changes rapidly in space due to the comprehensive influence of climate, provenance, and topography, which directly leads to strong reservoir

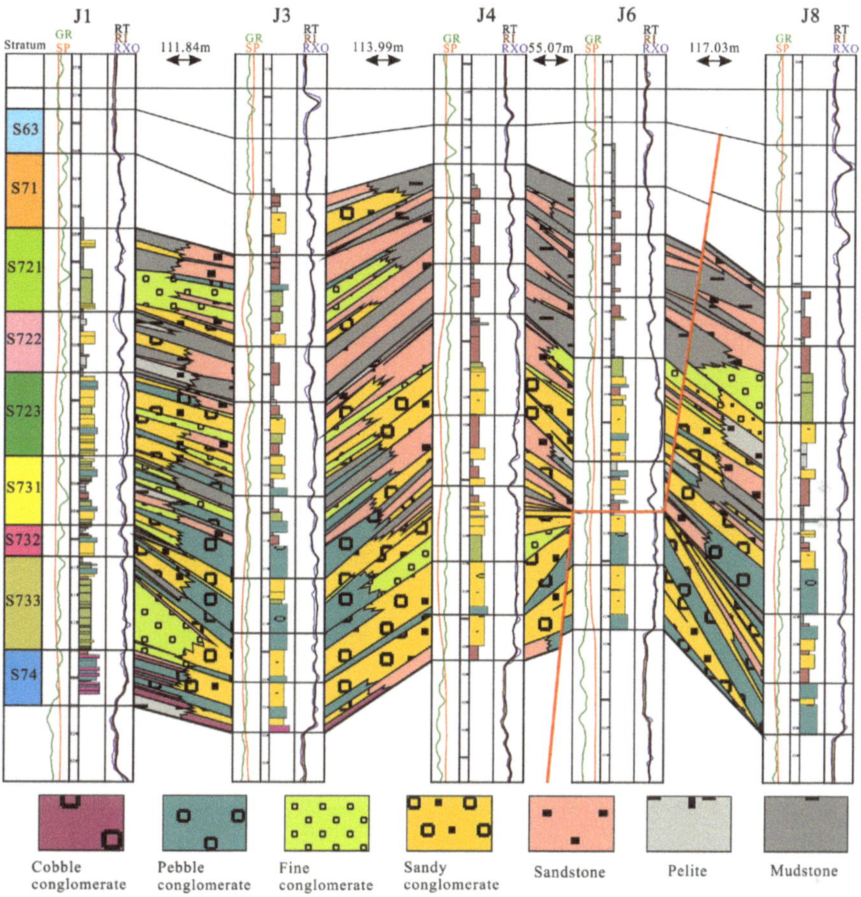


**Figure 3.45** Lithology profile of conglomerate in Lower Karamay Formation in an area at the northwestern margin of the Junggar Basin (vertical to the provenance direction).

heterogeneity. In countries abroad, to study reservoir heterogeneity, various experiments are conducted to detect the physical properties (porosity and permeability) of the reservoir and the impacts on fluid flow pattern, and the outcrop data are used for geological modeling and simulation to quantitatively analyze the reservoir heterogeneity. In China, the geological genesis (e.g., sedimentation and diagenesis) is analyzed to identify the reservoir heterogeneity, and the reservoir physical properties (e.g., permeability) derived from fine logging interpretation are adopted to quantitatively evaluate the reservoir heterogeneity.

**Table 3.7** Summary of reservoir heterogeneity analysis methods.

| Method | Contents | Advantages | Disadvantages |
|---|---|---|---|
| Geological genesis analysis | Heterogeneity genesis, macroscopic heterogeneity, and microscopic heterogeneity | It can help deeply understand the heterogeneity from the genetic perspective | It is slightly inadequate in quantitative analysis |
| Sequence stratigraphy | Interlayer heterogeneity, plane heterogeneity | It can describe macroscopic heterogeneity to guide oil and gas exploration | Its accuracy in the research on oil and gas development heterogeneity needs to be improved |
| Experiment | Microscopic heterogeneity, the effect of heterogeneity on oil and gas development | It can obtain the most intuitive information about heterogeneity | It is costly and may generate the conclusions to occasionally deviate from the actual situation due to the limitation of experimental conditions |
| Geostatistics | Intralayer heterogeneity and microscopic heterogeneity, etc. | It can obtain a large amount of quantitative information about heterogeneity, based on the accessible parameters | The accuracy is restricted by the reliability of reservoir physical property interpretation |
| Mathematical calculation | Spatial distribution of heterogeneity | It can provide quantitative results | It should be used together with geological analysis method |
| Logging interpretation | Spatial distribution of heterogeneity, intercalation feature | It can provide quantitative results | The accuracy is restricted by the reliability of reservoir physical property interpretation |
| Geological modeling | Spatial distribution of heterogeneity | It can describe the quantitative 3D distribution of reservoir heterogeneity | Accuracy and precision of the model need to be improved |
| Seismic prediction | Spatial development of intercalation | It can quantitatively describe the macrocharacteristics of heterogeneity | The accuracy is limited due to method, techniques, and data |
| Numerical simulation | Effect of heterogeneity on oil and gas development | It realizes the prediction by combining various data | The results are affected by human factors |
| Production performance analysis | Development of intercalation | Achievements can be used directly to guide production practice | Some nonheterogeneous factors may be considered as heterogeneous factors, leading to deviations |

### *3.5.1.2 Methods*

There are various methods for reservoir heterogeneity analysis. Maximov (1980) summarized the methods as mathematical statistics, parameter comparison, random function, and pressure build-up curve. This book presents a discussion of the methods such as geological genesis analysis, sequence stratigraphy, experiment, geostatistics, mathematical calculation methods, logging interpretation, geological modeling, seismic prediction, numerical simulation, and production performance analysis, together with the advantages and disadvantages of each method (Table 3.7).

#### 3.5.1.2.1 Geological genesis analysis method

Geological genesis analysis is the most commonly used method in the study of reservoir heterogeneity. Based on the geological genesis affecting reservoir heterogeneity, this method can reveal the main influencing factors of heterogeneity and describes the degree of heterogeneity. Generally, there are diverse factors influencing reservoir heterogeneity; nevertheless, structural factors, sedimentary factors, and diagenetic factors are dominant. Structural factors include faults and fractures. Sedimentary factors refer to the changes in sedimentary facies or sedimentary microfacies, which may lead to changes in reservoir properties (especially reservoir heterogeneity). Diagenetic factors mainly refer to compaction, cementation, and dissolution, which may result in differences in reservoir properties that correspond to the change of reservoir heterogeneity. Fan et al. (2006) analyzed the heterogeneity of Donghe sandstone on the east slope of Lunnan by diagenetic facies analysis, and their study results indicated that the extensively developed carbonate dense cementation facies and non-dense cementation facies led to strong reservoir heterogeneity in early diagenesis, and the carbonate non-dense cementation—weak dissolution facies, residual primary pore—weak dissolution facies, and residual primary pore—dissolution facies developed by local dissolution are the result of improved reservoir properties in late diagenesis. Qin et al. (2007) analyzed the heterogeneity of clastic rock reservoirs of Paleogene Shahejie Formation in the Dongying Sag from macroscopic and microscopic perspectives and analyzed its sedimentary genesis. Shizhong et al. (2008) studied the "architecture-controlled 3D heterogeneity model" of single-channel sand body, and they established the "architecture-controlled 3D heterogeneity model" of meandering channel sand body with "crescent wedge-shaped lateral deposits whose permeability decreases toward convex bank and upward, overlapping obliquely one by one and separated by thin baffles of non-permeable lateral deposits." Daniela and Claiton (2010) studied the heterogeneity of fluvial–eolian reservoirs of the Upper Jurassic Sergi Formation in the Reconcavo Basin, Brazil, based on the study of sedimentary architecture, and the results showed that the reservoir heterogeneity could be further understood through the analysis of sedimentary architecture and petrophysics. Fanhao et al. (2011) analyzed the diagenetic heterogeneity of Chang 6 reservoir of Upper Triassic in Jiyuan area, the Ordos Basin, using core and testing data, and the results showed that the Chang 6 Member is a low-porosity ultra-low permeability reservoir dominated by feldspar sandstone, where the mechanical differentiation at the beginning of deposition led to the pore heterogeneity of original sediments, which made the reservoirs with different original pore conditions evolve differently during the diagenetic evolution, and that went through four evolutionary stages (early compaction, early acid water injection, restrained carbonation cementation, and organic acid and hydrocarbon injection), ultimately leading to the diagenetic heterogeneity of reservoirs. Philippe et al. (2012) analyzed the relationship between stratigraphic architecture and multiscale reservoir heterogeneity by taking the Barremian-Aptian carbonate platform in Walkluz, southeastern France, as an example. When studying the heterogeneity of Yulou reservoir in an area of the Western Sag, the Liaohe Basin, we divided the factors influencing reservoir heterogeneity into three types: structural factors, sedimentary factors, and diagenetic factors; the former two types are dominant, while the diagenetic factors can only be seen locally in the lower part of the target interval (Fig. 3.46) (Huanqing et al., 2015a,b,c,d). It can be seen that, at 1105 and 1115 m, calcite tips generated due to calcite cementation appear on logs, indicating that at the depths, the formation properties were affected by diagenesis, enhancing the formation heterogeneity. The geological genesis analysis method, essentially, is to study the influence of the abovementioned processes on reservoir heterogeneity through structural analysis, division of sedimentary facies or microfacies, and genesis analysis. Its operation is basically the same as the research on structure, sedimentation, and genesis, but its focus is on reservoir heterogeneity. This method can enable the researchers to have a profound understanding of the genesis of reservoir heterogeneity, but it is slightly inadequate in quantitative analysis.

#### 3.5.1.2.2 Sequence stratigraphy method

Sequence stratigraphy has been applied tentatively in reservoir heterogeneity analysis. This method relies on the theory of sequence stratigraphy

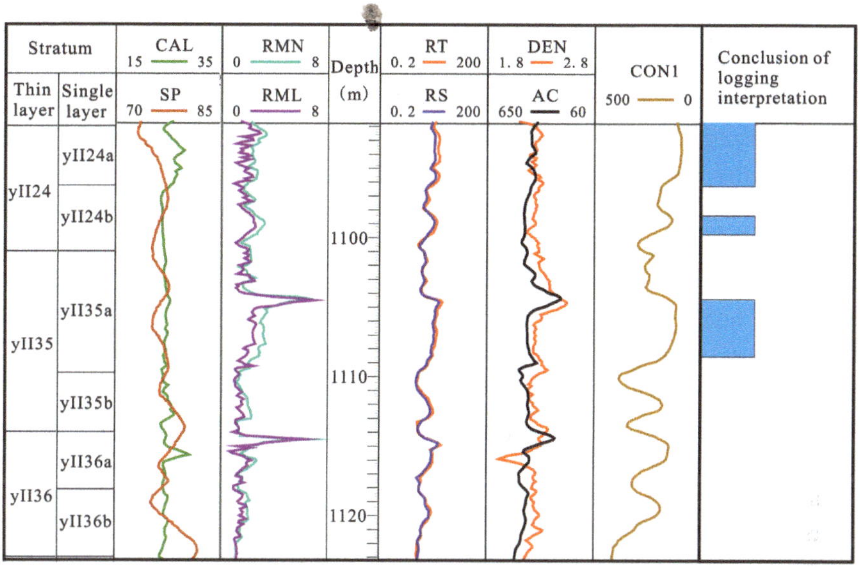


**Figure 3.46** Lithoelectric characteristics of diagenetic intercalation in a well in an area of the Western Sag, the Liaohe Basin. *Note*: depth: 1104–1105 m and 1114–1115 m. Blue represents the water layer interpreted. *From Huanqing, C., Chao, D, Yijing, D, Jue, W., 2015a. Advances of reservoir evaluation researches. Geological Science and Technology Information 34(5), 66–74. Huanqing, C., Chengfang, S., Chen, C., 2015b. Advances in the study of fine reservoir description. Geological Review 61(5), 1135–1146. Huanqing, C., Shuxian, L., Peng, J., et al., 2015c. Flow barrier characteristics of thermal rescovery of heavy oil reservoir: taking Yulou oil-bearing sets in some experimental area in the south end of westward slope of West depression in Liaohe Basin as an example. Acta Sedimentologica Sinica 33(3), 616–624. Huanqing, C., Shuxian, L., Zhirui, S., et al., 2015d. Caracteristics of conglomerate reservoir architecture of alluvial fan and its controlling effects to reservoir development: taking alluvial fan reservoir in some area of Northwest margin of Junggar Basin as an example. Journal of Jilin University (Earth science edition) 45(1), 13–24.*

to analyze the formation and distribution of reservoir heterogeneity from the perspective of genesis. Leyuan et al. (2005) analyzed the high-resolution sequence stratigraphic framework of Paleogene Shahejie Formation in North Dagang structural belt of the Huanghua Depression and its controlling effect on reservoir heterogeneity, and the results showed that short-term base-level cycles control the internal heterogeneity model of the sand body; and the medium- to long-term base-level cycles control the interlayer heterogeneity of reservoir sand body. Shiguang et al. (2009) took the Fuyu reservoir in Blocks Chao 1–Chaoqi 3 in Chaoyanggou Oilfield, the Songliao Basin, as examples to discuss the application of high-resolution sequence stratigraphy in the study on macroscopic reservoir heterogeneity. According to the sedimentary dynamics analysis of the origin of base-level cycles of different orders and the quantitative analysis of core test and statistical data for characterizing the macroscopic reservoir heterogeneity, it is stated to have been found that with the rise of long-term base level, the intralayer heterogeneity of short-term base-level cycles is enhanced, the interlayer heterogeneity of medium-term base-level cycles is weakened, and the plane heterogeneity of short-term base-level cycles is weakened; with the decline of long-term base level, these trends are just the opposite. It is pointed out that the variations of sedimentary environment caused by base-level cycles and its accommodation space changes are the decisive factor for the difference in macroscopic reservoir heterogeneity. The research on reservoir heterogeneity by sequence stratigraphy can: (1) provide a high-precision isochronal stratigraphic framework; and (2) enable a deeper understanding of the origin of reservoir heterogeneity (e.g., the spatial distribution of sand bodies and vertical depositional cycle and rhythm) through analyzing different orders of base-level cycles. The sequence stratigraphy method fully embodies the genetic mechanism of heterogeneity, but its accuracy needs to be further improved. Currently, it is mainly used to describe the macroscopic heterogeneity of reservoirs to guide oil and gas exploration. Its applicability to oil and gas field development needs to be analyzed.

#### 3.5.1.2.3 Experiment methods

Experiments are important foundations and tools for geologic study and, of course, for research on reservoir heterogeneity. First-hand data of reservoir properties can be obtained through various basic experiments. In recent years, especially with the advancement of relevant experimental instruments/devices and testing techniques, the experiment method has played an increasingly important and prominent role in reservoir heterogeneity analysis. Through these experiments, a series of characteristic parameters reflecting reservoir heterogeneity can be acquired, and then the distribution and variation of these parameters are analyzed to support the study on reservoir heterogeneity. Shehadeh (2002) used the centrifuge separation technology to judge how the wettability heterogeneity affects the capillary pressure curve, and the results showed that the recovery degree of reservoir wettability decreases with the increase of oil saturation. Zhongfeng et al. (2005) made 3D heterogeneity physical models by using the artificial physical model making technology and used the constant-rate method to test these models for water flooding so as to analyze the effect of reservoir heterogeneity on the oil displacement efficiency. Linyu et al. (2009) used the real sandstone micromodel to study the microscopic heterogeneity of Chang 3 reservoir in the Baibao area, the Ordos Basin, and the results showed that the target reservoir has a strong microscopic heterogeneity, which is mainly controlled by diagenesis and pore structure. Ming et al. (2013) conducted a 3D physical simulation of oil displacement to quantitatively study the variation of oil saturation and the sweep coefficient of positive-rhythm high water-cut reservoir geological model under different injection volumes and displacement pressures. Jesús et al. (2013) established the criteria for distinguishing bathyal mudstones from abyssal mudstones based on the heterogeneity of deep-water reservoirs with the support of SEM scanning. When studying the heterogeneity of volcanic reservoir in Ying 1 Member, Xudong area, the Songliao Basin, Huanqing et al. (2013a, b,c,d) used the CT scanning technology to analyze the microscopic heterogeneity of reservoir pore structure. The volcanic rocks in the target strata were formed by multistage eruptions, and these rocks are diverse in types, which are identified as volcanic lava and pyroclastic rocks, with ten types of lithologies. According to the CT images (Fig. 3.47), the lithologies vary greatly in physical properties (e.g., porosity), and even any one lithology may exhibit non-uniform physical properties in a very short distance.

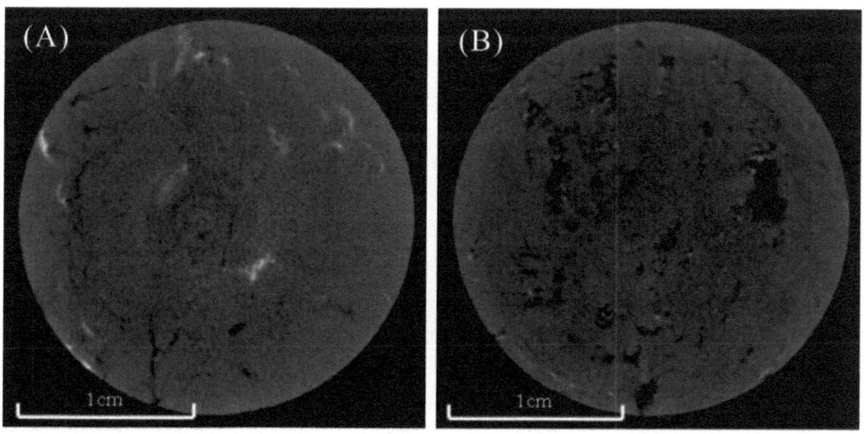


**Figure 3.47** CT images showing the pore development in different rock types in Ying 1 Member, Xudong area, the Songliao Basin (Huanqing et al., 2013a,b,c,d). (A) Green rhyolitic tuff, Well W21, Ying 1 Member, with porosity of 9.2%, the density of 2.39 g/cm$^3$, the permeability of 0.437 mD, and better-developed pores; (B) Green-gray rhyolite, Well W14, Ying 1 Member, with porosity of 14.3%, the density of 2.26 g/cm$^3$, the permeability of 1.097 mD, and well-developed pores.

Overall, due to the differences in rock properties and structures, the intralayer heterogeneity of sedimentary volcanic breccia is obviously stronger than that of tuff, and the intralayer heterogeneity is very different among volcanic lithologies. The experiment methods focus on the microscopic reservoir heterogeneity or on the influence of reservoir heterogeneity on oil and gas development. These methods can obtain the most intuitive information about reservoir heterogeneity, but the experiments are usually costly, aggravating the research costs. Moreover, some experiments are inoperable since sampling is limited by objective conditions.

#### 3.5.1.2.4 Geostatistical method

The geostatistical method mainly collects and calculates the reservoir parameters such as permeability to analyze the degree of reservoir heterogeneity. It also includes the statistics and analysis of information such as the thickness and lateral length of intercalation inside reservoirs. He Yan et al. (2001) studied the geostatistical method for reservoir heterogeneity description, and they found that this method is efficient and practical to a certain extent. Peter et al. (2006) quantitatively analyzed the reservoir porosity heterogeneity using the geostatistical method, which can acquire the quantitative information of reservoir heterogeneity based on accessible parameters, but the accuracy of the results is restricted by the reliability of reservoir physical property interpretation data. Huanqing et al. (2012a,b,c) adopted the geostatistical method to successfully identify the intralayer heterogeneity of volcanic reservoir in Ying 1 Member in Xudong area, the Songliao Basin, by comparing the ratio of vertical permeability to horizontal permeability inside the reservoir (Table 3.8). As the main criterion,

**Table 3.8** Ratio of vertical permeability to horizontal permeability of volcanic reservoir in Ying 1 Member in Xudong area, the Songliao Basin (Shenghe and Qihua, 1998).

| Bed | Ratio of vertical permeability to horizontal permeability | | | | | |
|---|---|---|---|---|---|---|
| | Kv/Kh1 | | | Kv/Kh2 | | |
| | Max. | Min. | Avg. | Max. | Min. | Avg. |
| YC1I1 (35) | 4.33 | 0.03 | 0.69 | 0.03 | 15.62 | 1.17 |
| YC1I2 (3) | 1.17 | 0.35 | 0.84 | 1.40 | 0.89 | 1.19 |
| YC1II1 (9) | 8.51 | 0.02 | 1.58 | 1.72 | 0.18 | 0.67 |
| YC1II2 (9) | 0.60 | 0.00 | 0.26 | 0.50 | 0.01 | 0.34 |

*Note:* The number in the brackets is the sample number.

low Kv/Kh indicates that the vertical permeability of fluid is relatively low; otherwise, the vertical permeability of fluid is high (Shenghe and Qihua, 1998). Peter et al. (2015) studied reservoir heterogeneity using the geostatistics method on the basis of a literature review. The geostatistical method is one of the most widely used methods in reservoir heterogeneity analysis. It should be noted that the study on reservoir heterogeneity based on core analysis and testing data requires a certain number of statistical samples that can cover, laterally and vertically, the target strata in the study area so as to ensure they can fully represent the heterogeneity characteristics of the target strata in the study area.

#### 3.5.1.2.5 Mathematical calculation methods

Mathematical calculation methods have always been a powerful tool to solve reservoir geology problems, so it does for the study on reservoir heterogeneity. Through specific mathematical calculation and analysis of various characteristic parameters that can reflect the reservoir heterogeneity, together with the certain experience, the degree of reservoir heterogeneity can be judged to support the study on reservoir heterogeneity. Bi Yanbin et al. (2003) used the variogram method to analyze the plane reservoir heterogeneity, where a mathematical model was established to quantitatively characterize the plane reservoir heterogeneity, and the results demonstrated that this method is superior to the traditional mathematical statistics methods. Zubing and Minggao (2007) took the lower V pay zone of Shuanghe Oilfield as an example to study the heterogeneity of glutenite reservoir using the method of heterogeneity index, with seven parameters (i.e., sedimentary microfacies, sand thickness, porosity, permeability, top and bottom microstructures of the reservoir, current composite water-cut, and flow zone indicator) selected through principal factor analysis to reflect the reservoir heterogeneity; the field practice proved that the results truly reflect the underground distribution of reservoir heterogeneity, with the coincidence more than 90%. Shi Dong et al. (2009) studied the reservoir heterogeneity using the GIS-based gray comprehensive evaluation system, achieving an integrated evaluation of reservoir heterogeneity. Shaochun et al. (2016) conducted the data envelopment analysis to characterize the heterogeneity of the clastic reservoir in the second member of the Dainan Formation in Zhenwu Oilfield, Yangzhou City, Jiangsu Province, and the results showed that the heterogeneity index of sandstone in the second member of the Dainan Formation in Zhenwu Oilfield ranges between 0.5 and 0.9, indicative of moderate to strong heterogeneity, and also that the distribution of heterogeneity index has a good correlation with the distribution of sedimentary microfacies; besides, the heterogeneity index has a direct impact on the development effect of water injection. The mathematical calculation method can realize quantitative study on reservoir heterogeneity on the premise of an in-depth geological analysis; otherwise, it will only be a digital game that cannot reflect the real information of reservoir heterogeneity.

#### 3.5.1.2.6 Logging interpretation method

Logging interpretation method is a way, being based on fine logging interpretation, to quantitatively describe the interlayer and intralayer heterogeneities through fine classification of correlation of sand and mud layers/beds to determine the positions of barriers and baffles between layers or between beds. As log data are most widely used for reservoir characterization, so are they for reservoir heterogeneity analysis. In the study on reservoir heterogeneity through logging interpretation, the permeability data derived from logging interpretation are used to calculate the variation coefficient, breakthrough factor, and differential so as to quantitatively compare the reservoir heterogeneity. Alternatively, the threshold physical properties of reservoirs and intercalation are determined using core test data, and the intercalation is classified by log data so as to study the reservoir heterogeneity. In this study,

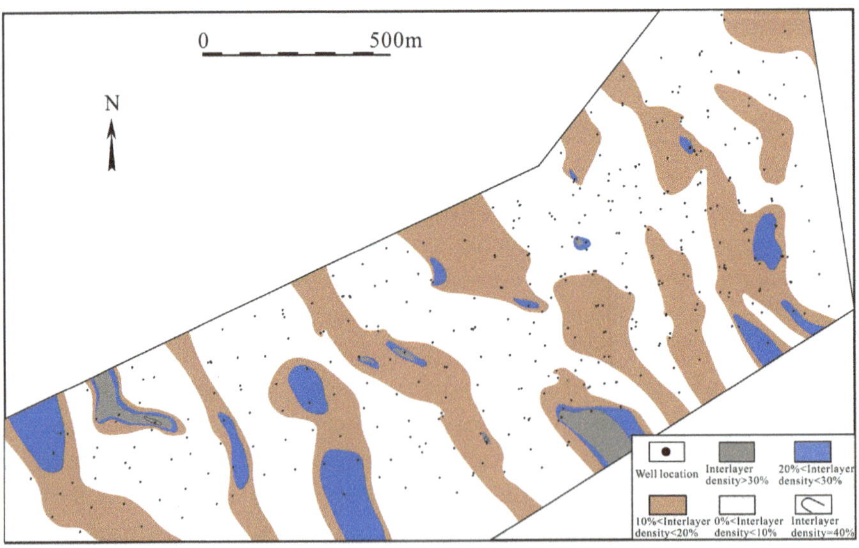


**Figure 3.48** Density distribution of baffles inside $yl_1^{1a}$ in an area of Western Sag, the Liaohe Basin.

we used this method to identify the intercalations between and within beds, and we drew the planar distribution maps of barriers between beds and baffles within beds so as to characterize the heterogeneity of Yulou reservoir in an area of Western Sag, the Liaohe Basin (Fig. 3.48).

#### 3.5.1.2.7 Geological modeling method

The geological modeling method is seldom used in reservoir heterogeneity analysis, but it has been discussed in some studies. The study on reservoir heterogeneity based on geological modeling includes several contents, such as establishing the geological models of reservoir intercalation and also establishing the models of reservoir permeability differential, variation coefficient, and breakthrough factor. Similar to the sedimentary microfacies modeling method, the study on reservoir heterogeneity based on geological modeling is extremely advantageous in the interwell prediction and 3D display of reservoir heterogeneity. Zhengyan (2003) took Niu 20, Chun 41, Zhuang 74, and Daluhu low-permeability sandstone reservoirs in Shengli Oilfield as examples to systematically discuss the heterogeneity analysis method based on the geological model, which makes full use of 3D seismic, geological, drilling, well logging, and testing data to achieve quantitative characterization of reservoir heterogeneity. Matthew et al. (2005) adopted the reservoir modeling method to investigate the petrophysical heterogeneity of dolomite reservoir on the Madison outcrop in Mississippi River, Sheep Canyon, Wyoming, USA, and the results showed that the outcrop modeling could provide quantitative information for describing the changes of reservoir petrophysical properties. Shaochun et al. (2006) established a 3D model of heterogeneity index, together with the variogram, to derive the changes of heterogeneity in the early, middle, and later development stages of Dongying Formation in Zone 2 of Shengtuo Oilfield, Dongying Sag, Jiyang Depression. At present, the study on reservoir heterogeneity based on geological modeling is still dependent on the basic data like seismic or log data, by which the data of intercalation or permeability variation coefficient, breakthrough factor, and differential that can reflect the reservoir heterogeneity are obtained, coupled with the calculation of variogram to establish the reservoir heterogeneity model for predicting the reservoir heterogeneity in 3D space. The geological modeling method can provide intuitive and quantitative information of reservoir heterogeneity, but the accuracy and precision of the model are restricted to a certain extent by the influence of the database and modeling algorithm. At present, it is still at the nascent stage of growth, with many pending issues.

#### 3.5.1.2.8 Seismic prediction method

Seismic prediction is seldom used in reservoir heterogeneity analysis. This method mainly describes the reservoir heterogeneity by seismic inversion and interpretation of reservoir sand/mud content or porosity, permeability, and oil/gas saturation. Lifeng et al. (2009) described the heterogeneity of gas reservoir that caused the difference of gas composition in Dongfang Gas Field, by identifying the vertical and horizontal distribution of reservoir in the mud flow gully zone based on the seismic data, in view of the model of gravity flow scouring after deposition, and simultaneously they also selected multi-dimensional linear regression of seismic attributes and porosity to predict the average porosity of each gas zone laterally and then make a semiquantitative prediction of reservoir heterogeneity characterized by gradual change of physical properties. For the study on reservoir heterogeneity based on seismic prediction, the reservoir physical properties are mainly predicted by ascertaining the corresponding relationship between reservoir porosity or permeability and seismic attributes. It focuses on the information of various mud intercalations, while the ordinary reservoir seismic prediction method stresses the information of reservoirs such as the sand bodies. The study on reservoir heterogeneity based on seismic prediction has been a relatively unique attempt in recent years. However, due to the limitations of relevant techniques and data accuracy, this study is still insufficient and needs to be improved.

#### 3.5.1.2.9 Numerical simulation method

Numerical simulation can also be used to effectively analyze the reservoir heterogeneity. This method is mainly used to simulate the changes of reservoir heterogeneity under different development modes, and it is of great practical significance. According to the results of numerical simulation, the corresponding measures can be adjusted in time to improve the development effect of oil and gas fields. Fuqing (2005) used the numerical simulation method to diagnose how the reservoir heterogeneity would affect the polymer flooding performance in Gudong Oilfield, and the results showed that the vertical heterogeneity, plane heterogeneity, and rhythm have different effects on polymer flooding and interlayer connectivity and macropore channels are also influential during polymer flooding. Due to the human factors, different parameters and simulation methods may be selected, which limits the application of this method in practice. The simulation results need to be verified by geological static data and production performance data.

### 3.5.1.2.10 Production performance analysis method

The production performance analysis method provides an effective way for static reservoir heterogeneity analysis from a dynamic point of view. Various production performance information, such as injected water volume, injected steam volume, oil production, and liquid production, can directly reflect the degree of reservoir heterogeneity. We used production data to analyze the development degree of reservoir intercalations in an area of Western Sag, the Liaohe Basin, thereby ascertaining the reservoir heterogeneity. The final success provided a solid geological basis for

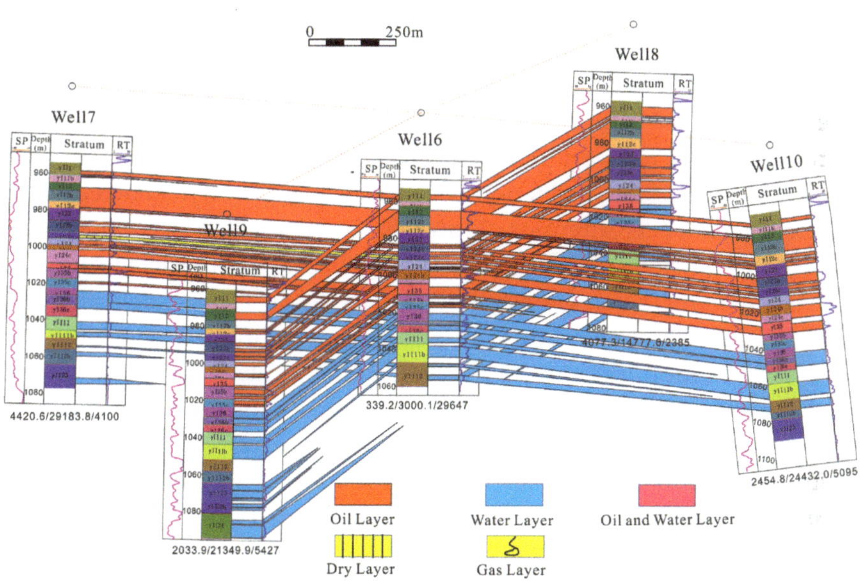


**Figure 3.49** Development characteristics of Yulou reservoir in the Well6 cluster in an area of Western Sag, the Liaohe Basin. *Note*: the numbers below the well name are: oil production ($10^4$t)/water production ($10^4$t)/injected steam volume ($10^4$m$^3$).

**Table 3.9** Information of Yulou reservoir in the Well6 cluster in an area of Western Sag, the Liaohe Basin ($yl_1{}^{1a}$–$yll_1{}^{2a}$).

| Well item | Well7 | Well6 | Well9 | Well8 | Well10 |
|---|---|---|---|---|---|
| Formation thickness (m) | 101.091 | 106.986 | 97.793 | 104.478 | 106.37 |
| Intercalation thickness (m) | 43.76 | 47.78 | 61.43 | 33.51 | 57.49 |
| Oil production ($\times 10^4$ t) | 4420.6 | 339.2 | 2033.9 | 4077.3 | 2454.8 |
| Injected steam volume ($\times 10^4$ m$^3$) | 4100 | 29647 | 5427 | 2385 | 5095 |

studying the heterogeneity of heavy oil thermal recovery reservoirs and the conversion from steam stimulation to steam flooding. Taking the Well6 cluster in the study area as an example (Fig. 3.49, Table 3.9), which includes five wells, the oil layers are found to be well-developed and connected in yI pay zone. Comparing the formation thickness, intercalation thickness, oil production, and injected steam volume, it is found that the intercalations in all five wells are mainly of sedimentary origin. Well7 and Well8 have the smallest intercalation thickness and injected steam volume, but the most oil production, while the other three wells present the opposite performance. The reason is that with the increase of intercalation thickness, the thermal efficiency of steam flooding decreases gradually, and the correspondence between injectors and producers becomes worse. The production practice of this cluster demonstrates that intercalation plays an important role in controlling and influencing the production of heavy oil thermal recovery reservoirs. As an effective method of reservoir heterogeneity analysis, production performance analysis has attracted more and more researchers' attention. However, it has some shortcomings. For example, production performance data are affected and restricted not only by static reservoir heterogeneity but also by many factors such as water injection intensity, water injection pressure, and surface engineering. In practical application, the production performance data should be analyzed comprehensively to avoid attributing some non-geological and reservoir effects to reservoir heterogeneity factors, which may result in erroneous results. Production performance data itself cannot directly describe the reservoir heterogeneity. For example, it cannot be used to identify intercalations. However, production performance data can be adopted to verify the reservoir heterogeneity, such as intercalations identified by well logging and other methods. Heterogeneity is a core reservoir property, and it directly decides the reservoir quality. Therefore the degree of reservoir heterogeneity can be demonstrated by the corresponding production performance of reservoirs. Of course, in this process, the influence of engineering technology should be excluded.

#### 3.5.1.3 Problems

Many researchers have recognized the importance of reservoir heterogeneity and carried out a lot of effective studies on it. However, there are still a lot of problems in reservoir heterogeneity analysis, mainly in eight aspects of research strength.

1. The quantitative methods to study reservoir heterogeneity are relatively simple. Typically, the intralayer heterogeneity is mainly investigated by calculating the permeability variation coefficient, breakthrough factor, and differential. At present, the less-accurate permeability derived from logging interpretation has seriously lowered the accuracy and precision of reservoir heterogeneity analysis. In this regard, we should learn from the studies abroad that integrate various quantitative methods into reservoir heterogeneity analysis. Benjamin et al. (2014) combined the methods of nuclear magnetic resonance, stable isotope geochemical analysis of small cores, and 3D geological modeling to analyze the genesis of heterogeneity of neritic carbonate reservoirs in the Middle Jurassic of the Paris Basin, France.
2. Fluid heterogeneity has been rarely handled, but it is worthy of further consideration. Yin Wei et al. (2001) studied the fluid heterogeneity of reservoir in Qian-12 block of Liaohe Oilfield, providing a reliable basis for the formulation and adjustment of the oilfield development plan.
3. Due to the limitation of data and methods/techniques, microscopic reservoir heterogeneity is far less studied than macroscopic reservoir heterogeneity. Chen Huanqing used the methods of geostatistical analysis, core description, and microscopic thin-section observation to analyze the microscopic heterogeneity of volcanic reservoir in Ying 1 Member in Xudong area, the Songliao Basin.
4. The heterogeneity of carbonate reservoirs has been studied maturely abroad, but it is compromised by a lot of problems in China, where the carbonate reservoirs are characterized by complex geologic genesis and limited distribution.
5. The study on reservoir heterogeneity is mostly based on the fine reinterpretation of log data but rarely relies on seismic data, which results in the inaccurate characterization of macroscopic reservoir heterogeneity.
6. Basic experiments have not been efficiently applied in reservoir heterogeneity analysis due to the limitations of sampling, the precision of instruments/devices, testing cost, and other factors.
7. Reservoir heterogeneity is a result of many geological factors, which have not been considered comprehensively enough.
8. The changes of reservoir heterogeneity in the development process have not been investigated, and they may impede the adjustment of development strategy and the recovery of remaining oil in the middle–late stage. As the oil and gas field development advances, the

reservoir properties (porosity, permeability, etc.) may change along with the velocity-, water- and salt-sensitive operations, which will ultimately alter the reservoir heterogeneity. Such changes should be considered carefully.

#### 3.5.1.4 Trend

The study on reservoir heterogeneity is considered to move forward in eight aspects.

1. The study on reservoir heterogeneity in view of geological factors, such as sequence stratigraphy, sedimentology, and diagenesis, should be strengthened since these factors are always the root causes for reservoir heterogeneity. Jingshan et al. (2007) summarized the genetic types and intralayer heterogeneity models of low-porosity and low-permeability sand bodies in the Fuxian exploration area. According to the vertical variations of reservoir parameters (e.g., rain size, porosity, and permeability) in a single sand body and its logging responses, they distinguished four basic patterns of intralayer heterogeneity, that is, upward-poorer porosity and permeability, upward-better porosity and permeability, composite, and complex, and pointed out that the first three patterns are mainly controlled by the factors such as sand microfacies type, sedimentary hydrodynamics, sediment grain size, and sedimentary sequence, while the last pattern is mainly related to diagenetic reworking factors such as differential cementation and differential burial dissolution in sand bodies. Samantha et al. (2010) studied the reservoir heterogeneity in the high-precision isochronal stratigraphic framework based on the sedimentary outcrop of Cedar Mesa sandstone and Jurassic Page sandstone in southern Utah, USA.
2. Field outcrop can be observed to establish the geological knowledge base for the purpose of quantitatively characterizing the reservoir heterogeneity. This practice is popular abroad, while it should be further considered in China. Matthew et al. (2007) conducted the outcrop simulation to study the heterogeneity of mesoscale fluvial point-bar sandstone reservoirs in Williamson Fock Formation, the Pishens Basin, Colorado, USA.
3. Fluid heterogeneity, which is an important part of reservoir heterogeneity analysis, should be further studied in China, while it has been frequently investigated abroad through flow field analysis. Chiyin (2004) analyzed the factors controlling the fluid heterogeneity in the Qianmiqiao buried-hill condensate gas reservoir and found that the fluid heterogeneity was mainly caused by bidirectional hydrocarbon supply,

multistage hydrocarbon accumulation, late reservoir formation, and presence of buried-hill faults. Zhan et al. (2010a) used deep electromagnetic and pressure monitoring equipment to dynamically monitor the reservoir fluids to study the reservoir heterogeneity. Fluid heterogeneity is bound to be a critical orientation for reservoir heterogeneity analysis.

4. The application of seismic data to reservoir heterogeneity analysis should be further considered. Seismic data are extremely beneficial for lateral reservoir prediction.
5. Complex lithology reservoirs (e.g., carbonate rock and volcanic rock) and tight reservoirs are expected to become one of the focuses of studies on reservoir heterogeneity in China. Dali et al. (2005) successfully studied the reservoir heterogeneity and remaining oil distribution of the reef limestone reservoir in Liuhua 11−1 Oilfield. Jingwei et al. (2013) investigated the heterogeneity and oil content lower limit of the Chang 7 tight sandstone reservoir in the Triassic, the Ordos Basin, and indicated that calcite cementation had caused the heterogeneity of tight sandstone reservoir, the residual hydrocarbon content presents a difference up to six times, and the tight sandstone mainly contains nanoscale throats (diameter $<1$ m).
6. Various experimental techniques such as digital cores should be adopted to characterize the micropore structure and heterogeneity of reservoirs, enabling the reservoir heterogeneity analysis to be more accurate and quantitative.
7. The application of advanced mathematical statistics methods should be enhanced in the quantitative study on reservoir heterogeneity, which is currently based on the statistical calculation of permeability data derived from logging interpretation. Qiang and Juanru (2009) adopted the fractal geometry method to identify the micropore structure of reservoir in an area of Gaoyou sub-sag, the Subei Basin, and then satisfactorily analyzed the microscopic heterogeneity of the reservoir.
8. The effect of reservoir heterogeneity on oil and gas development, one of the key objectives of reservoir heterogeneity analysis, will certainly be an orientation of the reservoir heterogeneity analysis. Ali et al. (2012) diagnosed how the heterogeneity of fractured reservoir would affect the steam assisted gravity drainage (SAGD) and concluded that the oil displacement efficiency of reservoirs with high fracture permeability at the top is lower than that of reservoirs with high fracture permeability at the bottom. Congjun et al. (2013) studied the reservoir heterogeneity of Quan 4 Member in Fuyu Oilfield and its control on

remaining oil distribution, and they demonstrated that the reservoir heterogeneity is strong, with three types of baffles inside: mud, calcite, and mud-gravel; they also found that interlayer sand bodies present in five patterns of vertical connection, and the lateral distribution of porosity, permeability, and heterogeneity coefficient are obviously influenced by sedimentary microfacies and sand body distribution. Meysam et al. (2015) studied the effect of reservoir heterogeneity on the productivity of vertical highly-deviated horizontal gas well produced by water flooding, and they found that the spatial distribution of permeability is the major driver for the properties of water flooded reservoirs.

## 3.5.2 Heterogeneity of Yulou reservoir in an area of Western Sag, the Liaohe Basin

In this study, we discussed the heterogeneity of Yulou reservoir, a heavy oil thermal recovery reservoir of fan delta front facies, which is characterized by frequent diversion of underwater distributary channels and fast variation of physical properties, based on the data of core, thin-section, log, analysis, and test to provide a geological basis for the transformation of development mode. The geology of Yulou reservoir has been introduced in Section 3.1, which will not be repeated here.

### *3.5.2.1 Characteristics of reservoir heterogeneity*

Reservoir heterogeneity, as a core element of reservoir development, has always been the focus of studies. For continental sedimentary reservoirs, which correspond to fast facies variation and limited data, their heterogeneity is difficult to accurately identify. In China, researchers often analyze reservoir heterogeneity by directly calculating the permeability variation coefficient, breakthrough factor, and differential. In this study, in addition to statistical analysis and calculation of reservoir permeability derived from fine logging interpretation, we specifically introduced the method of sedimentary microfacies analysis to understand the reservoir heterogeneity from the well, sectional, and plane perspectives.

#### 3.5.2.1.1 Single-well heterogeneity

Understanding the heterogeneity by well data is fundamental and also essential. Cores provide first-hand data of reservoir heterogeneity. According to the core observation (Fig. 3.50), the Yulou reservoir is highly heterogeneous, with the physical properties changing dramatically

**Figure 3.50** Core heterogeneity of Yulou reservoir in an area of Western Sag, the Liaohe Basin. (A) Well J2, trough bedding, 953.95–954.05 m; (B) Well J2, calcified animal shells, 965.04–965.14 m; (C) Well J2, fracture, 983.75–983.9 m; (D) Well 22–10, oil-cut fine sandstone, 1005.58–1005.7 m.

in a very short distance. Such heterogeneity is controlled by three factors, namely, sedimentation, diagenesis, and tectonism. Sedimentation, as the dominant factor, influenced the reservoir heterogeneity in two scenarios. In one scenario, depending on the sedimentary environment, the grain size of deposits changes, and the lithology changes accordingly. In the Yulou reservoir, which is a fan delta front deposit, the frequent redirection of underwater distributary channel causes the rapid variation of various lithologies (e.g., pebbly sandstone, medium sandstone, fine sandstone, siltstone, and mudstone) in space, thereby resulting in a strong spatial heterogeneity of reservoir. In another scenario, with the change of hydrodynamic conditions in the course of sedimentation, many bedding structures, such as trough cross bedding, tabular cross bedding, ripple cross bedding, flaser bedding, and oblique bedding, are formed (Fig. 3.50A). These bedding structures may be limited in thickness and scale, but they have huge impacts on the reservoir properties: they result in rapid variation of the reservoir properties in space. Diagenesis also influences reservoir heterogeneity to some extent and in many patterns. For example, the calcification of abundant cephalopod fossils in the Yulou reservoir worsened the physical properties (Fig. 3.50B) and enhanced the heterogeneity. The carbonization of fauna fossils deposited on different bedding interfaces aggravated the reservoir heterogeneity. Moreover, the overgrowth of feldspar and the existence of clay minerals made the reservoir's physical properties poorer and the heterogeneity stronger. Besides, fractures can be seen locally. Late filling of fractures or other factors altered the properties of reservoir beds near the fractures, thereby enhancing the heterogeneity (Fig. 3.50C). Heterogeneity decides the oil content of the reservoir. Core observation indicates that the oil-cut reservoir interval exhibits a weak heterogeneity, mostly consisting of pebbly fine sandstone or fine sandstone (Fig. 3.50D), which is in sharp contrast to the reservoir interval with strong heterogeneity that has a poor oil content (Fig. 3.50A). Therefore heterogeneity plays a very important role in controlling the reservoir's oil content. It should be considered carefully when steam flooding is applied to the heavy oil thermal recovery reservoir.

On the lithoelectric profile of a single well, the intralayer heterogeneity is represented as the rhythm of vertical permeability, location of the highest permeability zone, degree of heterogeneity, the macro ratio of vertical to horizontal permeability at the scale of the single sand bed, and distribution of baffles inside the sand beds. It directly controls the vertical sweep thickness of agents injected in a single sand bed. By analyzing the sedimentary rhythm patterns of the single well (Fig. 3.51), it is found that the Yulou reservoir is controlled by sedimentation, with a strong vertical heterogeneity. Three sedimentary rhythm patterns can be identified, that is, positive rhythm, negative rhythm, and complex rhythm, of which positive rhythm and complex rhythm are dominant. These rhythm patterns change in the vertical direction, leading to the strong intralayer heterogeneity of the reservoir. Under these three patterns, the effect of steam flooding is quite different. In steam flooding, perforation is often conducted at 1/2 in the lower part of the oil reservoir. In the case of positive rhythm, the

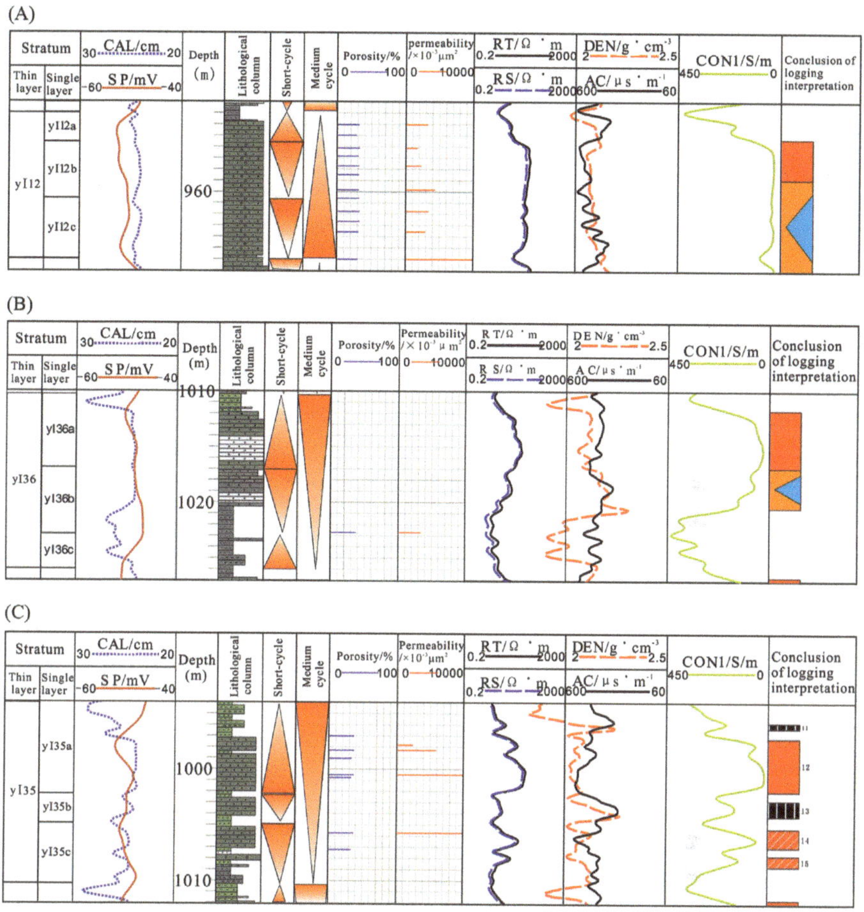


**Figure 3.51** Sedimentary rhythm patterns of Yulou reservoir in an area of Western Sag, the Liaohe Basin. (A) Positive rhythm; (B) reverse rhythm; and (C) compound rhythm.

steam overlap may cause a large part of the remaining oil to remain unrecovered, suggesting a poor flooding performance. In case of negative rhythm, the steam will advance upward due to the steam overlap, thereby displacing the heavy oil in the upper part with better physical properties, demonstrating the best flooding performance. In the case of complex rhythm, the steam moves in a complex law in the reservoir, so the thermal recovery layers should be divided carefully to ensure the ideal flooding effect. Generally, the production effect of steam flooding in negative rhythm and complex rhythm reservoirs is better than that in positive-rhythm reservoirs.

### 3.5.2.1.2 Sectional heterogeneity

The Yulou reservoir is monoclinic, with weak diagenesis, so that the reservoir heterogeneity is mainly controlled by the development of sedimentary facies. The sedimentary microfacies analysis shows that the fan delta front subfacies can be subdivided into five types of sedimentary microfacies: underwater distributary channel, estuary bar, underwater distributary interchannel sand, underwater distributary interchannel mud, and front sheet sand. The underwater distributary channel, estuary bar, and distributary interchannel sand are dominant, while the front sheet sand is occasionally observed near the southwest or south of the study area (Figs. 3.52 and 3.53). On the section of sedimentary microfacies (Fig. 3.52), laterally, most of the sand bodies extend to a limited distance, with small scales; vertically, the sedimentary microfacies change rapidly, with well-developed underwater distributary interchannel mud. Limited sand body scale laterally and rapid facies variation vertically lead to the dramatic changes in reservoir lithology and physical properties in space and ultimately aggravate the reservoir heterogeneity. The interlayer reservoir heterogeneity is mainly reflected in the section. It can be seen that barriers exist between different beds, with thickness more than 2 m and distributing stably in a lateral direction. According

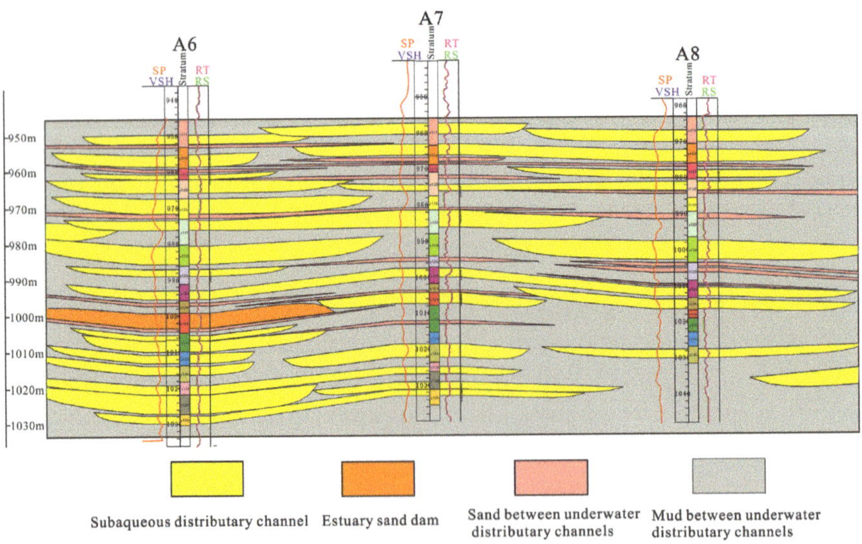


**Figure 3.52** Section of sedimentary microfacies in Yulou reservoir in an area of Western Sag, the Liaohe Basin. *Note*: nearly east-west, a direction perpendicular to the provenance.

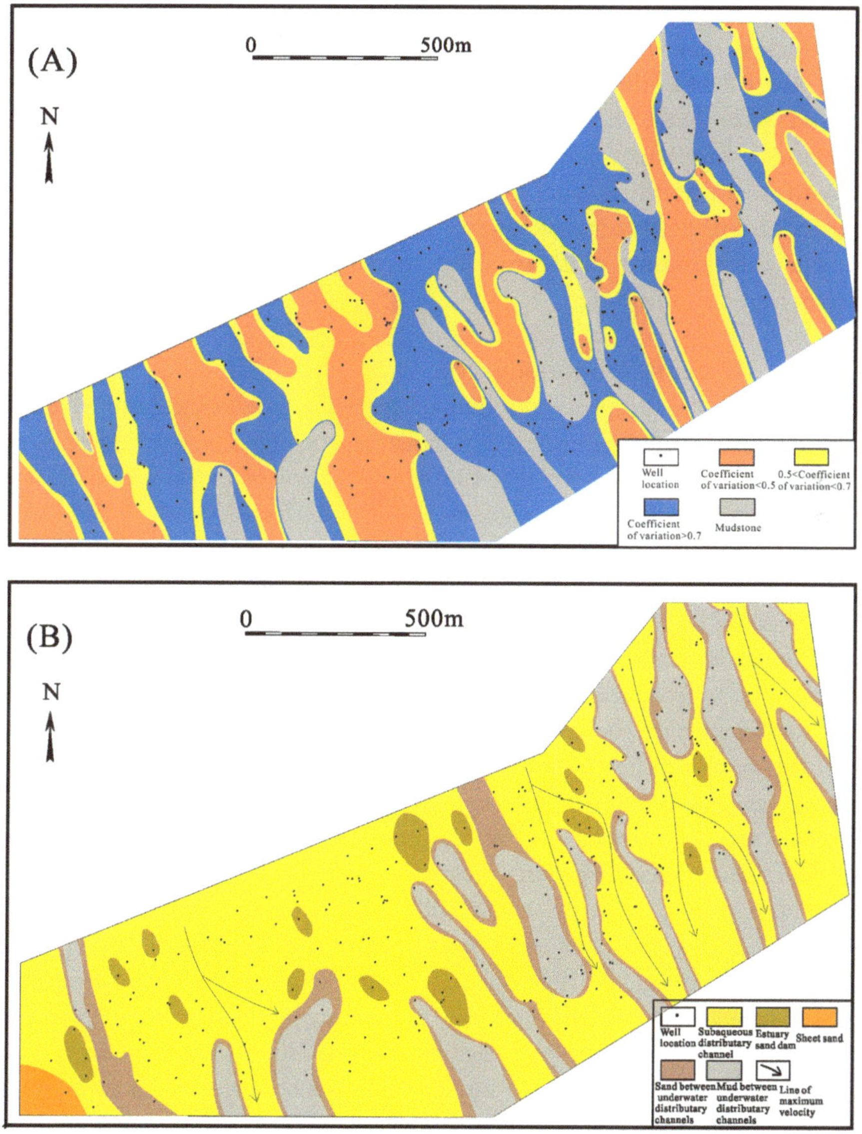


**Figure 3.53** Comparison between planar distribution map of permeability variation coefficient and sedimentary microfacies map of $yl_3^{6c}$ of Yulou reservoir in an area of Western Sag, the Liaohe Basin. (A) Planar distribution of permeability variation coefficient; (B) plane distribution map of sedimentary microfacies.

to the field experience in Liaohe Oilfield, for argillaceous rocks, a barrier thicker than 1 m can effectively shelter the steam. Therefore the influence of these thick barriers on steam flooding should be fully considered. Moreover, the existence of underwater distributary interchannel mud with a short lateral extension also has a certain impact on the path of steam flooding, thereby complicating the advancing path of the steam. The underwater distributary interchannel mud can also absorb a considerable amount of steam heat to partly reduce the thermal efficiency of steam flooding.

#### 3.5.2.1.3 Plane heterogeneity

Based on the physical properties derived from the fine logging reinterpretation, we calculated the permeability variation coefficient at the scale of bed (Table 3.10). It can be seen that the heterogeneity of the target interval is strong as a whole. The heterogeneity is moderate for $yI_1^{2a}$, $yI_1^{2b}$, and $yI_1^{2c}$, and it's strong for other beds. This is also consistent with the current production practice. We drew the plane distribution map of permeability variation coefficients of beds based on the calculated permeability heterogeneity parameters so as to analyze the heterogeneity within each bed. Taking $yI_3^{6c}$ as an example, by comparing the plane distribution map of permeability variation coefficient with the sedimentary microfacies map (Fig. 3.53), it is found that the reservoir heterogeneity is obviously controlled by sedimentary microfacies, and the zones with strong, moderate, and weak heterogeneity generally spread in stripes in NW-SE direction, which is basically consistent with the provenance direction. This bed, as a whole, presents strong reservoir heterogeneity, with the zones with weak heterogeneity accounting for about 1/3 of the study area. Among the rest of the 28 beds, $yI_1^{2a}$, $yI_1^{2b}$, and $yI_1^{2c}$ exhibit the weakest heterogeneity, with the zones with the permeability variation coefficient less than 0.5 accounting for about 1/2 of the test area; $yI_2^{3a}$, $yI_2^{4a}$, $yI_2^{4b}$, and $yI_2^{4c}$ exhibit relatively strong heterogeneity, with the zones with the permeability variation coefficient less than 0.5 accounting for about 1/3 of the test area; the other beds all have strong heterogeneity, with sporadic zones with the permeability variation coefficient less than 0.5, which is basically consistent with the distribution of porosity and permeability. These findings fully prove that reservoir heterogeneity is closely related to the physical properties; the former has a very important impact on the latter. According to the current effect of oilfield development, $yI_1^{2}$ ($yI_1^{2a}$ + $yI_1^{2b}$ + $yI_1^{2c}$), $yI_2^{3}$ ($yI_2^{3a}$ + $yI_2^{3b}$ + $yI_2^{3c}$), and $yI_2^{4}$ ($yI_2^{4a}$ + $yI_2^{4b}$ + $yI_2^{4a}$) are all major oil beds, and their reservoir heterogeneity is obviously weaker

**Table 3.10** Permeability heterogeneity parameters of yl of Yulou reservoir in an area of Western Sag, the Liaohe Basin.

| Heterogeneity parameter bed | Breakthrough factor | | Permeability differential | | Variation coefficient | |
|---|---|---|---|---|---|---|
| | Range | Mean | Range | Mean | Range | Mean |
| $yI_1^{1a}$ | 1−55.019 | 4.136 | 1−2016.243 | 132.144 | 0−6.012 | 0.944 |
| $yI_1^{1b}$ | 1−14.827 | 3.084 | 1−1865.889 | 74.372 | 0−2.197 | 0.804 |
| $yI_1^{2a}$ | 1−9.534 | 2.367 | 1−1890.263 | 55.508 | 0−2.177 | 0.619 |
| $yI_1^{2b}$ | 1−12.349 | 2.411 | 1−1937.565 | 60.981 | 0−2.55 | 0.59 |
| $yI_1^{2c}$ | 1−14.622 | 2.471 | 1−2009.335 | 113.937 | 0−2.356 | 0.639 |
| $yI_2^{3a}$ | 1−19.905 | 3.104 | 1−1995.416 | 143.156 | 0−2.732 | 0.792 |
| $yI_2^{3b}$ | 1−21.765 | 3.745 | 1−1999.619 | 462.418 | 0−3.237 | 1.007 |
| $yI_2^{3c}$ | 1.011−29.011 | 3.573 | 1.104−2016.496 | 379.088 | 0.034−3.781 | 0.922 |
| $yI_2^{4a}$ | 1−39.34 | 3.534 | 1−2015.274 | 215.589 | 0−4.606 | 0.84 |
| $yI_2^{4b}$ | 1−18.622 | 3.118 | 1−1963.716 | 147.499 | 0−3 | 0.809 |
| $yI_2^{4c}$ | 1−31.418 | 3.381 | 1−2017.05 | 196.971 | 0−4.193 | 0.835 |
| $yI_3^{5a}$ | 1−10.325 | 3.197 | 1−2012.191 | 303.551 | 0−2.847 | 0.876 |
| $yI_3^{5b}$ | 1−19.214 | 3.086 | 1−1972.341 | 276.334 | 0−2.511 | 0.816 |
| $yI_3^{5c}$ | 1−12.749 | 2.827 | 1−2013.072 | 247.945 | 0−2.523 | 0.778 |
| $yI_3^{6a}$ | 1−11.371 | 2.821 | 1−2010.65 | 199.597 | 0−2.784 | 0.74 |
| $yI_3^{6b}$ | 1−12.903 | 2.863 | 1−1878.689 | 124.074 | 0−2.288 | 0.78 |
| $yI_3^{6c}$ | 1−34.729 | 3.181 | 1−1821.182 | 104.367 | 0−3.94 | 0.771 |

than other beds, which also suggests that the reservoir heterogeneity directly affects the development effect: that is, better development effect is realized in zones with weaker reservoir heterogeneity. Therefore the study on reservoir heterogeneity can provide a useful reference for the implementation of effective reservoir development measures.

#### *3.5.2.2 Effect of reservoir heterogeneity on thermal recovery of heavy oil reservoir by steam flooding*

The degree of reservoir heterogeneity directly affects productivity, water injection efficiency, and ultimate oil recovery. The study on reservoir heterogeneity can help reveal the fluid behaviors in the reservoir, contributing to the division of development layer series, selection of injection-production system, prediction of productivity and production performance, and, finally, the improvement of oilfield development (Jinliang and Jun, 2011). As described above, the target interval in the study area has a strong heterogeneity as a whole. So, it is rare for the injected steam to distribute evenly in the target interval and on the plane. Following the principle of seeking the pathway with the lowest resistance, the steam preferentially enters the high-permeability pathways, such as the main channel of the underwater distributary channel, which may easily lead to steam channeling. If the flow resistance is small, the steam will quickly break through the production well. This will bring greatly negative impacts on the production: (1) a part of steam is directly produced from the production well; and (2) the oil displacement efficiency reduces to below the expected production performance, resulting in local enrichment of the remaining oil in the target interval. The influence of reservoir heterogeneity should be fully considered in thermal recovery by steam flooding. On the one hand, the injector−producer pattern should be designed to ensure the injection and production wells are placed in the zones with weak reservoir heterogeneity, such as the zones with identical sedimentary microfacies so that a good injection-production relationship can be achieved to maximize the steam flooding effect. Meanwhile, the zones with weak reservoir heterogeneity contain fewer intercalations so that the heat loss in the process of steam flooding can be minimized, and the economic benefits of thermal recovery by steam flooding can be maximized. On the other hand, in the zones with weak reservoir heterogeneity, the front of steam can advance evenly to enlarge the swept volume, so that the oil displacement efficiency is the highest. Taking the Well K1 cluster as an example (Table 3.11, Fig. 3.54), which includes five wells,

**Table 3.11** Comparison of evaluated heterogeneity with production effect of the Well K1 cluster in Yulou reservoir in an area of Western Sag, the Liaohe Basin.

| Well | Formation thickness (m) | Evaluated heterogeneity (number of beds) | | | Oil production ($\times 10^4$ t) | Water production ($\times 10^4$ $m^3$) | Injected steam volume ($\times 10^4$ $m^3$) |
|---|---|---|---|---|---|---|---|
| | | Strong | Moderate | Weak | | | |
| K1 | 84 | 5 | 2 | 3 | 126.1 | 1759.6 | 23,237 |
| W1 | 56 | 2 | 3 | 5 | 11159 | 55,278.26 | 31,502 |
| W2 | 49.6 | 5 | 4 | 2 | 5565.1 | 35,479.91 | 25,700 |
| W3 | 105.6 | 8 | 2 | 4 | 7955.3 | 18,784.69 | 25,621 |
| G2 | 79.7 | 8 | 4 | 2 | 9246.4 | 31,372.84 | 24,864 |

*Note*: The evaluated heterogeneity is determined through a comprehensive analysis of permeability variation coefficient, breakthrough factor, and differential derived from fine logging interpretation for corresponding horizons.

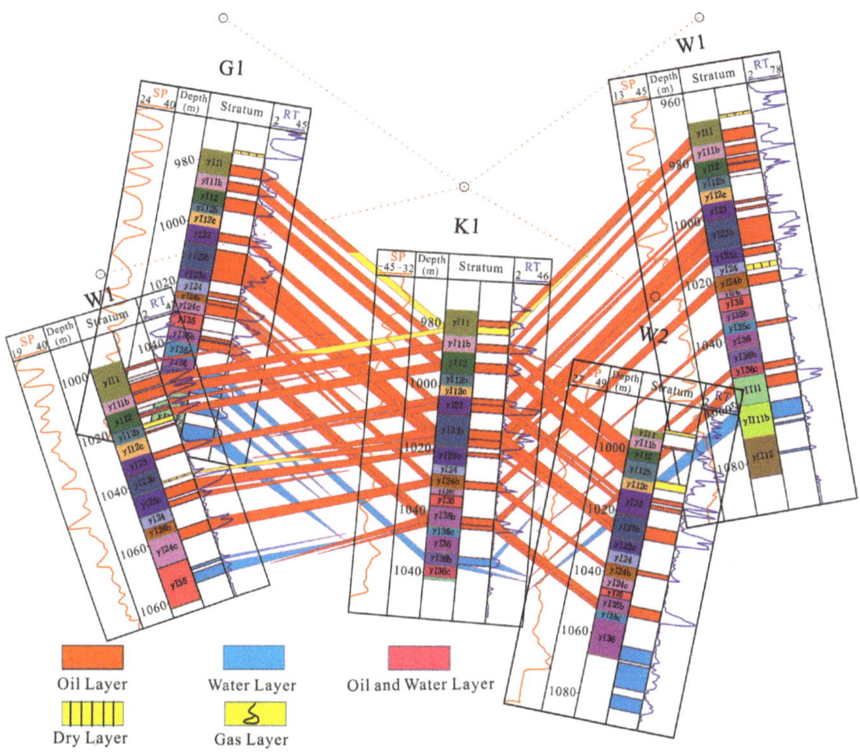


**Figure 3.54** Effect of reservoir heterogeneity on development of the Well K1 cluster in Yulou reservoir in an area of Western Sag, the Liaohe Basin.

that is, G2, K1, W1, W2, and W3, and for the purpose of comparing the production data such as injected steam volume and oil production, given G2 is not considered since its drilling and production time is different from other wells and K1 is a steam injector while W1, W2, and W3 are producers, we find that all the wells have experienced two production stages: steam stimulation, and steam flooding, and they are now in the stage of steam flooding. The analysis shows that W1, W2, and W3 are not distinct in injected steam volume but greatly different in oil production. Among them, W1 shows the largest oil production in spite of small formation thickness, while W2 and W3 achieve relatively small production. The main reason is that the overall reservoir heterogeneity of W1 is weak. Although W3 has a stronger heterogeneity than W2, it demonstrates larger formation thickness and reservoir thickness than W2; so, W3 achieves a higher production than W2.

## 3.6 Comprehensive quantitative reservoir evaluation based on geologic genesis analysis

Reservoir evaluation is to assess the capacity of the reservoir to store oil and gas, and its results can effectively guide the development of oilfields (Honghua et al., 2006). As a core in the research on exploration and development, reservoir evaluation has always been the focus of studies in China and abroad (Luo et al., 1991; Hongwen et al., 1997; Ma et al., 2000; Jiandong et al., 2003; Dubost et al., 2004; Jiyu et al., 2005; Xiwen et al., 2006; Saemi et al., 2007; Xudong et al., 2009; Takahashi and Kovscek, 2010; Al-Ghamdi et al., 2011; Wuguang et al., 2011; Zhonghong et al.,2012; Jiajia et al., 2012; Shafiei, 2013). Many researchers applied different methods to evaluate reservoirs from varying perspectives.

### 3.6.1 Current status

The methods for reservoir evaluation are transforming gradually from qualitative to quantitative, with more and more factors considered, given the comprehensive analysis of multiple factors is preferred rather than simply porosity and permeability. McCaffrey and Kneller (2001) studied the process control for the development of marginal stratigraphic traps in narrow turbidite sedimentary systems and its assistance to reservoir evaluation. Jiandong et al. (2003) used the analytic hierarchy process to evaluate the reservoirs in the plot area with dense well pattern in the eastern part of the North 2 zone of Saertu Oilfield, Daqing. Xiwen et al. (2006) used the reefs and beaches of the Late Permian Changxing Formation in the Jiannan Structure as an example to evaluate the reef-beach reservoirs. Saemi et al. (2007) adopted the genetic algorithm to design a neural network to evaluate the permeability of the reservoir; this method has been well applied in the South Pars gas field, the Persian Gulf. Chaoli et al. (2008) took the volcanic rocks in the Xingcheng gas field as an example, and the effective permeability, matrix effective porosity, matrix air permeability, average pore-throat radius, and rock density as indicators to quantitatively evaluate the low-permeability volcanic gas reservoir. Wilkes et al. (2008) evaluated the biodegradation of crude oil in the reservoir by using the stable carbon isotope of specific mixtures. Xudong et al. (2009) systematically evaluated the carbonate reservoirs in Yuroboqin Oilfield using the logs of the Russian logging series. Rabi Bastia et al. (2010) evaluated the potential of deep-water reservoirs in the Bengal fan at the northeastern margin of the Indian continent through analyzing the deposition of the middle and upper strata before and after plate collision. Takahashi and Kovscek (2010) evaluated the wettability of low-permeability siliceous shale using surface tension. Al-Ghamdi et al. (2011) used the carbonate reservoir in the Middle East as an example to evaluate the naturally fractured reservoirs based on an improved tri-porosity model. Wuguang et al. (2011) evaluated the reservoirs in Yangjiaba Oilfield through variogram analysis. Chalmers and Marc Bustin (2012) carried out a geological evaluation of the Triassic Halfwaye-Montneye-Doig shale gas and tight gas mixed reservoir in Groundbirch, northeastern British Columbia, Canada, covering the reservoir thickness, structure, total porosity, TOC, maturity of organic matter, pore size distribution, surface area, mineral characteristics, and permeability. Zhonghong et al. (2012) proposed a method to comprehensively evaluate the reservoirs in low-permeability zones by combining the mainstream throat radius, percentage of movable fluid, pseudo-start-up pressure gradient, crude oil viscosity, and clay mineral content as the evaluation parameters. Jiajia et al. (2012) used the wave impedance AlgR overlap method, petrophysical inversion method, and seismic multiattribute prediction method to evaluate the organic carbon content and oil content of oil shale. Shafiei et al. (2013) developed a screening tool for evaluating the performance of steam flooding in naturally fractured carbonate reservoirs based on an artificial neural network and particle swarm optimization.

At present, the data used in the research on reservoir evaluation mainly include core, test, log, and seismic data. The research methods are transforming gradually from qualitative to quantitative, with more and more factors considered: the comprehensive analysis of multiple factors is preferred rather than simply rock porosity and permeability. The research objects are diverse, ranging from clastic rocks carbonate rocks to volcanic rocks and even oil shale. In countries abroad, various mathematical methods, geological methods, and experimental methods are mainly applied to evaluate the physical properties (e.g., porosity and permeability) and the fractures. In China, based on optimal evaluation parameters, geostatistical or mathematical methods are adopted to comprehensively classify the reservoirs and then analyze the relationship between different types of reservoirs and hydrocarbon distribution, thereby guiding the oil and gas exploration and development. From the review of previous studies, it is found that all research methods for reservoir evaluation are less comprehensive. For some methods, the evaluation parameters are directly selected on the basis of geologic study experience, while the geostatistical analysis is missing. For other methods, just a classification by purely mathematical method is followed, but the geological basis is inadequate. The combination of qualitative geological empirical research and quantitative classification and evaluation has been rarely reported.

### *3.6.1.1 Research methods*

The selection of the reservoir evaluation method is inevitable for all researchers. The existing methods include geological empirical method, weight method, analytic hierarchy process method, fuzzy mathematics method, artificial neural network method, fractal geometric method, variogram method, cluster analysis method, gray relational analysis method, logging methods, and seismic methods. Here, we give a brief introduction to these methods and summarize their advantages and disadvantages (Table 3.12).

#### 3.6.1.1.1 Geological empirical method

The geological empirical method is widely used in reservoir evaluation. Specifically, based on the experience accumulated over the target interval in the study area, the corresponding parameters are selected, and the parameter threshold values of different reservoir types are determined so as to make the reservoir classification and evaluation. Zhiyong et al. (2000) established the logging-based standards for reservoir classification based on the experience, with the buried hill in Shu 103 block in Western Sag of the Liaohe Basin as an example. Mingyi et al. (2006)

**Table 3.12** Characteristics of different reservoir evaluation methods.

| Reservoir evaluation method | Rationale | Applicability | Advantage | Disadvantage |
|---|---|---|---|---|
| Geological empirical method | Based on geological experience, the parameters such as lithology and physical properties are selected to comprehensively classify and evaluate the reservoirs | Applicable to mature areas that have been studied sufficiently | It covers a deep geological genesis analysis and has been well applied in the study area | It is insufficient in quantitative analysis, and the method and relevant criteria can hardly be promoted |
| Weight method | It realizes the weight evaluation by analyzing, collectively, the relevant information of the research object's all original variables and determining the weight of different factors and the corresponding evaluation criteria | Applicable to the areas that correspond to rich basic data and experienced researchers | It combines the qualitative geological study and quantitative statistical calculation | The weights determined by different researchers may be very different |
| Analytic hierarchy process method | The simple mathematical tools are combined with operational ideas to decompose complex problems into various components, which are grouped by dominance relations to form a hierarchical structure, and then the relative importance of each component is clarified according to the interaction between components and their role in the system | Applicable to the areas where the main controlling factors of reservoir properties have been clearly understood | It simplifies the complex problems | The hierarchical division is random, with its reasonableness impossible to being verified |

(*Continued*)

**Table 3.12** (Continued)

| Reservoir evaluation method | Rationale | Applicability | Advantage | Disadvantage |
|---|---|---|---|---|
| Fuzzy mathematics method | Based on the mathematical system established according to the concept of membership function, an arbitrary value in the interval [0, 1] is selected to describe whether an object belongs to the set; on this basis, the classification and evaluation of the reservoir is realized | Applicable to areas with rich data and mathematically-skillful researchers | It can more objectively evaluate the reservoirs by overcoming the uncertainties in the conventional evaluation of oil content and quality | The process to establish the fuzzy system of geological factors is sensitive to many influencing factors, with intense uncertainty which will affect the accuracy of reservoir evaluation |
| Artificial neural network method | It mainly uses the known learning sample set, trains with error back propagation algorithm, and builds the network to achieve the goal of reservoir description and evaluation | Applicable to areas which have been studied to a certain extent and where the basic reservoir properties have been partly understood | It is more time-saving, more objective, and more accurate | It is limited in application, because the neural network is modeled and predicted with known prior knowledge |
| Fractal geometric method | From the fractional dimension perspective, mathematical method is adopted to describe and study the objective things, namely, the mathematical tools of fractal dimension are used to describe the objective things | Researchers are required to have knowledge in the basic levels of computer technology and mathematics | It can quantitatively evaluate the reservoir distribution and heterogeneity | It requires the object with self-similarity and self-inversion |

| | | | | |
|---|---|---|---|---|
| Variogram method | It characterizes the properties of regionalized variables by the parameters such as range, nugget constant, and sill and the variograph | Applicable to areas with reservoirs containing simple, scale-stable, and less-variable microfacies types | By statistically calculating the scale and regularity of reservoirs of different properties, the reservoirs in unknown zones can be evaluated and predicted by means of variogram tool | The research results are greatly influenced by the representativeness of the samples |
| Cluster analysis method | It is a multivariate statistical analysis technique for quantitative analysis of object according to the nature or origin of the object | Applicable to areas with rich well data, and reservoirs characterized by strong heterogeneity, great variation in space, and absence of large pores | The qualitative analysis of geological genesis and the quantitative classification of clustering algorithms are organically combined | The parameter selection, weights of parameters, and clustering algorithm need to be adjusted frequently |
| Gray relational analysis method | It is a method to study and analyze the correlation between things according to the similarity of time series curves among factors | Applicable to areas where the reservoir properties are only controlled by one or two factors | The complex reservoir evaluation process is transformed into a systematic mathematical operation process, thus, making the reservoir evaluation more quantitative | No systematic and intensive analysis of the genetic factors of reservoir may lead to deviations in establishing the evaluation criteria and determining the indicator weight, thereby, impeding the final evaluation results |

(*Continued*)

**Table 3.12** (Continued)

| Reservoir evaluation method | Rationale | Applicability | Advantage | Disadvantage |
|---|---|---|---|---|
| Logging methods | It realizes the reservoir classification and evaluation through characterizing the reservoir quality based on the lithoelectric, AC, and nuclear magnetic logging responses | Applicable to areas with rich log data, complete logging series, and high coincidence of logging interpretation of physical properties | The data are accessible and can be processed in batches, and the evaluation results are quantitative. The methods can also be combined and supplemented with other methods | Limited by the logging tools and the interpreter's experience, the evaluation results may be somewhat discrepant from actual geological conditions. Moreover, in view of costs and geological conditions, some logging methods are limited to not being applied |
| Seismic methods | It realizes the reservoir classification and evaluation through characterizing and predicting the reservoir and describing the physical properties and oil content based on the seismic reflections of lithology, porosity, permeability, and fluid content | Applicable to areas where the seismic data are good and logging–seismic calibration is accurate | It can describe the macroscopic development of reservoirs with different properties in space through reservoir prediction | Seismic data must be calibrated with well data. Limited by the technical resolution, the methods fail to be efficiently applied in evaluating the reservoirs with thin sand-mud interbeds |

defined the criteria for reservoir classification and evaluation according to the previous studies and field practices, taking the tight sandstone reservoir of the Upper Triassic Xujiahe Formation in the Western Sichuan foreland basin as an example. Chunjun and Yanbin (2011) conducted a genesis analysis and evaluation of low-permeability reservoirs in Da'niudi Gas Field, and they described the characteristics of different types of reservoirs in terms of rock type, physical properties, pore structure parameter, and pore types based on the geological experience. As the biggest advantage, the geological empirical method can fully reflect the genetic factors of the reservoir, and it is based on the evaluation criteria derived from the exploration and production practices. This method has been well applied in the study area. However, this method is insufficient in quantitative analysis and statistical calculation, and it, together with the relevant criteria, can hardly be promoted satisfactorily to different oilfields/blocks which are characterized by greatly variable geological characteristics.

#### 3.6.1.1.2 Weight method

The weight method is a relatively simple semiquantitative reservoir evaluation method (Yi et al., 2012a). It realizes the weight evaluation by collectively analyzing the relevant information of the research object's all original variables and by determining the weight of different factors and the corresponding evaluation criteria. Keqi et al. (2005) used the weight method to evaluate the reservoir in the fourth member of Shahejie Formation in the Weicheng 81 fault block, the Dongpu Sag, which laid the foundation for the adjustment of the development plan. Yi et al. (2012a) elaborated the role of the weight method in evaluating the Carboniferous volcanic reservoir in Zhongguai 58 block, the Ordos Basin. The weight method is the much known quantitative method in contrast to the geological empirical method. However, its evaluation results may vary greatly since the determination of weight for parameters is susceptible to human factors; namely, the weights determined by different researchers for evaluating the same reservoir may be very different.

#### 3.6.1.1.3 Analytic hierarchy process method

Analytic hierarchy process (AHP) was proposed in the mid-1970s by Saaty T. L., a well-known American operations researcher and Professor at the University of Pittsburgh. As to the AHP reservoir evaluation, the simple mathematical tools are combined with operational ideas to decompose complex problems into various components, which are grouped by dominance relations to form a hierarchical structure, and then the relative importance of each component is clarified according to the interaction between components and their role in the system; on this basis, the reservoir quality is evaluated (Jiandong et al., 2003). Wang Jiandong et al. (2003) used the AHP method to evaluate the reservoir in the test area with a dense well pattern in the eastern North 2 zone, Saertu Oilfield, Daqing. Lingyun and Binggao (2009) used AHP to quantitatively evaluate the tight reservoirs in the Baise Basin. The biggest advantage of the AHP method is that it decomposes a complex system structure into several levels or subsystems to determine the relative importance of each component in the system. Disadvantageously, however, the hierarchical division is random, with its reasonableness impossible to being verified.

#### 3.6.1.1.4 Fuzzy mathematics method

Fuzzy mathematics is a mathematical system established according to the concept of membership function. The membership function can use an arbitrary value in the interval [0, 1] to describe whether an object belongs to the set or not, just like the exact function which chooses 1 (belonging to the set) or 0 (not belonging to the set). Clearly, the membership function has entitled the capacity to describe the gradual transition of things. Fuzzy mathematics approximates ambiguity while admitting mathematical precision (Wei et al., 2013). The fuzzy mathematics comprehensive evaluation method is used to determine the reservoir evaluation factor set, that is, the characteristic parameters of reservoir classification and evaluation so as to comprehensively and quantitatively classify and evaluate the reservoir. Chunying et al. (2008) used the fuzzy mathematics method to evaluate the Chang 4 + 5 oil reservoir in the Baiyushan area, the Ordos Basin. Wei et al. (2013) adopted the fuzzy mathematics method to evaluate the terrigenous clastic reservoirs in the Triassic–Jurassic, the southeastern Pre-Caspian Basin, Kazakhstan. The fuzzy mathematics method is most advantageous given the fact that it can more objectively evaluate the reservoirs by overcoming the uncertainties in the conventional evaluation of oil content and quality. When fuzzy mathematics is used for reservoir evaluation, a fuzzy system of geological factors must be firstly established, and the membership function is given depending on the researcher's geological experience, thereby determining the membership of different geological factors. The process is sensitive to many influencing factors, with intense uncertainty, which will affect the accuracy of reservoir evaluation.

#### 3.6.1.1.5 Artificial neural network method

The artificial neural network is a computational model used to simulate the thinking process of the human brain. It mainly uses the known learning sample set, trains with error back propagation algorithm, and builds the network to achieve the goal of reservoir description and evaluation (Yanwu et al., 2001). Saemi et al. (2007) used the genetic algorithm to design a neural network to evaluate the reservoir permeability; the method has been applied successfully in the South Pars gas field, the Persian Gulf. Peigang et al. (2010) successfully identified the reservoir lithology and fluid based on the structure and principle of a self-organizing neural network. Shafiei et al. (2013) developed a screening tool for evaluating the steam flooding performance of naturally fractured carbonate reservoirs based on the ANN and particle swarm optimization. The Artificial neural network (ANN) method is more time-saving, more objective, and more accurate. However, it is limited in application because the neural network is modeled and predicted with known prior knowledge.

#### 3.6.1.1.6 Fractal geometric method

Fractal geometry was first proposed in 1975 by the famous French mathematician Mandelbrot, and it provides an effective mathematical tool for describing irregular, nonlinear, and complex things in nature (Lei et al., 2012). For the reservoir evaluation by fractal geometry, the pore structure or properties like fractures of the reservoir are described depending on the fractal dimension study, and then the development degree of the reservoir is evaluated. Liqiang et al. (1998) calculated the fractal dimension and fractal lower limit of pore structure of the Jurassic reservoir in the Bogda piedmont according to the basic principles of fractal geometry and the mercury intrusion data, and they discussed the application of fractal geometry in reservoir evaluation. Lili et al. (2008) used the variable-scale fractal technique to predict the distribution of natural fractures in the Chang 4 + 51 + 2 reservoirs in Wellblock Yuan48, Tiebiancheng Oilfield, Changqing, and comprehensively classified and evaluated the reservoir depending on the distribution characteristics and other formation coefficients in the zone where fractures are relatively developed, revealing the results in a good agreement with practical production. This method can quantitatively evaluate the distribution law and heterogeneity of reservoir, but it requires the object with self-similarity and self-inversion.

#### 3.6.1.1.7 Variogram method

The variogram is mainly used in geostatistics to describe the spatial geometry of regionalized variables. It is the basis of various prediction and simulation algorithms, and it can also characterize the properties of regionalized variables by the parameters such as range, nugget constant, sill, and the variograph. Wuguang et al. (2011) discussed the application of variograms in the evaluation and development of reservoirs in Yangjiaba Oilfield. In Li's study, the variogram range of different layers was calculated by sedimentary microfacies types and the use of permeability data; through the influence range of variables reflected by range, the reservoir was finely described. The variogram method integrates the geostatistics to the reservoir evaluation. By statistically calculating the scale and regularity of reservoirs of different properties, the reservoirs in unknown zones can be evaluated and predicted by means of the variogram tool; thus qualitative geologic study and quantitative statistical prediction are organically combined in reservoir evaluation. However, its research results are greatly influenced by the representativeness of the samples.

#### 3.6.1.1.8 Cluster analysis method

The cluster analysis method is one of the most widely and successfully used methods in the classification and evaluation of reservoirs. Cluster analysis, also called point group analysis, is a multivariate statistical analysis technique for quantitative analysis of an object according to the nature or origin of the object (Hanlin, 1998). Tang Jun et al. (2012) successfully used the Q-type cluster analysis and discriminant analysis to comprehensively evaluate the Chang 81 reservoir in the Jiyuan area, the Ordos Basin. In the study, we adopted the method of cluster analysis in the quantitative evaluation of Yulou reservoir in an area of Western Sag, the Liaohe Basin. Firstly, through the geological genesis analysis of reservoir, the major controlling factors for reservoir development were determined, and then the characteristic parameters that could best reflect these factors were selected for statistical analysis via the SPSS cluster analysis platform; finally, the reservoir evaluation and classification were achieved. As the biggest advantage of the cluster analysis method for reservoir evaluation, the qualitative analysis of the geological genesis and the quantitative classification of clustering algorithms are organically combined. However, the classification is strenuous because the parameter selection, weights of parameters to be determined, and clustering algorithm need to be adjusted frequently.

#### 3.6.1.1.9 Gray relational analysis method

Gray relational analysis is an important part of the gray system theory. It is a method to study and analyze the correlation between things according to the similarity of the time series curves among factors. The purpose is to find the main relations among factors in the system and identify the key factors affecting the target value, thereby mastering the primary features of matters (Jiafan et al., 2003). Jiyu et al. (2005) conducted a comprehensive evaluation of the reservoirs in the North 2 zone of Saertu Oilfield, Daqing, by using the gray relational analysis method, thus avoiding the problem of overlapping and nonunique evaluation results obtained by a single factor. Yi et al. (2012b) used the gray relational analysis method to comprehensively evaluate the clastic rock reservoirs in the Qingdong Sag. The gray relational analysis method is widely applied in reservoir evaluation now. The biggest advantage is that the complex reservoir evaluation process is transformed into a systematic mathematical operation process, thus making the reservoir evaluation more quantitative. However, the disadvantage is that no guarantee can be given that a systematic and intensive analysis of the genetic factors of the reservoir may not lead to deviations in establishing the evaluation criteria and determining the indicator weight, thereby impeding the final evaluation results.

#### 3.6.1.1.10 Logging methods

Logging is a very effective tool for evaluating and describing reservoirs. In particular, FMI logging and formation dip logging are extremely superior to conventional techniques for evaluating fractured-vuggy carbonate reservoirs. Through the fine logging interpretation of reservoir physical properties such as porosity and permeability, or the characterization of pore space such as fractures, pores, and vugs, the development degree of the reservoir is evaluated, and the vertical and lateral development of different types of reservoirs is analyzed. Xiaoyu and Qiang (1995) conducted a comprehensive evaluation of the Sinian dolomite reservoirs in the Ziyang area by using the log-based pore-cavity hybrid probability method. Yang Bin et al. (2010) used the logging method to evaluate the reservoirs in the Carboniferous of Shaguanping area, the eastern Sichuan Basin. Xiaoming (2012) adopted the natural gamma ray spectrum logging to comprehensively evaluate the carbonate reservoir in the Yubei area. The logging methods have been widely used in reservoir evaluation. Advantageously, the data are accessible and can be processed in batches, and the evaluation results are quantitative. The methods can also be combined and supplemented with other methods. Nevertheless, there are some defects. For example, limited by the logging tools and the interpreter's experience, the evaluation results may be somewhat discrepant from actual geological conditions. Moreover, in view of the costs and geological conditions, some logging methods are limited as to their unconstrained applicability.

#### 3.6.1.1.11 Seismic methods

Seismic methods/techniques are always regarded by the researchers as an important tool for characterizing and predicting reservoirs but have rarely been applied in reservoir evaluation. Seismic reservoir evaluation involves multiple aspects, including a description of reservoir structure (especially the development degree of fractures), inversion prediction of reservoir sand/mud content or fluid (e.g., oil, gas, and water) development, and change monitoring during development. The seismic methods are best for standing out for their quantitative characterization of macroscopic reservoir properties. Liu Zhen et al. (1995) adopted the modified Differential Interlayer Velocity Analysis method to evaluate the reservoirs in the LD structure. Xiaodong et al. (2011) used the amplitude spectrum gradient of seismic data for reservoir evaluation. MacBeth et al. (2006) used four-dimensional seismic data to monitor and evaluate the pressure drop of the Rotliegende low-porosity tight gas reservoir in the continental shelf of the gas-bearing basin, southern UK, where the seismic responses of reservoirs containing fractures were improved, and the measures proposed could help enhance the gas production. The seismic methods can describe the macroscopic development of reservoirs with different properties in space through reservoir prediction. However, seismic data must be calibrated with well data. Besides, limited by the technical resolution, the methods fail to be efficiently applied in evaluating the reservoirs with thin sand-mud interbeds. Reservoir evaluation is a systematic discipline. Due to the diversity of data and the difference in research purposes, the optional methods are variable as well. In practice, a variety of data and methods should be organically combined to optimize the reservoir evaluation.

### *3.6.1.2 Considerations*

The following three aspects should be considered specifically in reservoir evaluation.

#### 3.6.1.2.1 Selection of data for reservoir evaluation

The selection of data is fundamental for reservoir evaluation, and such data are diverse, including basic geology, log, seismic, test, and production performance (Zhihong et al., 2007; Yanzhong et al., 2008; Yingli et al., 2011; Likun, 2013). The type, abundance, and reliability of such data determine the scheme and methods of reservoir evaluation to a certain extent and even decide the correctness of evaluation results.

#### 3.6.1.2.2 Selection of parameters for reservoir evaluation

Another important aspect of reservoir evaluation is the selection of parameters. In a sense, the rationality of parameter selection directly determines the rationality and correctness of the evaluation results. Yang Zhengming et al. (2008) studied the evaluation parameters of low-permeability water-bearing gas reservoirs, and they revealed that the throat radius, irreducible water saturation, and critical pressure gradient could be used as the parameters for evaluating the low-permeability water-bearing gas reservoirs. The parameters selected may be very different, depending on the data, research objectives, and the researchers' targets and focuses, as well as the complex, diverse, and unique geological properties of the reservoir. In practice, the researchers should fully consider the research objectives and the available data so as to select the optimal combination of parameters. It is worth noticing that the case "the more, the better" is not true for reservoir evaluation parameters. Generally, the reservoir properties are mainly controlled by three geological processes, that is, tectonism, sedimentation, and diagenesis. These three factors are reflected by different parameters in reservoir evaluation (Table 3.13) (Huanqing et al., 2012a,b,c). If too many parameters are selected, one factor may be repeatedly reflected by several parameters; thus the weight of this factor is artificially increased, resulting in the distortion of reservoir evaluation results.

**Table 3.13** Evaluation parameters of the volcanic reservoir in Ying 1 Member, Xudong area, the Songliao Basin.

| Property class | Reservoir thickness (m) | Effective porosity (%) | Total permeability ($10^{-3}$ $\mu m^2$) | Gas saturation (%) | Variation coefficient |
|---|---|---|---|---|---|
| I | 8.8−164.1 | 5.83−14.48 | 5.01−17.44 | 35.68−70.46 | 0−1.01 |
| II | 11.2−97.7 | 4.74−9.21 | 0.06−4.11 | 43.71−64.08 | 0.52−1.71 |
| III | 5.9−240.8 | 4.97−8.14 | 0.13−7.12 | 2.20−50.42 | 0.80−5.24 |
| IV | 1.5−70 | 3.90−7.95 | 0.017−4.79 | 2.53−57.2 | 0−1.04 |

#### 3.6.1.2.3 Verification of reservoir evaluation results

Reservoir evaluation by cluster analysis is a systematic discipline. First of all, a detailed geological genesis analysis should be made. Then, the parameters that can best reflect the factors affecting the reservoir properties should be selected, and an appropriate calculation method should be selected during classification. Finally, the reservoir evaluation results should be verified. If the results are not rational, the evaluation parameters and weights of parameters should be adjusted. Through repeated modification in this way, a relatively rational reservoir classification and evaluation is finalized. Based on the practical evaluations of the Ying 1 Member reservoir in the Xudong area, the Songliao Basin, and the Yulou reservoir in an area of the Western Sag, the Liaohe Basin, we put forward the following principles for verifying the reservoir evaluation results:

1. The parameters that can fully reflect the factors affecting the reservoir properties should be selected.
2. The optimal calculation method and data normalization processing algorithm are selected to ensure that the hit ratio of the software reaches an empirical value.
3. The evaluation result for each class should be ensured with a certain proportion and by excluding the influence of minor singular values.
4. The relationship between parameters is compared from result to result so as to ensure that each parameter plays its due role in clustering, and the cluster analysis is made comprehensively and reasonably.
5. The plane distribution map of the reservoirs classified is compared with the distribution map of sedimentary microfacies, structure map, and diagenetic facies map (Fig. 3.55). If a good correlation and consistency is observed, the reservoir evaluation results are proved rational.

Verification of reservoir evaluation results is an integral part of reservoir evaluation. It should be further emphasized in practice.

### *3.6.1.3 Problems and trend*

#### 3.6.1.3.1 Problems

We summarize the problems in the reservoir evaluation as follows:

1. There are no recognized evaluation criteria. The evaluation criteria vary greatly from region to region, affected by data, geology, and research objectives, and no acceptable uniform criteria have been established. This is unfavorable for the promotion and citation of existing research results.

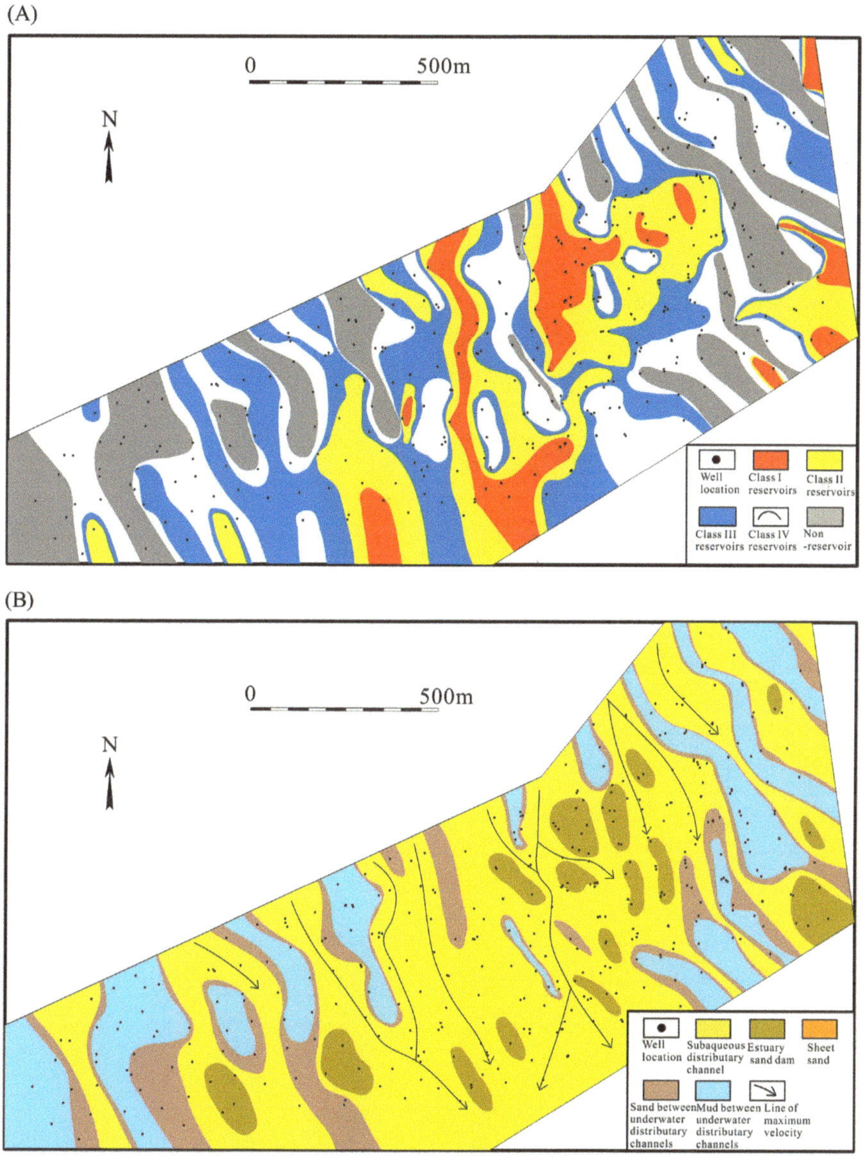


**Figure 3.55** Plane distributions of sedimentary microfacies and reservoir evaluation results of the Yulou reservoir in an area of Western Sag, the Liaohe Basin. (A) $yl_1^{2b}$ plane distribution diagram of reservoir evaluation; (B) Plane distribution of sedimentary microfacies of $yl_1^{2b}$.

2. There are no recognized valid standards for the evaluation of parameters. Currently, the most popular parameters for reservoir evaluation are porosity and permeability, and other parameters such as oil saturation, effective thickness, and shale content are also used. The parameters are very different from region to region. So far, no uniform standards for parameter evaluation have been recognized.
3. The research methods are not comprehensive enough, with less combination of qualitative and quantitative methods. Most studies focus on mathematical methods such as fuzzy mathematics and gray relational analysis, but there are rare reports on the combination of qualitative analysis of geological genesis with the quantitative calculation of mathematical statistics.

**Table 3.14** Physical properties-based evaluation of pore structure of Ying 1 Member volcanic reservoir in Xudong area, the Songliao Basin.

| Parameter class | Porosity $\Phi$/% | | | Permeability $K$/ $10^{-3}\ \mu m^2$ | | | Median pore throat radius $R_{50}$/μm | | |
|---|---|---|---|---|---|---|---|---|---|
| | Max. | Min. | Avg. | Max. | Min. | Avg. | Max. | Min. | Avg. |
| I | 14.4 | 10.5 | 13.0 | 1.19 | 0.04 | 0.20 | 0.299 | 0.062 | 0.161 |
| II | 9.9 | 7.7 | 8.8 | 1.12 | 0.02 | 0.10 | 0.251 | 0.021 | 0.113 |
| III | 7.5 | 4.9 | 6.3 | 1.35 | 0.01 | 0.05 | 0.128 | 0.018 | 0.06 |
| IV | 4.8 | 2 | 4.0 | 1.21 | 0.01 | 0.04 | 0.086 | 0.016 | 0.03 |

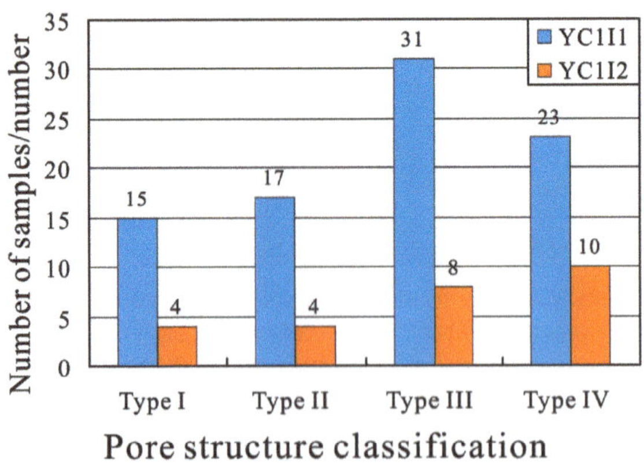


**Figure 3.56** Pore structure classification of YC1I1 and YC1I2 in the Yingcheng Formation, Xudong area, the Songliao Basin.

4. The overall reservoir is often evaluated, but rarely have studies been conducted that deeply look into a particular property of reservoir. Huanqing et al. (2013a,b,c,d) used the porosity, permeability, and median pore-throat radius to evaluate and classify the pore structure of volcanic reservoirs in the Xudong area, the Songliao Basin, providing a basis for the effective development of volcanic gas reservoirs (Table 3.14, Fig. 3.56). Porosity, permeability, reservoir heterogeneity, and fractures, as integral parts of reservoir properties, have significant impacts on reservoir properties. However, each of these items has never been thoroughly studied in its own capacity.
5. The reservoir evaluation mainly targets conventional oil and gas reservoirs such as sandstone and carbonate rocks but rarely involves tight oil and shale gas. The tight oil and shale gas reservoirs are difficult to evaluate since they are unique in pore structure and hydrocarbon flow mechanism.
6. The mathematical-statistical methods have been widely and maturely applied in reservoir evaluation, but the test methods have not been used adequately. This is extremely unfavorable for clarifying the development characteristics and genesis mechanism of microscopic reservoir properties.
7. Seismic techniques/methods, as one of the powerful tools for reservoir characterization, have been rarely used and considered in reservoir evaluation.
8. The verification of reservoir evaluation results has not attracted the researchers' adequate attention. The verification method is relatively simple, and no uniform standards for evaluating results are available.

#### 3.6.1.3.2 Trends

We believe that the reservoir evaluation trends have the following aspects:

1. The data for reservoir evaluation are more abundant and more comprehensive. Baofeng et al. (2009) combined well testing curve and geophysical data to evaluate the carbonate reservoir in the central Tarim Basin, China.
2. The methods of reservoir evaluation have gradually evolved from qualitative or semiquantitative to quantitative.
3. Geological genesis analysis has gradually become attractive for researchers. It makes the selection of reservoir evaluation parameters more scientific and representative, and it also enables the selection of evaluation methods and the evaluation results to be more reasonable and more consistent with the underground geology.

4. Reservoir evaluation is increasingly fine. In addition to the evaluation of the overall reservoir, the reservoir properties such as heterogeneity and pore structure have begun to emerge in more and more studies. Dubost et al. (2004) used the core analysis test and reservoir description techniques to evaluate the permeability heterogeneity of the reservoir, and they numerically simulated the production results and fitted the production history.
5. As the technology advances and the exploration and development of conventional reservoirs face increasingly rigorous situations, unconventional reservoirs such as tight sandstone and shale gas become the frontier and hotspots of reservoir evaluation. Aghighi and Rahman (2010) discussed how the lateral permeability anisotropy of tight sandstone gas reservoirs would affect the evaluation and design of hydraulic fracturing and hydraulic re-fracturing.
6. With the continuous development of related disciplines and technologies, new methods/techniques for reservoir evaluation have emerged. For example, core CT scanning imaging technology and constant-rate mercury intrusion technology enable the reservoir evaluation to be more microscopic and quantitative. Some researchers are working on this aspect. Shi Dong et al. (2004) adopted the GIS technology in the reservoir evaluation for Sulige Gas Field.
7. Seismic methods/techniques should be further considered in reservoir evaluation. Seismic data can help quantitatively describe the macroscopic reservoir development characteristics, thereby contributing to a more quantitative reservoir evaluation.
8. Verification of evaluation results will be one of the key orientations. Through the verification, any evaluation mistakes can be avoided, and the selection of evaluation parameters and methods can be improved, thereby greatly aggregating the level of reservoir evaluation.

### 3.6.2 Classification and evaluation of Yulou reservoir in an area of Western Sag, the Liaohe Basin

In this study, we made a comprehensive and quantitative evaluation on the Yulou reservoir in an area of Western Sag, the Liaohe Basin, with the SPSS cluster analysis software, based on the analysis of factors affecting the genesis of fan delta front reservoir and the characteristic parameters, and according to the fine logging interpretation results of 400 wells. The qualitative genesis analysis and quantitative classification and evaluation are closely combined to ensure the rationality of parameter selection and the reliability of evaluation

Figure 3.57 Core photos showing the lithology of the Yulou reservoir in an area of Western Sag, the Liaohe Basin. (A) Well A10, 944.73–944.78 m, greyish brown microconglomerate; (B) Well A2, 945.39–945.53 m, greyish brown fine sandstone; (C) Well A2, 948–948.1 m, siltstone; (D) Well A22, 1040.8–1041 m, mudstone.

results, providing a geological basis for the transformation of development mode for the fan delta front reservoir. The geology of Yulou reservoir has been introduced in Section 3.1, which will not be repeated here.

### *3.6.2.1 Reservoir characteristics*

#### 3.6.2.1.1 Lithology

According to the detailed observation had from and description made of the 668 m cores from six coring wells (Fig. 3.57) and according to the grain-size analysis of 233 samples, the Yulou reservoir is composed of diverse rock types, including microconglomerate, glutenite, coarse sandstone, medium sandstone, fine sandstone, siltstone, argillaceous siltstone, silty mudstone, and mudstone. The rock colors are usually gray, greyish black, and greyish green. The rocks are mostly mixed with gravel and have a high shale content. The medium-to-fine sandstones are dominant.

#### 3.6.2.1.2 Sedimentary microfacies

Based on the geologic, well log, and core data, and based on the analysis of the well, and the sectional and plane sedimentary facies distribution, the Yulou reservoir is defined as a fan delta front deposit, which can be subdivided into five sedimentary microfacies, that is, underwater distributary channel, estuary sand bar, underwater distributary interchannel sand, underwater distributary interchannel mud, and front sheet sand (Fig. 3.58). Among them, the underwater distributary channel and estuary sand bar are dominant.

#### 3.6.2.1.3 Pore structure

The pore structure of the reservoir refers to the geometry, size, distribution, and connection of pores and throats in rocks. The study on the pore structure to reveal the internal structure of the reservoir is of great significance for the exploration and development of oil and gas fields (Huanqing et al., 2013a,b,c,d). From the mercury intrusion curve (Fig. 3.59), it can be seen that the mercury intrusion curve extends to the lower left, indicating that the pores and throats are well sorted, with coarse skewness, and that the pore throats are well-developed, with coarser skewness and good sorting. In this study, based on the core and thin-section data, we divided the pore structure of the Yulou reservoir into two categories: primary pores and secondary pores, which are subdivided into ten sub-classes: intergranular pore, intragranular pore, matrix micropore, cleavage fracture, intergranular dissolved pore, intragranular

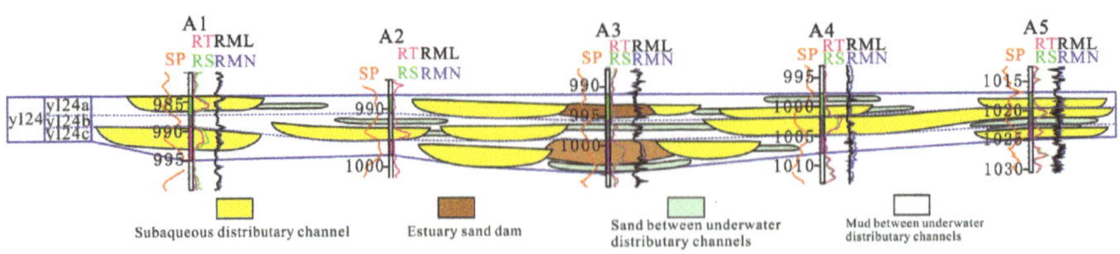


**Figure 3.58** Section of sedimentary microfacies in the Yulou reservoir in an area of Western Sag, the Liaohe Basin.

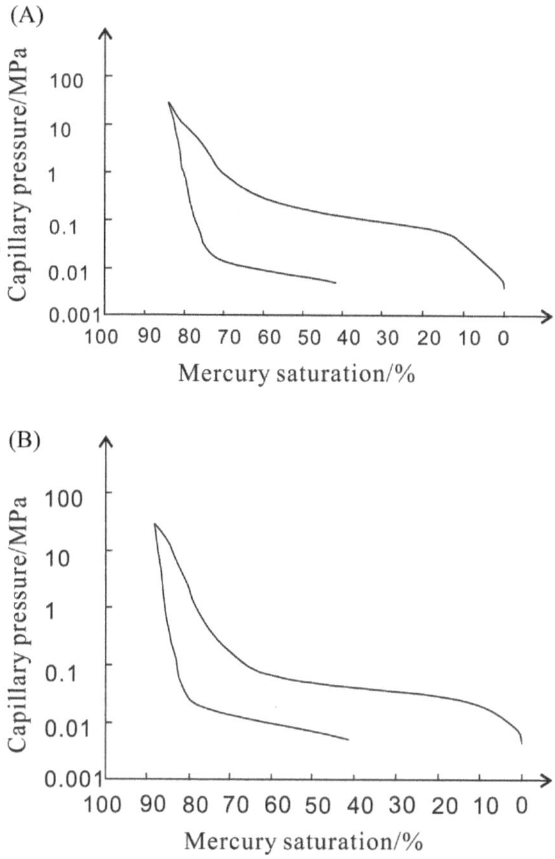


**Figure 3.59** Mercury intrusion curve of Well J1 in the Yulou reservoir in an area of Western Sag, the Liaohe Basin. (A) yI, depth: 923.18 m; (B) yII, depth; 1001.99 m.

dissolved pore, moldic pore, superlarge intergranular dissolved pore, structural fracture, and dissolved fracture. Each pore type shows unique behaviors on thin-sections (Fig. 3.60). Generally, the intergranular pore and intergranular dissolved pore are dominant.

#### 3.6.2.1.4 Physical properties

The test results of 150 core samples show that the porosity is 25%–40%, with an average of 31.25%, while the permeability varies greatly in a range of 1–5,000 mD, with an average of 1829.3 mD, suggesting that the Yulou reservoir is a high-porosity, high-permeability reservoir.

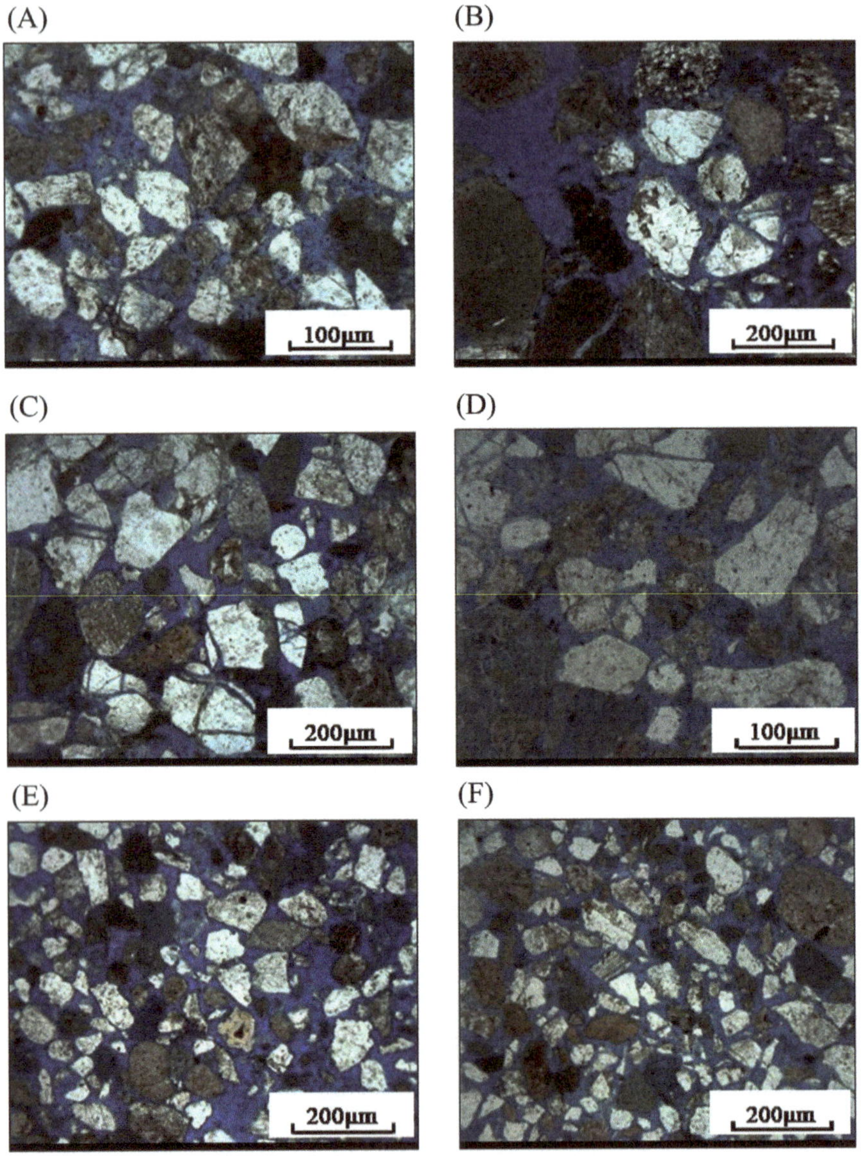


**Figure 3.60** Pore structure of the Yulou oil reservoir in Well A2 in an area of Western Sag, the Liaohe Basin. (A) Intergranular pores, 945.3 m; (B) Superlarge intergranular dissolved pores, 965.72 m; (C) Structural fractures, 974.56 m; (D) Matrix micropores, 960.32 m; (E) Moldic pores, 1022.72 m; (F) Cleavage fractures, 1005.72 m.

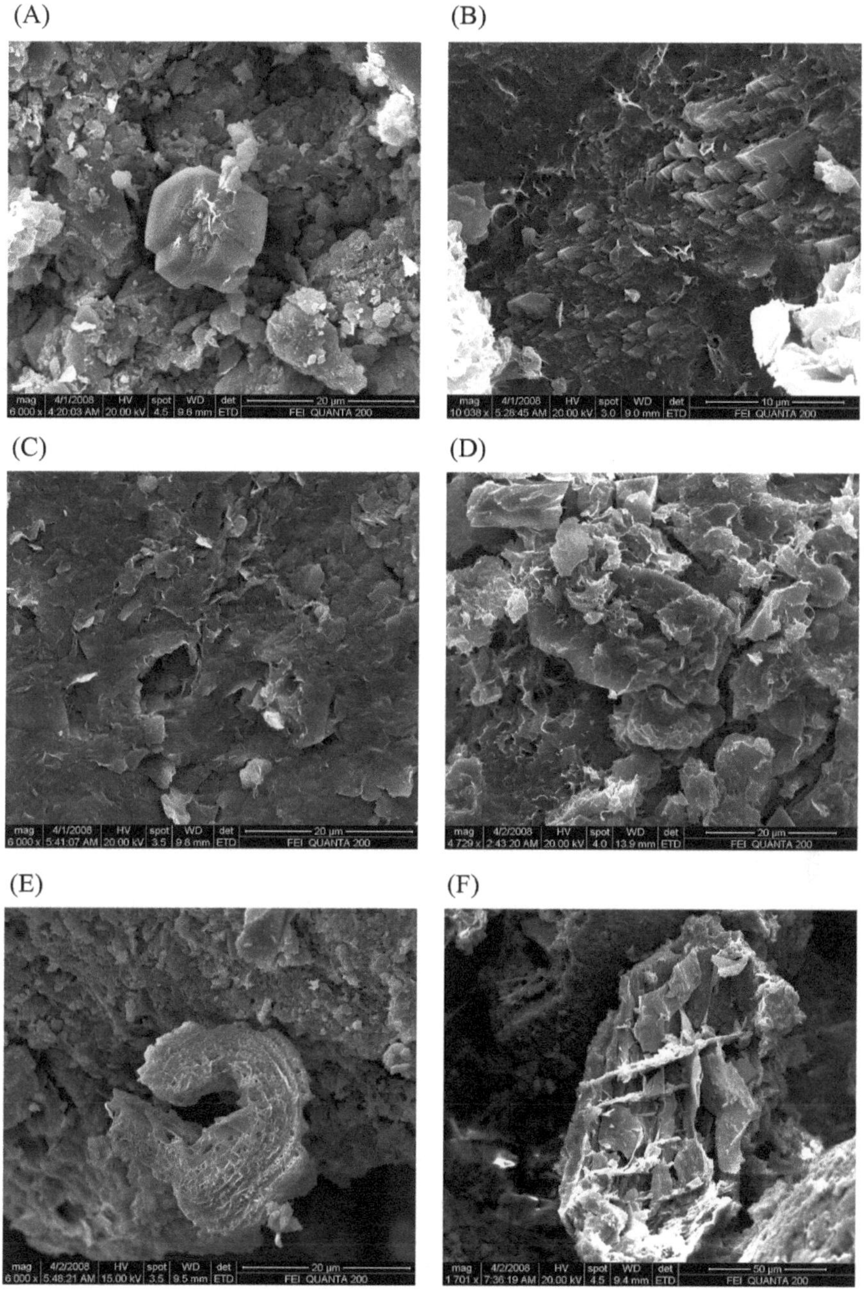

**Figure 3.61** Diagenesis of the Yulou reservoir in Well A2 in an area of the Western Sag, the Liaohe Basin. (A) Carbonate minerals, 965.72 m, sandstone; (B) overgrowth of feldspar to Grade I, 979.32 m, sandstone; (C) flaky illite, 984.35 m, sandstone; (D) smectite, 989.18 m, sandstone; (E) kaolinite, vermicular distribution, 1005.72 m, sandstone; (F) granular dissolution, 1022.72 m, sandstone.

#### 3.6.2.1.5 Diagenesis

Diagenesis is closely related to the reservoir properties (Huanqing et al., 2013a,b,c,d). The pore structure of the Yulou reservoir is also subjected to diagenetic processes, such as cementation and dissolution (Fig. 3.61). The cementation includes both carbonate cementation and clay cementation, with the latter in dominance.

#### 3.6.2.1.6 Heterogeneity

Reservoir heterogeneity is one of the main factors affecting underground oil, gas, and water movement and oil and gas recovery. The study on reservoir heterogeneity is of great significance for the effective development of reservoirs (Huanqing et al., 2012a,b,c). By analyzing the single-well sedimentary rhythm pattern, it is found that the Yulou reservoir is controlled by sedimentation, with a strong heterogeneity in the vertical direction. Three sedimentary rhythm patterns can be identified, that is, positive rhythm, negative rhythm, and complex rhythm, of which the positive rhythm and the complex rhythm are dominant. These rhythm patterns change in a vertical direction, leading to the strong intralayer heterogeneity of the reservoir. According to the plane distribution of permeability variation coefficient of yI beds, it is found that the reservoir heterogeneity is obviously controlled by sedimentary microfacies. Specifically, $yI_1^{2a}$, $yI_1^{2b}$, and $yI_1^{2c}$ exhibit the weakest heterogeneity, with the zones with the permeability variation coefficient less than 0.5 accounting for about 1/2 of the test area; $yI_2^{3a}$, $yI_2^{4a}$, $yI_2^{4b}$, and $yI_2^{4c}$ exhibit relatively strong heterogeneity, with the zones with the permeability variation coefficient less than 0.5 accounting for about 1/3 of the test area; the other beds all have strong heterogeneity, with sporadic zones with the permeability variation coefficient less than 0.5, which is basically consistent with the distribution of porosity and permeability.

### *3.6.2.2 Quantitative reservoir evaluation*

#### 3.6.2.2.1 Parameter selection

The selection of evaluation parameters is one of the most critical factors for the success of reservoir evaluation. There are many factors that influence the reservoir properties; therefore the parameter selection is variable depending on the researchers' prerogatives. As described above, the case "the more, the better" is not true for reservoir evaluation parameters; if too many parameters are selected, some factors may be overlapping, while the main factors controlling the reservoir quality fail getting reflected effectively. Based on the comprehensive analysis of reservoir genesis, the parameters of reservoir evaluation are selected. Tectonism mainly affects the reservoir properties through fractures, which can be reflected by the permeability. Sedimentation plays an important role in controlling reservoir properties. The frequent diversion and redirection of underwater distributary channels further exacerbate the spatial distribution of reservoir heterogeneity, which can be reflected by the permeability variation coefficient. Besides, the shale content can also reflect whether the target reservoir is an underwater distributary channel deposit or interchannel mud deposit, so it is also a very important indicator. The zones with high porosity mainly correspond to the underwater distributary channel or estuary sand bar, while the zones with small porosity correspond to the underwater distributary interchannel mud. Therefore porosity is also considered an important parameter. Thickness is a comprehensive representation of reservoir properties, and it reflects the lithology, lithofacies, and diagenesis to some extent. Based on careful consideration, five parameters that can fully reflect the influencing factors of the reservoir, that is, shale content, porosity, permeability, reservoir thickness, and permeability variation coefficient, are selected to comprehensively and quantitatively evaluate the Yulou reservoir.

#### 3.6.2.2.2 Method selection

In the clustering process, the Q-clustering in the pedigree clustering provided by the SPSS software is selected. The so-called "case Q clustering" is based on the characteristics of the variables, namely, the cases with similar characteristics are categorized into one class (Jianqiong et al., 2006). This study involves a Q-clustering. According to the five parameters of the reservoir (shale content, porosity, permeability, reservoir thickness, and heterogeneity reflected by permeability variation coefficient), we made a cluster analysis to classify the reservoir types. In the clustering process, the abovementioned five variables were first normalized to enable their size comparison and mathematical operation. The SPSS software provides various cluster analysis methods, such as Pearson correlation, Chebychev, Minkowski, Block, and Ward's method. According to the principles of effective classification and that the classification results by five parameters present consistency in trend and with the current development status, Ward's method was selected. The "effective classification" means that the number of results in none of the classified reservoir categories is significantly smaller than that of other categories (magnitude difference), thereby reflecting the rationality of parameters, method, and classification results. The "consistency in trend" means that the sampling points with high porosity, high permeability, and low shale content are categorized into the same class of good reservoirs (e.g., Class I or Class II), and the sampling points with low porosity, low permeability, and high shale content are categorized into the same class of poor reservoirs (e.g., Class III or Class IV). In this way, the reservoir classification and evaluation results can be prevented from the influence of some singular points such as low porosity or high permeability due to the presence of microfractures. The "consistency with the current development status" means that the Class I and Class II reservoirs are located in the zones with good development effects now, while the Class III and Class IV reservoirs are located in zones with poor development effects. Ward's method is finally determined through repeated experiments and also with consideration to the researchers' experience in cluster analysis. Different parameters and methods may be selected depending on the study area and data base. Ultimately, the reservoirs are classified according to the distance and interaction (Table 3.15, Fig. 3.62).

#### 3.6.2.2.3 Results of quantitative reservoir evaluation

In this study, the target reservoirs are divided into Class I, Class II, Class III, and Class IV. Among them, the Class I and Class II reservoirs have good physical properties and are currently the primary targets for development; the Class III and Class IV reservoirs have poor physical properties and cannot be developed economically under current technical conditions. All these reservoirs are obviously controlled by sedimentary facies. The Class I and Class II reservoirs are mostly located in underwater distributary river and estuary sand bar, while the Class III and Class IV reservoirs are mostly located in underwater distributary interchannel sand or front sheet sand.

#### 3.6.2.2.4 Reservoir development characteristics

Comparing the distribution of reservoirs in beds in the plane, we found that sedimentary factors play a leading role in the formation of reservoirs. In the plane, all four classes of reservoirs are generally distributed in stripes trending in NW-SE, extending in a direction basically the same as that of sedimentary facies. Vertically, with the migration and diversion of major underwater distributary channels of different depositional stages on the fan delta front, the distribution of areas corresponding to the physical

**Table 3.15** Parameters for reservoir classification and evaluation of Yulou reservoir in an area of Western Sag, the Liaohe Basin.

| Parameter class | Porosity/% | | | Permeability/$\mu m^2$ | | | Shale content/% | | | Effective thickness/m | | | Permeability variation coefficient | | |
|---|---|---|---|---|---|---|---|---|---|---|---|---|---|---|---|
| | Max. | Min. | Avg. | Max. | Min. | Avg. | Max. | Min. | Avg. | Max. | Min. | Avg. | Max. | Min. | Avg. |
| I | 39.000 | 35.170 | 38.590 | 6.057 | 4.618 | 5.520 | 39.980 | 0.660 | 17.650 | 7.260 | 0.000 | 2.870 | 0.522 | 0.000 | 0.194 |
| II | 38.240 | 23.730 | 36.580 | 4.615 | 2.709 | 3.594 | 44.990 | 1.240 | 21.520 | 8.450 | 0.000 | 2.920 | 1.030 | 0.147 | 0.635 |
| III | 37.110 | 23.140 | 34.350 | 2.748 | 1.001 | 1.751 | 49.970 | 1.160 | 24.710 | 10.980 | 0.000 | 2.560 | 1.780 | 0.000 | 0.917 |
| IV | 35.030 | 6.180 | 30.050 | 1.346 | 0.000 | 0.355 | 54.840 | 0.310 | 27.300 | 8.660 | 0.000 | 1.450 | 6.012 | 0.000 | 0.921 |

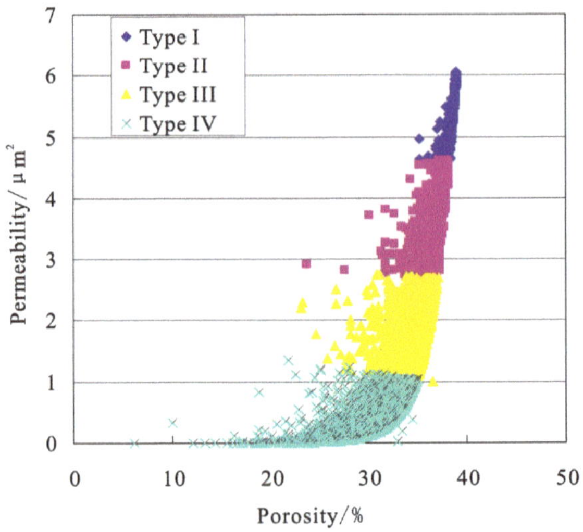


**Figure 3.62** Porosity–permeability relationship under the reservoir classification and evaluation of beds in Yulou reservoir in an area of Western Sag, the Liaohe Basin.

properties of the reservoir changes constantly. Comparing the reservoirs in the yI pay zone, we found that the Class I and Class II reservoirs are mainly in $yI_1^{2a}$, $yI_1^{2b}$, $yI_1^{2c}$, $yI_2^{3a}$, $yI_2^{3b}$, $yI_2^{3c}$, $yI_2^{4a}$, $yI_2^{4b}$, and $yI_2^{4c}$, especially in $yI_1^{2a}$, $yI_1^{2b}$, and $yI_1^{2c}$.

### 3.6.2.2.5 Control and verification of reservoir classification and evaluation results

The reservoir classification and evaluation based on cluster analysis is a systematic work. In this study, the classification and evaluation results are ensured rationally from the following aspects of planning for results:

1. The parameters that can fully reflect the factors affecting the reservoir properties are selected.
2. The optimal calculation method and data normalization processing algorithm are selected to ensure that the hit ratio of the software reaches 85%.
3. The evaluation result for each class should be ensured with a certain proportion and exclude the influence of minor singular values.
4. The relationship between parameters is compared from result to result so as to ensure that each parameter plays its due role in clustering and the cluster analysis is made comprehensively and reasonably.
5. The plane distribution map of the reservoirs classified is compared with the plane distribution map of sedimentary microfacies (Fig. 3.63),

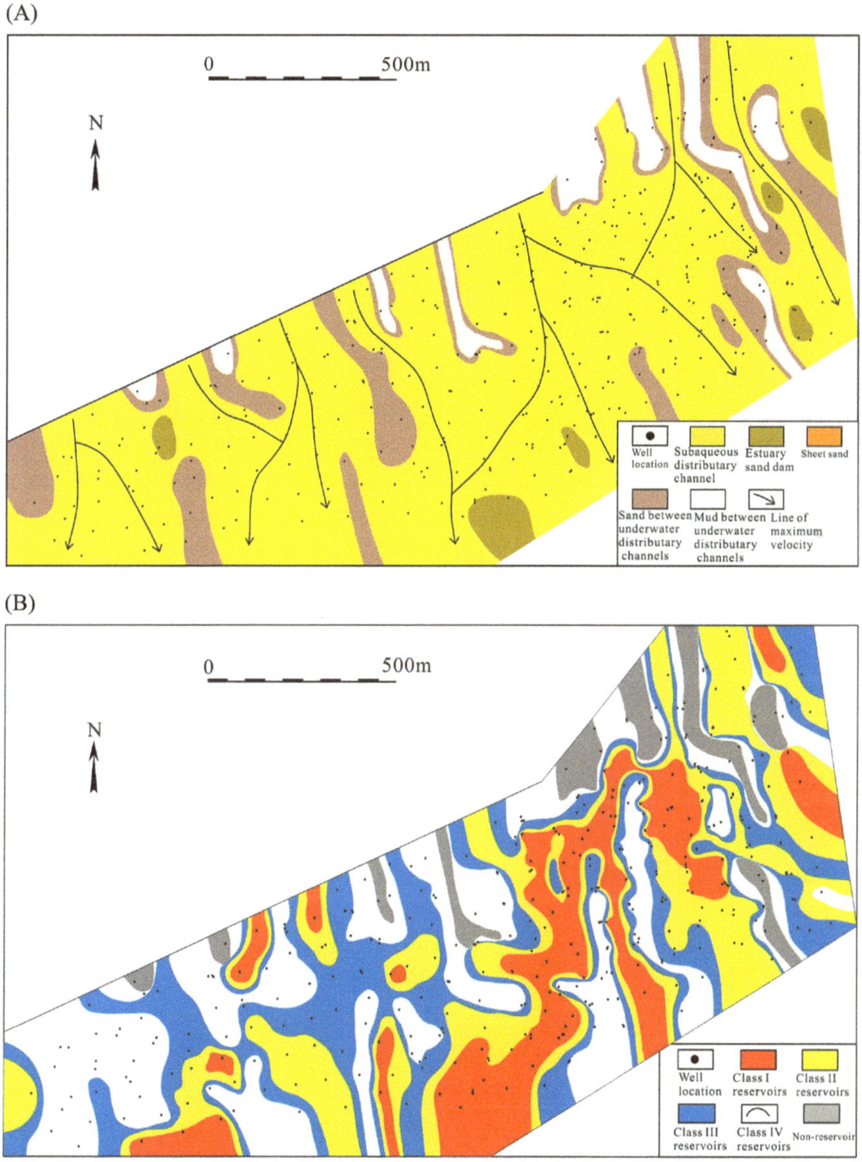


**Figure 3.63** Plane distributions of sedimentary microfacies and reservoirs classified in Yulou reservoir in an area of Western Sag, the Liaohe Basin. (A) Plane distribution of sedimentary microfacies of yl12b; (B) plane distribution diagram of yl12b reservoir evaluation results.

revealing a good correlation and consistency, which also indicates the rationality of the reservoir evaluation results and fully reflects the controlling effect of sedimentary factors on the reservoir properties.

#### *3.6.2.3 Influence of reservoir classification and evaluation results to the conversion of thermal recovery mode*

The results of reservoir evaluation can provide a solid geological basis for the conversion of heavy oil thermal recovery mode from steam stimulation to steam flooding for fan delta front reservoirs. The nature of the reservoir, as the material basis for steam flooding, directly determines the effect of steam flooding. The following points should be considered during steam flooding:

1. According to the results of reservoir evaluation, the steam injectors and producers should be placed in areas with similar physical properties to ensure the steam injectors inject steam effectively to drain hydrocarbons via the producers. As far as the study area is concerned, the steam injector and producers should be placed in the areas with Class I and II reservoirs; preferably, the reservoirs between the injector and the producer belong to the same class and are continuous and unchanged.
2. According to the reservoir properties, the injector—producer spacing should be reasonably designed to prevent the occurrence of steam channeling when the well spacing is too small. In the conversion of development mode, specific well patterns and well spacing should be designed depending on the class of reservoirs.
3. Since the Class I and Class II reservoirs are developed separately as two formation series in a vertical direction, the results of reservoir evaluation for beds in the two formation series should be carefully considered to ensure as many beds as possible are swept by steam flooding. Constrained by the current development status, most beds must be developed collectively for the purpose of cost-effective thermal recovery by steam flooding, and the reservoir properties are different from bed to bed, so most of the beds in one injection-production system should be ensured in the areas with Class I and Class II reservoirs in order to achieve the optimal development results.

## 3.7 Reservoir flow unit classification

As a typical transitional facies, the fan delta can be seen in many sedimentary basins, and it has always been one of the research focuses due to its rich abundance of hydrocarbon (Shouqing, 1993; Longxin and Ailin, 2003; Zaixing, 2003; Huanqing et al., 2017a,b). By definition, a fan delta is an alluvial fan propelled from the highland to the stable waters such as the sea and the lake (Zaixing, 2003). Since the fan delta reservoir is characterized by fast facies variation, strong heterogeneity, and complex oil–water relationship, the classification and evaluation of flow units in the fan delta reservoir for guiding the oil and gas field development is very important task. Among countries abroad, Hearn first defined flow unit as the accommodation consisting of vertically and laterally continuous reservoirs with similar permeability, porosity, and bedding characteristics (Dou, 2000; Yang and Jianmin, 2005). Aggoun et al. (2006) characterized the flow units in argillaceous sandstone reservoirs in Hassi R'mel Field, Algeria, through the conditional simulation with an optimal model selected after a detailed analysis of production history data. Mikes and Geel (2006) used a standard facies model to characterize the spatial shape, scale, direction of extension, and boundaries within and between facies of different sedimentary facies types, which incorporates the study on flow unit, providing the information basis for the accurate establishment of the final sedimentary facies model. Taghavi et al. (2007) classified the flow units in the geological modeling of the carbonate reservoirs in the Cretaceous Salvak Formation, southwestern Iran, and the results showed that flow units could be applied in reservoir characterization and geological modeling to describe the uncertainty and variation of the spatial distribution of porosity and permeability. In their study, the target reservoir was divided into eight types of flow units depending on the sedimentary environment, diagenetic process, and evaluation of porosity and permeability, each of which was divided by porosity, permeability, water saturation, and pore-throat distribution. Ghiasi-Freez et al. (2012) proposed two artificial intelligence models to improve the prediction accuracy of reservoir flow units by taking the South Pars gas field in the Persian Gulf basin as an example. Kadkhodaie-Ilkhchi et al. (2013) made cluster and mathematical analysis on the electric characteristics of flow units, from the sedimentological and geophysical perspective, in the tight sandstone reservoirs of the Willersby Formation in Whicher, the Perth Basin, Western Australia, and divided the flow units into Types A, B, C, D, and E, of which Types A and B flow units exhibit good reservoir properties, while Types C, D, and E flow units have poor reservoir properties. Enayati-Bidgoli et al. (2014)

studied the flow units in the Permian–Triassic carbonate reservoirs in the South Pars gas field, Iran, and they concluded that the characteristics of flow units are controlled by sedimentation and diagenesis. Roberto Aguilera (2014) systematically summarized the studies on flow units in conventional reservoirs and unconventional reservoirs such as tight oil/gas and shale oil/gas. Hatampour et al. (2015) studied the hydraulic flow units, sedimentary facies types, and pore structure characteristics of the Kangan and Dalan reservoirs in the South Pars gas field, Iran, and they classified the flow units through cluster analysis of core data into six types, of which the flow units with good reservoir quality are mainly composed of bioclastic limestone, oolitic limestone, intragranular grainstone, and agglutinated limestone, according to the bedding analysis of thin-section and cores. Jazi and Eckstein (2015) simulated the groundwater flow system through a case study of alluvial fan sediments and weathered bedrock belts with fractures in the Sand-Lick Watershed, Boone County, West Virginia, USA. Mahjour et al. (2016) studied the reservoir flow units in the Tabnack gas field, southern Iran, using the statistical zoning features and the improved Lorenz plots. Bidgoli and Bonab (2016) analyzed the characteristics of reservoir flow units when studying the geological conditions for the zonal distribution of reservoirs within the sequence stratigraphic framework for the offshore Carboniferous–Triassic gas-bearing reservoirs in Iran. Oliveira et al. (2016) measured the flow unit center to determine the dominant position of perforations in the well. Lopez Jimenez and Aguilera (2016) studied the flow units in shale condensate reservoirs, and they revealed that the ratio of permeability to porosity could be practically used as an index for classifying and evaluating the flow units in shale condensate reservoirs. In China, flow units have also been discussed frequently (Huanqing et al., 2010; Chen et al., 2011a,b,c). Yangquan et al. (1998) analyzed the differences in porosity and permeability between reservoir flow units when studying the heterogeneous hierarchical structure of clastic rock reservoirs, and they concluded that flow units are closely related to reservoir lithology and can present much lower physical properties if the effect of diagenesis is stacked. Jiyu et al. (1998) summarized the methods and significance of studies on flow units, and they proposed two research methods of flow units: core-based qualitative analysis and modified Kozeny-Carman equation. Wei Bin et al. (2000) combined the research results of reservoir flow units and the analysis of sedimentary facies to describe the distribution of the remaining oil in the high water-cut oilfields. Hanqing (2001) believed that the reservoir flow unit is a reservoir heterogeneous unit characterized by flow behaviors, which reflects a further refinement and quantitative representation of the reservoir structure model (deposition model). Hanqing (2002) discussed the approaches of research on reservoir heterogeneous system, on the internal architecture of sand bodies, and on flow units, and they indicated that the division of flow units is also a part of the research on reservoir heterogeneity, which must be based on the structural unit division. Yang (2003) characterized the distribution of the remaining oil in reservoirs based on the classification of flow units. Shaohua et al. (2007a,b) studied the GIS-assisted division of reservoir flow units, which achieved a classification of flow units with higher efficiency, authenticity, and objectivity. Song Ziqi et al. (2007) used the gray system to organically evaluate the flow units in the ultra-low permeability reservoirs in central Northern Shanxi slope, which are characterized by greatly different storage capacities and flow patterns and complex flow zones/belts. Songjiang et al. (2008) took the Mesozoic reservoirs in the Guan 142 fault block of the Huanghua Depression as an example to discuss the methods for studying the flow units in thick sandstone reservoirs. Haifa et al. (2009) finely divided the flow units in the thick conglomerate reservoir of alluvial fan facies, with a case study of the Badaowan Formation reservoir in Karamay Oilfield, using the flow zone indicator (FZI) histogram and the cumulative frequency curve plotting method. Qingsheng (2009) conducted a microscopic physical simulation experiment of the remaining oil in reservoirs using flow unit as constraints, and the results showed that the distribution of remaining oil is closely related to reservoir heterogeneity and displacement conditions. Zifei et al. (2014) analyzed the remaining oil distribution of carbonate reservoirs based on the flow unit characterization by taking the Γ-North oil reservoir in the Zanazor fractured-porous carbonate oilfield. Chao et al. (2016) classified and evaluated the reservoir flow units from the genetic perspective based on the study of high-resolution sequence stratigraphy, and they revealed that the plane distribution of sedimentary microfacies has a good agreement with the flow unit zone in the short-term base-level cycles.

This book comprehensively summarized the advantages and disadvantages of studies on reservoir flow units around the world—from the aspects of research content, methods, and trend (Table 3.16) (Yangquan et al., 1998;

**Table 3.16** Studies on reservoir flow units in China and other countries.

| Research needs | Maturity | Immaturity |
|---|---|---|
| Research contents | Classification and evaluation of flow units, flow unit analysis of genesis, distribution characteristics described of flow unit, heterogeneity of flow units | Definition and connotation of flow unit, geological modeling of flow unit, prediction of flow units in three-dimensional space, flow units and their influence on the development and production of oil fields such as distribution of remaining oil |
| Research object | Clastic rocks reservoir and carbonate reservoir | Shale condensate, tight oil, shale oil, and other unconventional reservoirs |
| Research methods | Outcrops, sedimentology, diagenesis analysis, numerical simulation, mathematical statistics and analysis, clustering analysis of various well logging data, and analysis and testing of physical properties of reservoir | Sequence stratigraphy, petrophysics, reservoir structure, geological modeling, artificial intelligence, gray system theory, microphysical simulation experiments |

Jiyu et al., 1998; Dou, 2000; Bin et al., 2000; Hanqing, 2001; Hanqing, 2002; Yang, 2003; Yang and Jianmin, 2005; Aggoun et al., 2006; Mikes and Geel, 2006; Taghavi et al., 2007; Shaohua et al.; Ziqi et al., 2007; Songjiang et al., 2008; Haifa et al., 2009; Qingsheng, 2009; Huanqing et al., 2010; Chen et al., 2011a,b,c; Ghiasi-Freez et al., 2012; Kadkhodaie-Ilkhchi et al., 2013; Enayati-Bidgoli et al., 2014; Aguilera, 2014; Zifei et al., 2014; Hatampour et al., 2015; Jazi and Eckstein, 2015; Mahjour et al., 2016; Enayatie Bidgoli and Rahimpoure Bonab, 2016; Oliveira et al., 2016; Lopez Jimenez and Aguilera, 2016; Chao et al., 2016)—and the analysis reveals that the study of reservoir flow unit can help quantitatively to describe the movement law of reservoir fluids, thereby providing references for oilfield development and production. In this study, the fan delta frontal flow units were comprehensively analyzed using the core, thin-section, and test data of five coring wells, together with the fine logging reinterpretation data of 400 wells and seismic data in the work area, to understand the flow behaviors of reservoir fluids, thereby providing a geological basis for conversion of development mode from steam stimulation to steam flooding for heavy oil thermal recovery reservoirs.

### 3.7.1 Geology and development of the Yulou reservoir in an area of Western Sag, the Liaohe Basin

The study area is structurally located at the southern end of the western slope of the Western Sag, the Liaohe Basin, and presents as a monoclinic structure dipping southeastward (Fig. 3.64) (Minggang et al., 2010; Huanqing et al., 2014a,b). The target reservoir is the Yulou reservoir in the first member of Shahejie Formation ("Sha 1 Member") of Paleogene, which is dominated by the fan delta front clastic deposits, and is mainly composed of thick inequigranular sandstone and medium to fine sandstone. The study area has generally experienced four development stages, that is, the steam stimulation test stage, the formal steam stimulation stage, the infilling adjustment stage, and the steam flooding test stage (in the middle–late steam stimulation stage). The Yulou reservoir has entered the middle–late steam stimulation stage, with worsening production effect. It is urgent to convert the development mode. Due to the frequent swinging of underwater distributary channels, the channel and interchannel sand bodies of different stages are superposed vertically and intertwined in the form of stripes laterally, resulting in strong reservoir heterogeneity, which restricts the further expansion of steam flooding in the block. In such circumstances, we made a sedimentary analysis to deeply understand the development law of reservoirs, thereby providing a reference for the effective development of the reservoir. The Yulou reservoir is divided into 29 beds, corresponding to 29 short-term base-level cycles (Huanqing et al., 2014a,b). According to the geological, core, log, seismic, and test data, the Yulou oil reservoir is believed as a fan delta front deposit, which can be subdivided into five types of sedimentary microfacies, that is, underwater distributary channel, estuary sand bar, underwater distributary interchannel sand, underwater distributary interchannel mud, and front sheet sand. Among them, the underwater distributary channel and estuary sand bar are dominant. The core analysis shows that the reservoir in the study area is of high porosity (31.25% averagely) and high permeability (1829.3 mD averagely). The frequent diversion of underwater distributary channels, coupled with diagenesis, leads to

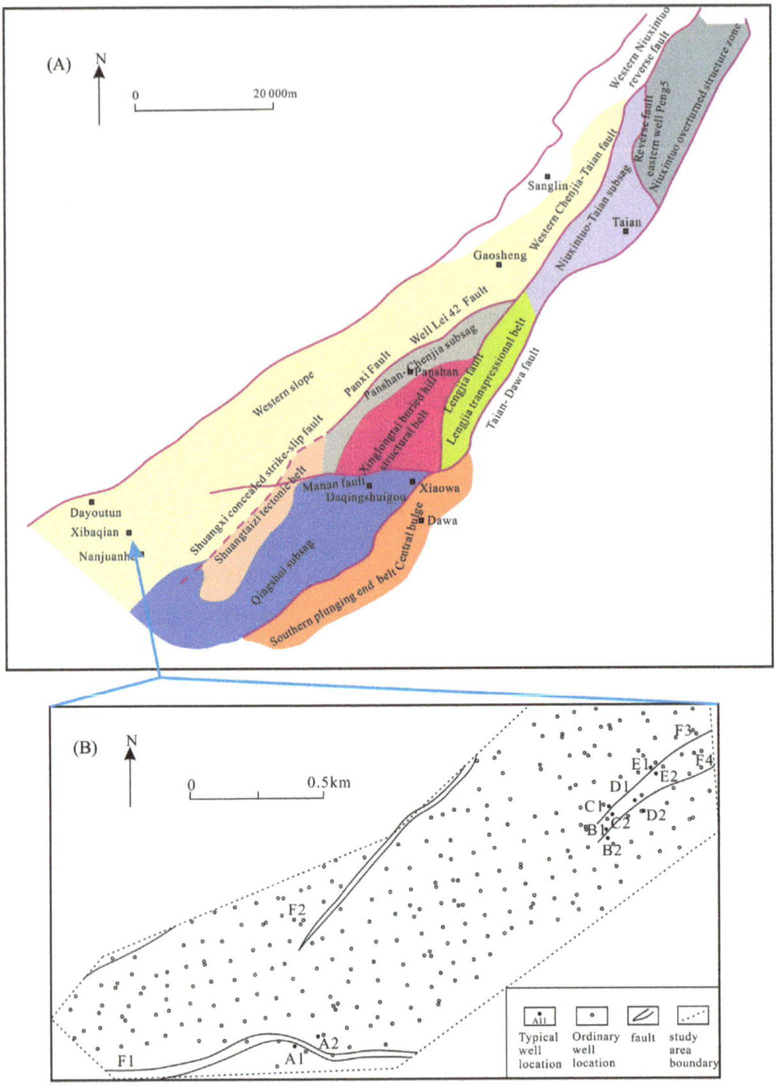


**Figure 3.64** Structural division of the Western Sag, the Liaohe Basin, and location of the study area. *From Minggang, L., Jiafu, Q., Hengmao, T., et al., 2010. Cenozoic fault structure and hydrocarbon accumulation in Western Sag, Liaohe Depression. Petroleum Exploration and Development 37(3): 281–288. Huanqing, C., Chao, D, Yijing, D, Jue, W., 2015a. Advances of reservoir evaluation researches. Geological Science and Technology Information 34(5), 66–74. Huanqing, C., Chengfang, S., Chen, C., 2015b. Advances in the study of fine reservoir description. Geological Review 61(5), 1135–1146. Huanqing, C., Shuxian, L., Peng, J., et al., 2015c. Flow barrier characteristics of thermal rescovery of heavy oil reservoir: taking Yulou oil-bearing sets in some experimental area in the south end of westward slope of West depression in Liaohe Basin as an example. Acta Sedimentologica Sinica 33(3), 616–624. Huanqing, C., Shuxian, L., Zhirui, S., et al., 2015d. Caracteristics of conglomerate reservoir architecture of alluvial fan and its controlling effects to reservoir development: taking alluvial fan reservoir in some area of Northwest margin of Junggar Basin as an example. Journal of Jilin University (Earth science edition) 45(1), 13–24.*

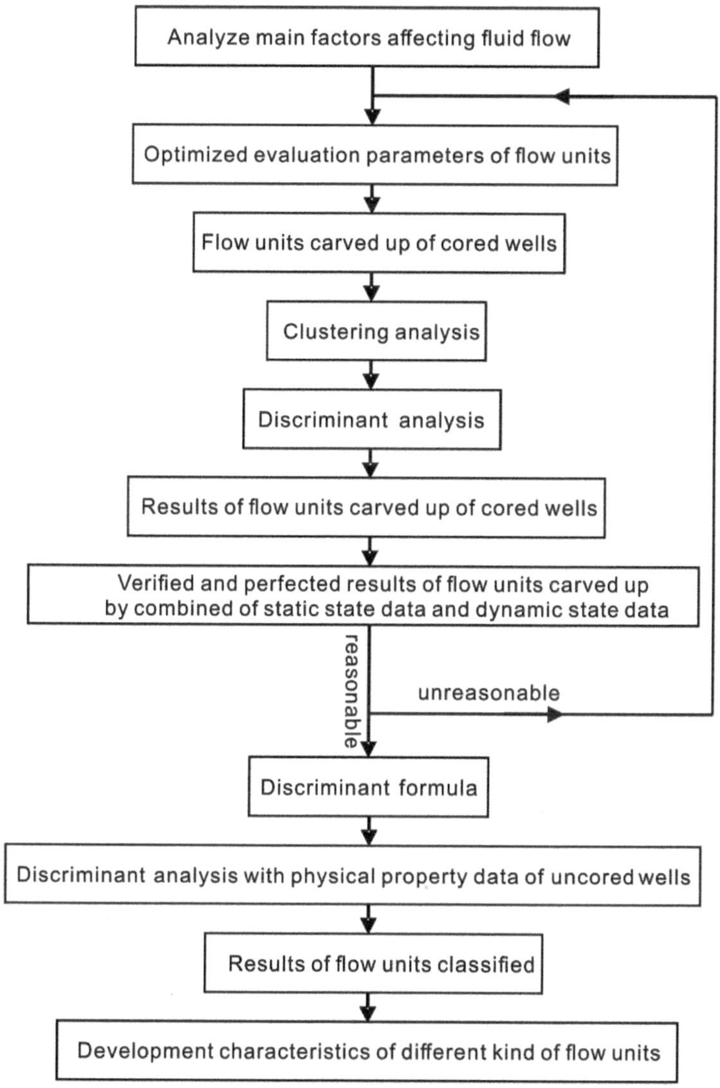


**Figure 3.65** Flowchart for classification and evaluation of reservoir flow units.

strong reservoir heterogeneity. The study area is in the late steam stimulation stage, so it is urgent to convert the development mode from steam stimulation to steam flooding.

The study area is located on the Xibaqian fan delta. Four faults were developed, of which two were larger, and two were smaller. The fault strike is from north-east to south-west and east to west.

### 3.7.2 Research approach

According to the research purpose and the data availability, the research approach is designed (Fig. 3.65). Firstly, the main factors affecting the fluid flow behaviors inside the reservoir are identified. On this basis, the optimal parameters for reservoir flow unit classification are selected. Then, cluster analysis and discriminant analysis are conducted on the test data of cores from the coring well. Based on the work experience, dynamic and static data are closely combined for repeated experiments until a satisfactory discriminant formula is obtained. The discriminant formula is used to calculate and discriminately analyze the physical property data derived from fine logging reinterpretation for noncoring well. Finally, the classification and evaluation of reservoir flow units in the whole study area are completed, and the spatial development of different types of reservoir flow units is characterized. This approach can efficiently and quickly derive a relatively accurate result by virtue of organically combining the qualitative genetic mechanism analysis with the quantitative mathematical operation, together with the operation analysis based on SPSS. The data of fine logging reinterpretation is the most important database for this study. Here, the data source is briefly introduced. The study area is an old area with a history of nearly 40 years of development being undertaken here. Now, a large amount of log data are available. These data were acquired by several logging companies with different logging tools in different time periods; so it is necessary to unify these data. Moreover, as more data of cores, dynamic monitoring, and oilfield development are acquired, these additional data should be integrated into logging interpretation to improve the original results; thus fine logging reinterpretation is needed. This approach inherits from the procedure of the additional test data being closely combined with the log-interpreted electrical data and the dynamic monitoring and oilfield development data to reinterpret the reservoir porosity, permeability, and oil/gas/water layers, and then the reinterpreted data being imported into this flow unit classification. Such data serve as a solid basis for accurate classification and discrimination of flow units.

### 3.7.3 Parameter determination for reservoir flow unit classification

Parameter determination is the focus of and a challenge in reservoir flow unit classification, and it directly controls the classification accuracy. In this study, a number of parameters affecting the fluid flow behaviors in the reservoir were analyzed and compared. Based on the analysis of the factors influencing reservoir fluid flow characteristics, the classification and evaluation parameters of flow units are optimized to ensure that these parameters can fully reflect the influence of various factors on reservoir flow characteristics. According to the analysis, tectonism, sedimentation, and diagenesis, all have an influence on the fluid flow characteristics of the reservoir, among which sedimentation plays a major role. The comparison can reflect the parameters such as porosity, permeability, mud content, sand body thickness, reservoir thickness, and FZI and, finally, determines the flow unit classification parameters. The tectonic action mainly affects the fluid flow characteristics of the reservoir through cracks, which can be reflected by permeability. Sedimentation plays a very important role in controlling the fluid flow characteristics of the reservoir, mainly through porosity and permeability. The influence of diagenesis on fluid flow characteristics of the reservoir is mainly reflected by FZI. In addition to porosity and permeability, the FZI of the reservoir hydraulic unit was used. The weights of the three parameters were 0.2, 0.4, and 0.4, respectively, determined by repeated tests, and after being verified according to the quantitative comparison of different classification results, the distribution of reservoir properties like porosity and permeability, and the previous working experience. In the process of cluster analysis and discriminant analysis, the parameters involved in the calculation were normalized empirically. It should be noted that, different from the reservoir classification and evaluation, the flow unit classification and evaluation don't necessarily generate better results when too many parameters are used. Instead, just the parameters that can best reflect the fluid flow behaviors are preferred.

Among the abovementioned parameters, porosity and permeability are obtained through core analysis and test, or from fine logging reinterpretation for noncoring wells. FZI is defined as follows (Xinghe, 2009):

$$\text{Hydraulic unit FZI} = \text{RQI}\ \varphi Z = 10 - 2 \cdot \pi \cdot K\phi \cdot 1 - \phi\phi \tag{3.1}$$

where,

$$\text{Reservoir quality index RQI} = 10 - 2 \cdot \pi \cdot K\phi \tag{3.2}$$

$$\text{Ratio of pore volume to particle volume } \phi Z = \phi/1 - \phi \tag{3.3}$$

$\varphi$ is the porosity, and K is the permeability, or mD, both of which can be obtained by refining logging reinterpretation of the reservoir.

Note that the data should be selected with full consideration to Eq. (3.1) the factors such as reservoir heterogeneity; Eq. (3.2) the location of coring well, namely, coring wells as scattered as possible should be selected on the plane to control the entire study area; and Eq. (3.3) a certain depth in the horizon, so that the selected sampling points can fully represent the physical characteristics of the target reservoir in the entire study area.

It should be pointed out that when selecting data, reservoir heterogeneity and other necessary factors must be taken into consideration. The locations of cored wells should be dispersed as far as possible laterally, and the reservoir should cover enough depth to ensure sample points can fully represent the target layer in the study area. Porosity and permeability reflect the reservoir seepage characteristics from the static perspective, while FZI reflects the reservoir seepage characteristics from the dynamic perspective. In this way, static and dynamic indexes are effectively combined when dividing flow units. It should be pointed out that although FZI is derived from porosity and permeability, FZI based on mathematical calculations with physical implications of reservoir quality has been found to be completely different from porosity and permeability, reflecting reservoir fluid flow characteristics from a dynamic perspective. Therefore there is no duplication in the choice of parameters.

Detailed well logging secondary interpretation data is the most important data basis of this study, so the source is briefly discussed and introduced. Because the research area is old, it has nearly 40 years of development history and has accumulated a lot of well logging data. These data come from different well logging companies of different ages and different well logging instruments, so the basis of these data needs to be unified. In addition, with the increasing enrichment of core well data, dynamic monitoring data, and oilfield development data, it is necessary to add the newly acquired data to the log interpretation results to constantly improve and perfect the original well logging interpretation results, so the well logging fine secondary interpretation can be carried out. The method is used to increase the analysis and test data, combining well logging data, at the same time reference for dynamic monitoring and oilfield development data, the reservoir porosity, permeability and oil and gas layer and so on to explain, the application of the data of these to explain to the flow unit points? The class study, for flow unit

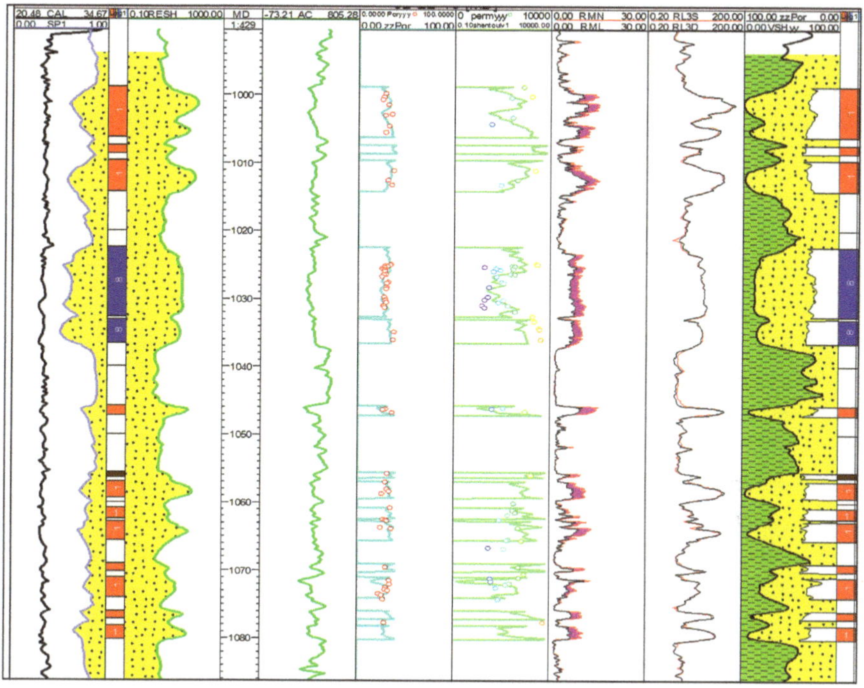


**Figure 3.66** W2 well logging fine secondary interpretation map.

classification criterion, accurately provides a solid material data foundation (Fig. 3.66). Columns five and six, in the Fig. 3.66, are coring well core porosity and permeability analytical test results compared with the well logging interpretation results, respectively. And the corresponding better that it prove that the fine secondary logging interpretation result believable. The right column is well logging oil–water layer interpretation results, including red for reservoir, blue for water, brown for dry layer. The results of oil–water layer interpretation can be considered as one of the factors for prediction of interwell connectivity of different types of flow units.

## 3.7.4 Results of reservoir flow unit classification

### *3.7.4.1 Single-well reservoir flow unit classification*

Based on the porosity, permeability, and FZI determined, the SPSS cluster analysis platform was adopted to analyze the 150 core samples from 5 coring wells. Specifically, the two-step cluster was used. It can accept both nominal variables and within-interval or above variables. For the two-step cluster, which takes Cases as the object of clustering (equivalent to the Q-clustering of Cases), the data not obvious in one set are clustered into several natural groups or classes (Jianqiong et al., 2006). In the cluster analysis, Ward's method was adopted for calculation, using the basic data with their quality as controlled. Particularly, different parameters were normalized by using the SPSS software to make them reach the same order of magnitude. In Ward's method, the sum of squared deviations is the sum of the squares of the difference between the terms and the average term, which, physically, is used to reflect the difference between different samples and the average term, thereby ensuring that the flow units with the same characteristics are assigned into the same category. The discriminant analysis was performed on the results of cluster analysis to obtain the discriminant formula. The discriminant formulas (Eqs. (3–4) to Eqs. (3–7)) were used to calculate and discriminate the physical property data derived from fine logging reinterpretation for noncoring wells. Finally, the classification and evaluation of flow units in the target reservoir were completed for the entire study area. The algorithms and data processing methods used in cluster analysis and discriminant analysis were obtained through repeated tests. The Yulou reservoir in an area of Western Sag, the Liaohe Basin, is divided into four classes of flow units, that is, Class A, Class B, Class C, and Class D, which respectively correspond to the best, good, moderate, and poor properties of the reservoir (Table 3.17, Fig. 3.67). The SPSS software showed a hit ratio of 98.7%, which proves that the results of the flow unit classification and evaluation are true and reliable. The discriminant formulas for different classes of flow units are as follows:

$$\text{Class A flow unit: } 2.133\,\phi - 0.346\,K + 0.449FZI - 103.222 \tag{3.4}$$

$$\text{Class B flow unit: } 1.879\,\phi - 0.284\,K + 0.294FZI - 68.035 \tag{3.5}$$

$$\text{Class C flow unit: } 1.787\,\phi - 0.23\,K + 0.225FZI - 59.105 \tag{3.6}$$

$$\text{Class D flow unit: } 1.66\,\phi - 0.193\,K + 0.181FZI - 50.557 \tag{3.7}$$

where $\phi$ is the porosity; K is the permeability, mD; FZI is the hydraulic unit flow zone indicator.

As mentioned above, the calculation method and data processing method were repeatedly tested to obtain the final results of flow unit

**Table 3.17** Results of flow unit classification for the Yulou reservoir of coring well in an area of Western Sag, the Liaohe Basin.

| Parameter classification | Porosity/% | | | Permeability/md | | | FZI | | |
|---|---|---|---|---|---|---|---|---|---|
| | maximum | minimum | average | maximum | minimum | average | maximum | minimum | average |
| A class flow unit | 37.9 | 28.4 | 33.3 | 7531.00 | 887.00 | 2780.12 | 8.69 | 3.34 | 5.53 |
| B class flow unit | 38.0 | 26.2 | 30.7 | 818.00 | 61.00 | 254.80 | 2.64 | 1.22 | 1.80 |
| C class flow unit | 40.1 | 25.4 | 30.8 | 201.00 | 16.00 | 70.27 | 1.06 | 0.73 | 0.86 |
| D class flow unit | 36.7 | 21.9 | 28.8 | 35.87 | 1.00 | 11.03 | 0.67 | 0.14 | 0.39 |

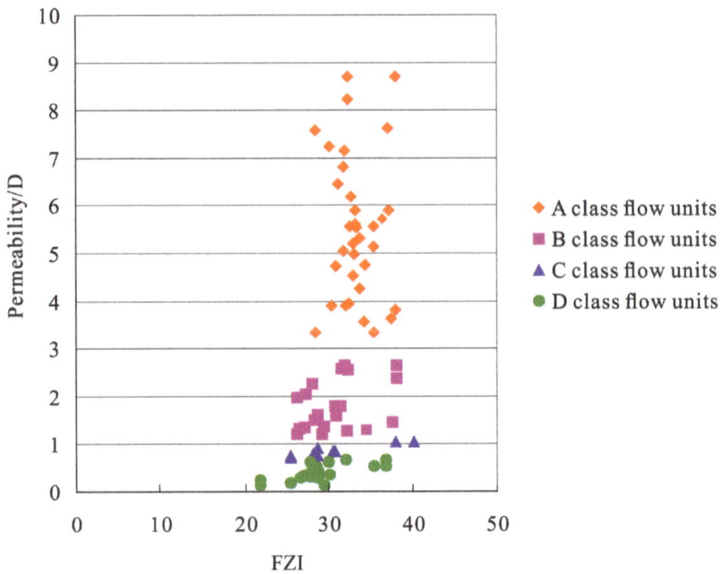


**Figure 3.67** The relationship between porosity and FZI of different classes of flow units in Yulou oil-bearing sets in cored wells in some areas of West Depression in Liaohe Basin.

classification and evaluation. Such results were verified according to the following principles:

1. If the number of one or two classes of flow units is significantly smaller than the rest of the classes, this is considered as a singular value in the physical property data of the coring well caused by the presence of fractures. Such singular value should be eliminated, and a reclassification should be made.
2. If there is a great contradiction between the results of flow unit classification and the results of sedimentary microfacies analysis, which can't be explained from the perspective of geological geneses such as tectonism and diagenesis, this may be caused by the incorrect selection of algorithm or data processing method for cluster analysis or discriminant analysis. In this case, the algorithm or data processing method should be reselected for calculation.
3. The results of flow unit classification and evaluation may be inconsistent with the current oilfield development and production practices. Wells corresponding to Class A or Class B flow units are inconsistent that have poor physical properties and low steam flooding volume and oil production.

4. The values of parameters may have overlapped for different classes of flow units, with no clear boundary. Each of the three parameters should be reflected in both cluster analysis and discriminant analysis.
5. The hit ratio of the SPSS software is less than 85%, which reveals the potential problems in the selection of parameters, cluster analysis, discriminant analysis, or parameter processing algorithms.

The author has done a study on the classification of flow units of volcanic rock reservoirs of the Member One of Yingcheng Formation in the Xudong Area of Songliao Basin in China. It was found in the work that the test was repeated in the discriminant analysis. When the discriminant positive rate of flow units was less than 85%, the classification results of flow units in different layers were quite different from the development status of volcanic rock gas reservoirs. When the value is greater than 85%, the research result of the flow unit agrees well with the actual development status of the gas reservoir (Chen et al., 2011). This 85% positive rate also meets the requirements of the project research contract for the accuracy of this study, so it is used as an empirical indicator for reference in this study. As it turns out, it works. The determination of this quantitative index also makes the identification of the classification results of flow units to have a quantitative index, rather than just a qualitative index, which promotes the development of flow unit research from qualitative to quantitative.

### 3.7.4.2 Sectional reservoir flow unit classification

In the study, a typical section was selected and the distribution of flow units was plotted. The results of four types of flow units based on the secondary fine interpretation of well logging in a single well were marked on the well first. Then the flow units types of sand bodies of different sedimentary origin were determined with reference to the results of a single well and section of sedimentary microfacies. As for the superposition relation and connectivity of different types of flow units on the plane, the results of oil and water layer interpretation and the size of the sand body, the shape of logging curves, and the results of single-layer partition are referred to. The algorithm of clustering analysis is introduced in detail in the fourth part of this paper, and arithmetic is Ward's method. Ward's method is the sum of the squared differences of each item and the average term, and its physical meaning is used to reflect the differences between different samples and the average term so as to ensure that the flow units with consistent characteristics are divided into the same category. Typical sections were selected to plot the distribution characteristics of flow units (Fig. 3.68). It can be seen

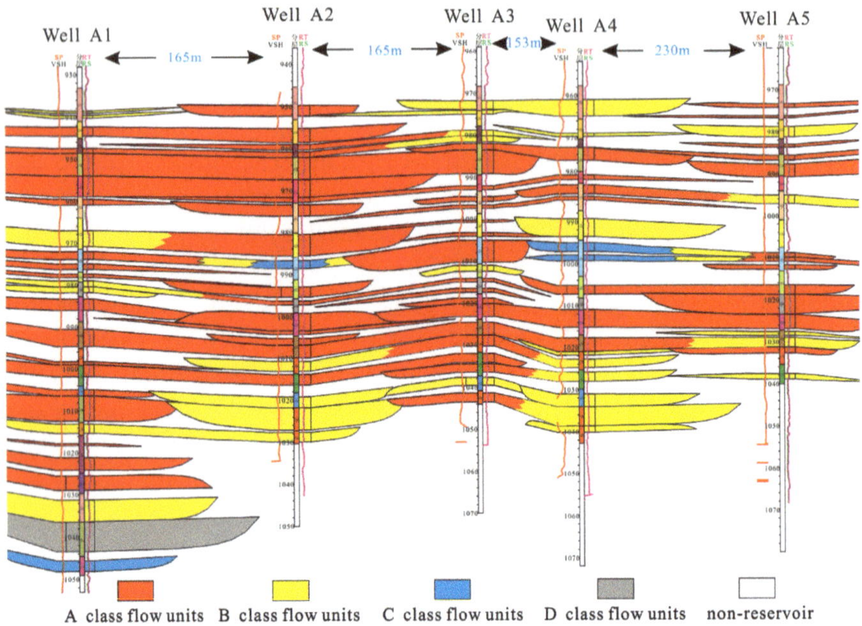


**Figure 3.68** Sectional distribution of flow units in the Yulou reservoir in an area of Western Sag, the Liaohe Basin (parallel to the provenance direction).

that the flow units distribute complexly in space. Generally, Class A and B flow units are dominant, while Class C and D flow units are less. Upward, Class A and B flow units decrease gradually, while Class C and D flow units increase gradually, which is consistent with the distribution of reservoir properties. According to the well spacing data, the flow units can change in the range of 200–300 m, suggesting that the underground fluid also has a strong heterogeneity. This implies a great challenge to the late conversion of development mode and the enhancement of oil recovery.

Based on the sectional distribution of flow units and the core observation and test data, the genesis of the flow unit was analyzed (Fig. 3.69). The underground distribution of flow units is mainly controlled by sedimentary microfacies but by tectonism and diagenesis locally, where the same class of flow units distributes in a less range. In other words, similar to the development of the sedimentary microfacies on the section, different classes of flow units show much better continuity in the direction parallel to the provenance direction (NW-SE) than perpendicular to the provenance direction. From the perspective of sedimentary microfacies, Class A and B flow units mainly correspond to the underwater distributary channel and estuary sand bar, while Class C and D flow units mainly correspond to

**Figure 3.69** Genetic factor analysis on flow units in Yulou oil-bearing sets in some area of West Depression in Liaohe Basin. (A) Well J2, gray-black fine sandstone; (B) Gray siltstone, ripple cross bedding; (C) Well J2, intergranular pore, 1022.72 m, 25 ×; (D) Well J2, intergranular pore, 1027.19 m, 25 ×; (E) Well J2, I/S, 991.02 m; (F) Well J2, grade I–II overgrowth of feldspar, 1000.76 m.

the underwater distributary interchannel sand. From the perspective of structure, the existence of some large-scale third-order faults, coupled with the development of fractures near the faults, optimizes the spatial connectivity of the reservoir to some extent, which is positive for the distribution of flow units. Besides, the existence of local microstructures further complicates the distribution of underground oil and water and the oil–water interaction in the development process. Moreover, diagenesis such as compaction, dissolution, and cementation greatly affects the spatial distribution of different classes of flow units. Overall, diagenesis plays a negative role in the distribution of flow units by making their connectivity worse in space.

Due to the influence of sedimentation and diagenesis, the sand body has poor continuity on the profile, resulting in poor continuity of different flow units on the profile. The underwater distributary channel mainly corresponds to A type of flow units and B type of flow units. Flow unit types change quickly, usually around 100 m.

Fig. 3.69A and B show that sedimentation has a controlling effect on flow units. Fig. 3.69C–F show that diagenesis has a certain influence on the flow units of the reservoir, in which dissolution is favorable for the distribution of flow units while cementation is unfavorable for the distribution of flow units.

#### 3.7.4.3 Plane reservoir flow unit classification

According to the single-well flow unit classification results, the plane development of flow units in each bed was plotted (Fig. 3.70). All flow units are roughly distributed in stripes along the NW-SE direction. In the 29 beds, Class A and B flow units are dominant, while Class C and D flow units are less. Comparing the flow unit distribution map and the sedimentary microfacies plane distribution map (Fig. 3.71), it is found that the flow units are mainly controlled by the sedimentary microfacies, which is consistent with the characteristics of flow units in section. However, different classes of flow units can occur in the sedimentary reservoirs corresponding to the same sedimentary microfacies. This is mainly due to the differences in the lithology and physical properties of the strata under different hydrodynamic conditions in the early stage and the different degrees of diagenesis in the late stage. At the positions where the hydrodynamic conditions are strong and the dissolution is strong, but the compaction and cementation are weak, Class A and B flow units are dominant; otherwise, Class C and D flow units are dominant. In view of the distribution range, the sum of Class A and B flow units

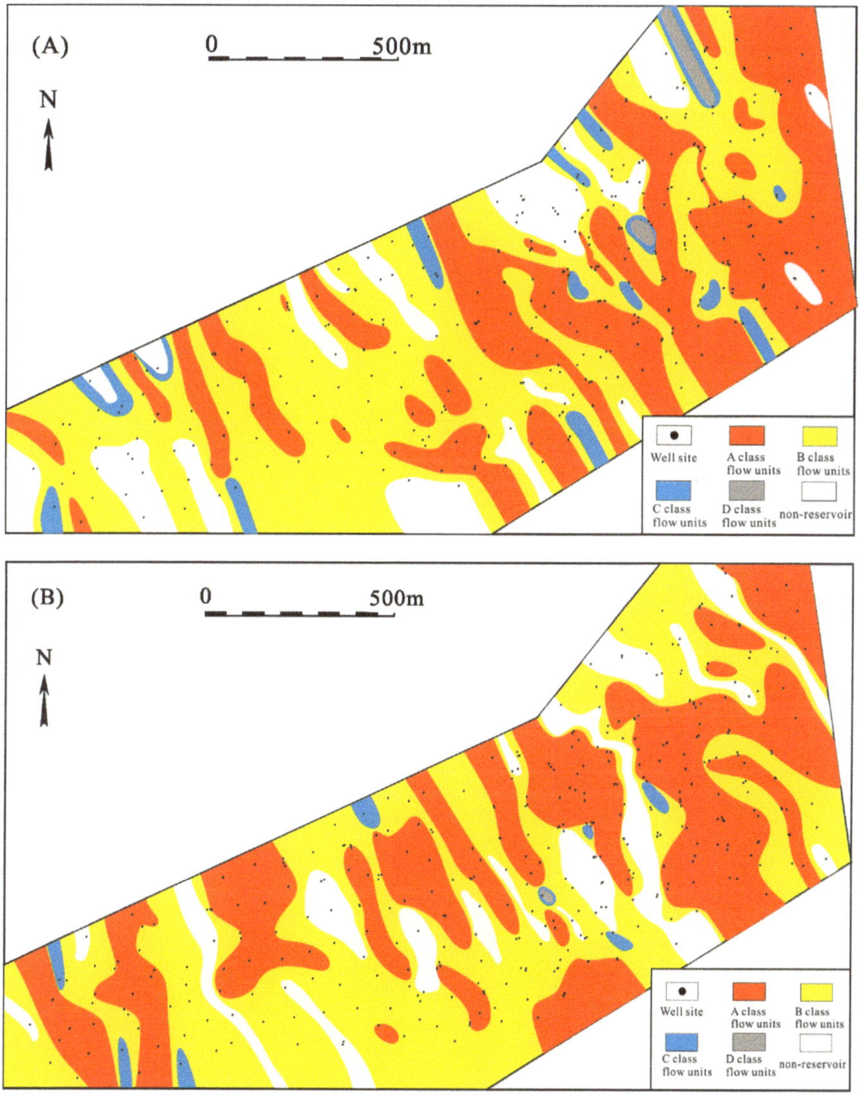


**Figure 3.70** Plane distribution of flow units in yl11a and yl11a of Yulou oil reservoir in some area of West Depression in Liaohe Basin.

exceeds 2/3 of the study area. All 29 beds present the development characteristics of flow units basically consistent with that of $yI_1^{1a}$ and $yI_1^{1b}$.

However, due to the limitations of data and research technology, no 3D model of flow units was established. Therefore, in this study, there are no drawings of three-dimensional flow units. In order to draw the 3D

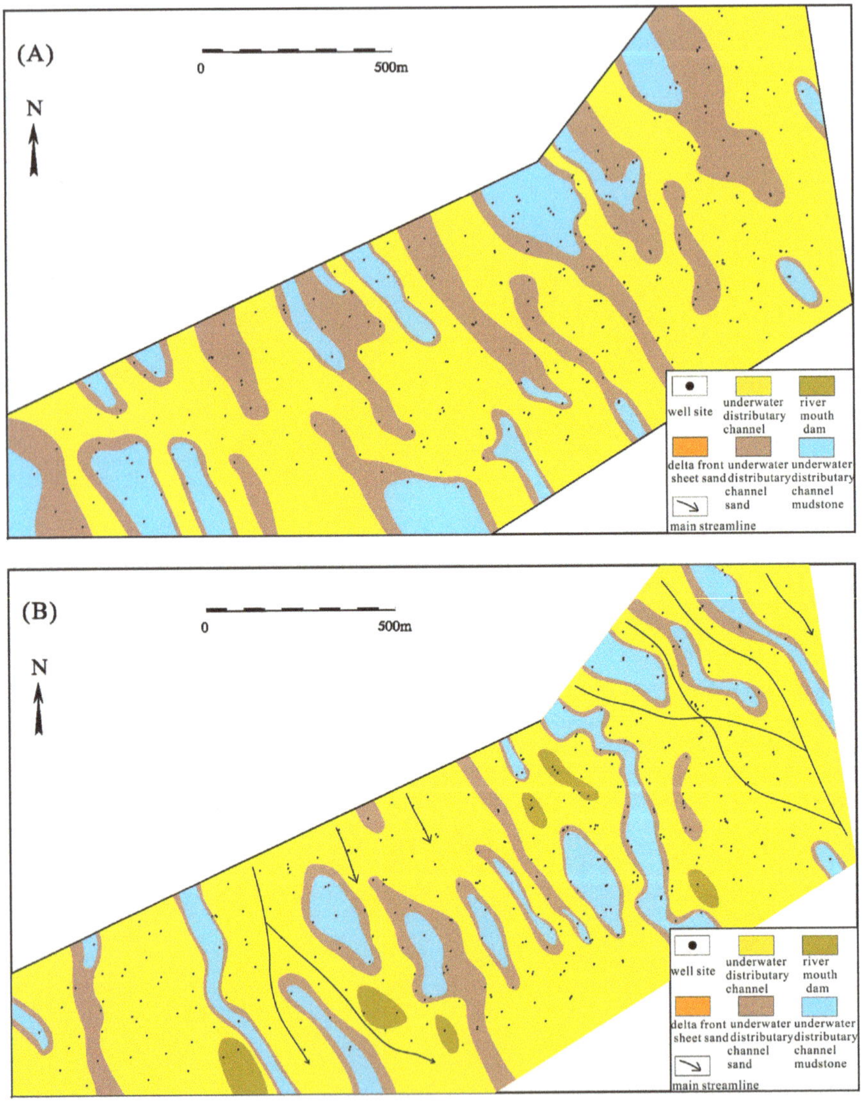


**Figure 3.71** Lateral distribution characteristics of microfacies in yl11a and yl11b in Yulou oil-bearing sets in some areas in the West Depression in Liaohe Basin.

flow units development feature map, it is necessary to develop appropriate calculation methods to complete interwell prediction of the flow units in 3D space, which will be a very important research direction for the author in the future. At present, the development characteristics of different monolayer flow units have been realized only by combining the results of sedimentary microfacies in each layer.

In the plane, the flow units in the target layers of the study area are controlled by sedimentary microfacies development features, mainly class A flow units and class B flow units, which are roughly distributed along the north-south and east-south directions, consistent with the characteristics of high porosity and high permeability reservoirs.

In the plane, the sedimentary microfacies are mainly distributed in place of subsurface distributary channels, while other types of sedimentary microfacies are less in the distribution. Controlled by the direction of the sedimentary source, different types of sedimentary microfacies are roughly distributed from northwest to southeast, and the estuary sand dam presents a thallous distribution feature with a small area.

### 3.7.5 Influences of flow units on oilfield development and countermeasures

For the fan delta front reservoirs, the frequent divergence of the underwater distributary channels made them strongly heterogeneous, and then influenced by the structure and later diagenesis, they became more complicated and more difficult to develop. In addition, influenced by development measures of long-term steam stimulation, the movement and distribution of the oil and water are more complicated. Through the classification and quantitative evaluation of the flow units, the distribution of the underground fluid mobility is further understood and provides a basis for the conversion of development mode from steam stimulation to steam flooding. Production practice shows that the classification results of the flow units are closely related to the present development performance. In general, the reservoirs corresponding to Classes A and B flow units are better in physical properties, while those corresponding to Classes A and B are poor. In this study, by taking Well group JK280 as an example, a grid diagram of the flow unit was drawn, and combined with the results of well logging interpretation, the influence of distribution characteristics of flow units on production was analyzed (Table 3.18). The well group includes five wells, of which Well JK280 is a steam injection well. The remaining wells are those of oil. Analysis indicates Classes A and B in Well J270C group are thickest if ignoring the effect of water flooding, and they contributed the most to oil production. Among the other three production wells, Class A flow units in Well JK32 have been flooded thoroughly, while Class B flow units are water or flooded layers; they

**Table 3.18** The thickness of the target layers (y11a-yll12a) and isolated layers, and production data of Well group of JK280 in the study area.

| Well name item | JK270 | J270C | JK280 | JK290 | JK32 |
|---|---|---|---|---|---|
| Layer thickness (m) | 123.00 | 86.50 | 107.00 | 123.50 | 122.3 |
| The thickness of A class flow units/thickness of water flooding layer (m) | 11.33/ 8.14 | 25.00/ 12.00 | 15.10/ 12.70 | 23.87/ 8.70 | 31.90/ 31.90 |
| Thickness of B class flow units/thickness of water flooding layer (m) | 24.68/ 15.46 | 18.6/5.30 | 29.7/ 14.00 | 34.63/ 7.90 | 22.09/ 4.70 |
| Oil production ($\times 10^4$T) | 4420.626 | 12356.148 | 339.200 | 3099.771 | 2400.757 |
| Steam injection volume ($\times 10^4$ $m^3$) | 4100.00 | 35564.00 | 29647.00 | 1810.278 | 4982.91 |

were injected with the most steam, but delivered the lowest production. Compared with Well JK270, Well JK290 was injected steam 44.2% of Well JK270, but its oil yield reached 70% of Well JK270, showing performance of the latter is better than the former's. This further proved that the development characteristics of the flow units had a very important influence and controlling effect on the production practice.

Based on the results of flow unit classification and evaluation, the implementation of measures to improve oil recovery should be considered from the following aspects in the conversion process of steam stimulation to steam flooding in the target layer of the study area. First, the flow units are mainly Classes A and B (Fig. 3.74), which shows that reservoir connectivity is good. However, in combination with the results of fine reinterpretation of well logging data, the sandbodies in the lower part of the target layer are mostly water layers. Therefore, during the development process, especially the injection process of steam flooding, special attention should be paid to the injection pressure control and the selection of injected sections to prevent the edge water and bottom water from channeling and to minimize the influence of flooding. Second, in the deployment and adjustment of infill wells, injection wells, and production wells should be located within the same flow unit to ensure good injection-production communication between injection and production wells. Third, the main targets by steam stimulation and steam flooding are Classes A and B flow units which are the major components of the remaining oil because Classes A and B flow units mainly correspond to good reservoirs. In the study, the development characteristics of flow units in the other well groups were also analyzed. Their characteristics are similar to those in Well JK280. Therefore the above problems and corresponding countermeasures are common in the study area. Fourth, when applying the research results of flow units, it should be used in combination with the results such as fine reinterpretation of well logging data and takes full account of the differences between oil and water layers. Because the results of flow unit research only fully demonstrate the development characteristics of the underground fluid, it is different from the result of reservoir evaluation. Classes A and B flow units may also correspond to water layers which are not development targets. In application, the limits and deficiencies of the research on flow units should be clearly recognized.

Flow unit types are closely related to production practice. The flow units are mainly controlled by sedimentary microfacies, which are affected by diagenesis. The flow unit types can change greatly at the level of 100 m, and it is very difficult to predict the flow unit interwell. The development of flow units of types A and type B is not only the main stratum of oilfield development at present but also the main targets of future remaining oil characterized (Fig. 3.72).

## 3.8 Multipoint geostatistical modeling

The final result of reservoir description is a quantitative reservoir geological model as the basis of reservoir simulation, reservoir engineering, and oil recovery process. A complete reservoir geological model includes three parts: structural model, reservoir model, and fluid model (Yinan and Ziqi, 1996). According to the model precision required in different development stages, the reservoir geological model can be divided into three categories: conceptual model, static model, and predictive model. The predictive model is highlighted in the fine reservoir description. Wu Shenghe et al. (1999) systematically introduced the concept, significance, and representation of reservoir geological modeling, as well as deterministic modeling, stochastic modeling algorithm, and geological modeling software (Table 3.19). Jiahua and Tuanfeng (2001) published a monograph entitled "Stochastic Reservoir Modeling," detailing the geological application of stochastic reservoir modeling, the spatial difference of geological variables, stochastic simulation and basic principles of geostatistics, and the Gaussian field model, indicator model, truncated Gaussian model, random walk model, marked point process model, and the likes, and they stated that spatial stochastic simulation of sedimentary

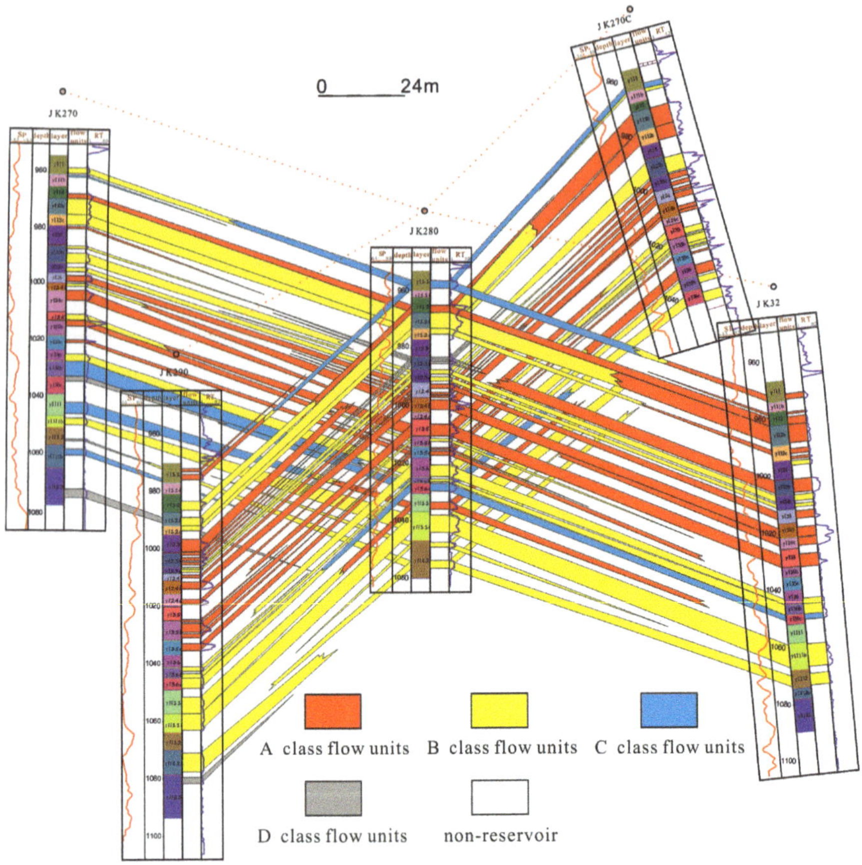


**Figure 3.72** Grid diagram of flow units in Yulou oil-bearing sets of a well group in some area of West Depression in Liaohe Basin.

facies is a core issue in the entire stochastic reservoir modeling process. Zhu Shijun et al. (2005) systematically introduced in the monograph "Quantitative Reservoir Modeling" how to construct the quantitative reservoir model by integrating the seismic data and related processing and interpretation techniques. Jia Ailin (2010) introduced the geological modeling technique in the fine reservoir description from three aspects: stochastic simulation technology, reservoir prototype model technology, and the establishment of a fine reservoir geological model by integrating miscellaneous geological information. Carls Jeff et al. (2014) concretely introduced the contents of petroleum geostatistics and its application in geological modeling. Carls Jeff et al. (2016) expounded on various techniques of uncertainty modeling in geoscience. Beginning with the significance and current status of

**Table 3.19** Main stochastic models, algorithms, and methods.

| **Algorithm and model** stochastic modeling technique **stochastic model and nature** | | Sequential simulation | Error simulation | Probability field simulation | Optimized algorithm (simulated annealing and iterative algorithm) | Model nature |
|---|---|---|---|---|---|---|
| Purpose-based stochastic model | Marked point process (Boolean model) | | | | Marked point process simulation (Boolean simulation) | Discrete |
| | Stochastic genetic model | | | | Sedimentary process simulation | Discrete |
| Pixel-based stochastic model | Gaussian domain | Sequential Gaussian simulation | Steering belt simulation | Probability field Gaussian simulation | (Simulation annealing can act as post-processing) | Continuous |
| | Truncated Gaussian domain | | Truncated Gaussian simulation | | (Simulation annealing can act as post-processing) | Discrete |
| | Indicator stochastic domain | Sequential indicator simulation | | Probability field indicator simulation | (Simulation annealing can act as post-processing) | Discrete/ Continuous |
| | Random fractal domain | | Fractal simulation | | (Applicable simulation annealing) | Discrete/ Continuous |
| | Markov random domain | | | | Markov simulation | Discrete/ Continuous |
| | Random walk | | | | Random walk simulation | Discrete |
| | Multipoint statistic | Multipoint statistic simulation | | | Multipoint statistic simulation | Discrete |

*Source*: From Shenghe, W., 2010. Reservoir Characterization and Modeling. Beijing: Petroleum Industry Press. 136–174.

geological modeling in fine reservoir description, this book summarizes the current status and principles of multipoint geostatistical modeling and its differences from traditional geological modeling methods. Moreover, some examples of multipoint geostatistical modeling are given, and the problems and trends of multipoint geostatistical modeling are proposed.

## 3.8.1 Current status

The reservoir geological model is a concrete manifestation of the final results of reservoir description and reservoir characterization, and it is also the core content and frontier of current oil and gas reservoir geology study (Xinghe, 2009). This book investigates and summarizes the geological modeling in China and abroad.

### *3.8.1.1 Geological modeling abroad*

1. Content

   The geological modeling abroad mainly includes modeling of structural and fault systems, modeling of sedimentary reservoirs, modeling of reservoir pore structure, and the likes, involving the reservoirs such as clastic rocks and carbonate rocks. Escuder-Viruete et al. (2004) used the three-dimensional geological modeling techniques of fault zones to study the geological, geophysical, and geochemical, structural features of the granite fault zone. P. Leroy et al. (2007) conducted geological modeling of the composition of pore water in the Callovian–Oxfordian claystone strata in France, based on the calculation of the flow pressure and the ion composition of pore water from a microscopic perspective. Caumon et al. (2009) conducted a three-dimensional modeling study based on surface geological structures, which provided key concepts, principles, and procedures of quality-controlled geological modeling. Maesano et al. (2013) derived the slip rate of thrust fault from geological modeling with examples from the Mache coastal and offshore contraction belt, Northern Apennines, Italy.

2. Methods

   The geological modeling methods include deterministic modeling and stochastic modeling. Caumon (2010) proposed the stochastic time-varying geological modeling based on the optimal method that satisfies both random and time-shift conditions and also stated that the application of inversion should rely not only on random field models but also on geological concepts and parameters. Lillah and Boisvert (2013) presented the stochastic distance-based geological boundary modeling with curvilinear features through a case study of a gold deposit in Northern Canada. Pyrcz and Deutsch (2014) introduced the geological modeling principles, prerequisites, methods, and applications in the monograph "Geostatistical Reservoir Modeling (Second Edition)."

3. Data source

   Data sources include field outcrops, logs, seismic, tests, and others. Hynes and Dixon (2005) conducted geological mapping and simulation modeling for a northwestern part, Riyadh, Kotaneelee, and Tlogotsho, Canada. Li et al. (2015) used the seismic imaging of subsalt reservoirs as an example to evaluate the seismic uncertainty velocity model disturbances and image registration by geostatistics so that the uncertainty of seismic data can be largely ignored in the geological modeling process.

4. Application of geological modeling in oil and gas field development

   Rezvandehy et al. (2011) integrated seismic attributes in the accurate modeling of geological structures and determined the storage of the gas reservoir in Gorgan Plain, North of Iran. Deschamps et al. (2012) assessed the heavy oil recovery through the SAGD process by using numerical models of a reservoir analog of a meander belt through (1) a very fine description of the outcrops in terms of architecture and geological heterogeneities, (2) the evaluation of the impact of genesis on the heterogeneity distribution in the reservoir based on different scales of data, and (3) the reservoir SAGD simulations to assess the impact of heterogeneity on heavy oil production.

### 3.8.1.2 Geological modeling in China

1. Content

   The geological modeling in China mainly includes structural modeling, sedimentary facies modeling, and reservoir architecture modeling, involving clastic rocks or sometimes carbonate rocks and volcanic rocks. Huanqing and Xiaomin (2008) systematically summarized the sedimentary microfacies modeling in fine reservoir description as the object- and pixel-based stochastic modeling methods using geological, geophysical, and oilfield development performance data; architecture analysis method; and cross-well seismic methods. Wu Shenghe (2010) expounded the processes of reservoir interpolation modeling; object-based geological modeling that effectively integrates prior geological knowledge; multipoint geostatistical modeling; and modeling based on the deposition process as the future orientations of content.

2. Methods

Stochastic modeling is the focus. Dianxing et al. (2012) presented a geomodeling technology under multifactor control with examples of sandy mudstone reservoirs, volcanic reservoirs, and carbonate reservoirs. Yuming and Jiagen (2016) summarized the 3D geological modeling methods for fractured-vuggy carbonate reservoirs by taking the Ordovician reservoir in Tahe Oilfield as an example.

3. Algorithms

Shaohua et al. (2012) summarized the reservoir modeling algorithms, including de-clustering, variogram, Kriging, sequential Gaussian simulation, sequential indicator simulation, truncated Gaussian simulation, simulated annealing, target-based channel simulation, Snesim simulation, and phase model post-processing. Donghui et al. (2014) applied the multipoint geostatistical modeling to simulate the lithofacies in the Dongsheng gas field, thus truly reproducing the spatial distribution pattern of channel sand bodies with high prediction accuracy. Zhaowei et al. (2017) used target-based modeling, sequential indicator simulation, and multipoint geostatistics modeling methods to establish different sedimentary facies models for Sulige Gas Field. We used the sequential indicator modeling method in the geological modeling of the Yulou fan delta front sedimentary reservoir in an area of Western Sag, the Liaohe Basin (Fig. 3.73), and we constructed the models of sedimentary microfacies, porosity, and permeability of the reservoir. In our work, we used the variogram fitting to model the sedimentary microfacies, and we built the reservoir physical property models under the control of sedimentary facies. This is also the most popular method in China.

4. Data source

Log data are mainly used, and seismic data are also used sometimes. However, field outcrop and test data are rarely used. Wenjie et al. (2015a,b) proposed a characterization method based on the presimulated geological vector information and constructed the sedimentary microfacies model by using the multipoint geostaticstical modeling based on vector information, according to the modern sedimentary features and the characterization of reservoir sedimentary microfacies, with a case study of the Karamay Formation alluvial fan reservoir of the Lower Triassic in Karamay Oilfield.

Obviously, the studies of geological modeling abroad mostly involve fault systems and have made great progress in pore structure modeling and

(A)

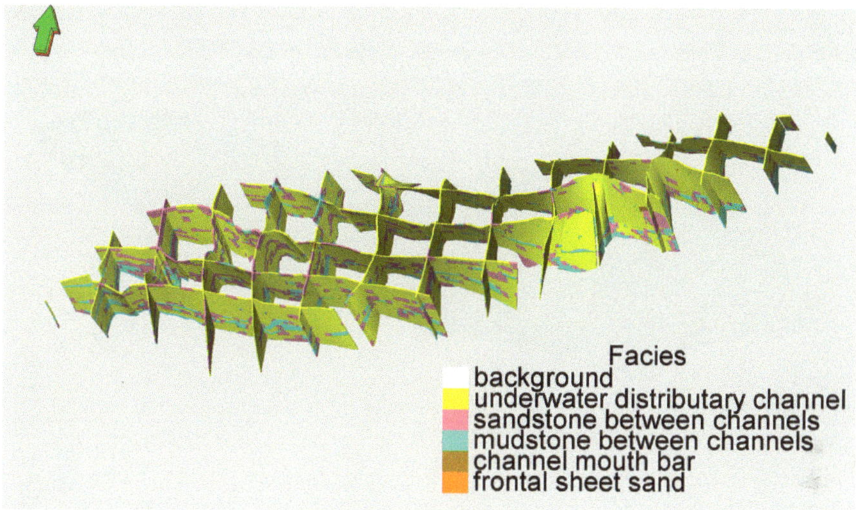


(B)

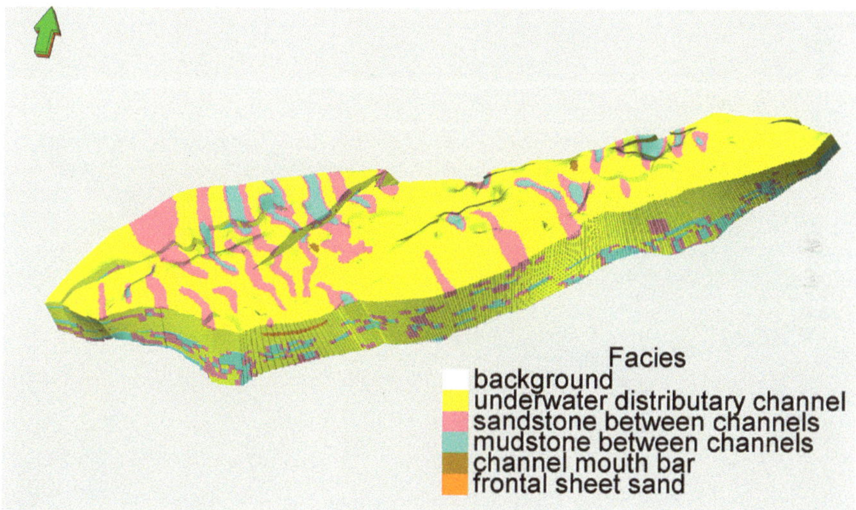


**Figure 3.73** Geological model of the Yulou reservoir in an area of Western Sag, the Liaohe Basin. (A) Grid diagram of sedimentary microfacies model; (B) Sedimentary microfacies model; (C) porosity model; (D) permeability model.

seismic-based modeling. Moreover, the geological modeling of carbonate fractured-vuggy reservoirs is well established abroad. All these are worthy of reference for the researchers carrying out studies in China.

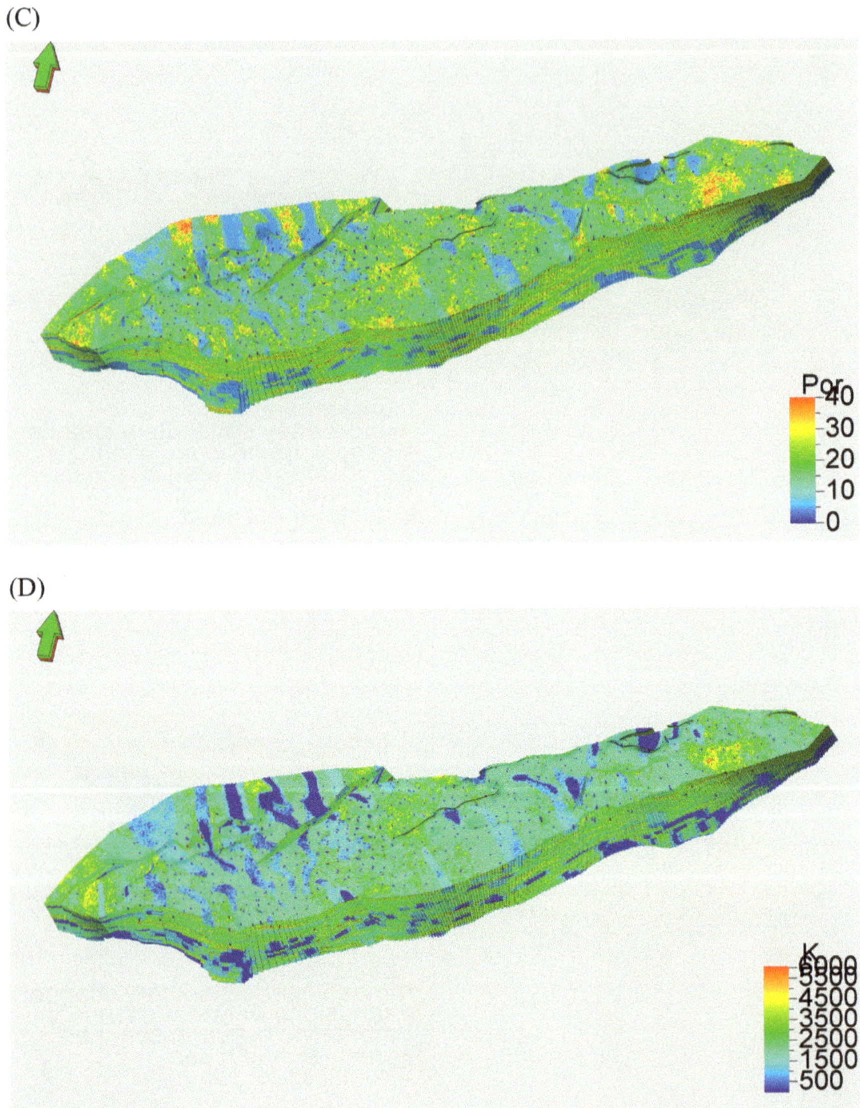


**Figure 3.73** (*Continued*)

## 3.8.2 Current status and principle of multipoint geostatistical modeling and comparison with traditional geological modeling methods

### *3.8.2.1 Current status*

The multipoint geostatistical modeling was first proposed in the 1990s. This method, combining the advantages of pixel-based and target-based methods, uses the training image to describe the relationship between points in space and its prospective for application (Shenghe, 2010). Many foreign scholars have reported on this method. Yamada and Okano (2007) successfully combined the multipoint geostatistics and probability perturbation theories into the integrated facies distribution modeling and history matching of a complex pressure system for volcanic reservoirs. Ezequiel F. González et al. (2008) presented a novel seismic inversion technique that combines rock physics and multiple-point geostatistics to describe the geological information of reservoir, which is demonstrated as valid and applicable. Ravalec-Dupin et al. (2008) reconstructed the geological models using the multipoint geostatistical method to fit the production performance data. Babak and Deutsch (2009) proposed a new geostatistical modeling method based on the improved CoKriging method and integrating the secondary variables (mainly the multiple seismic attributes, geological trend, and structure control), which is proved to produce results identical to those obtained when several secondary variables are used. Comunian et al. (2011) discussed the 3D multiple-point statistics simulation using 2D training images, which is proved practically feasible. Mariethoz and Caers (2015) explained, in their book "Multiple-In Point Geostatistics: Stochastic Modeling with Training Images," the concepts and methods of the multipoint geostatistics in stochastic modeling, and the applications of the method, such as the reservoir forecasting—the West Coast of Africa (WCA) reservoir, the geological resources modeling in mining, and the climate modeling application—the case of Murray-Daling Basin. In China, more and more researchers are working on the multipoint geostatistics. Shaohua et al. (2007a,b) pointed out that the popular Snesim and Simpat algorithms used abroad are inadequate, and they proposed a multipoint geostatistical stochastic modeling technique based on reservoir skeleton, which effectively reproduces the channel morphology and distribution. Jiahua and Jian (2007) proposed to build a reservoir modeling process system framework based on the workflow technology by studying the multipoint statistical reservoir modeling process, which realizes the multipoint geostatistical modeling with higher efficiency and lower cost. Yanshu et al. (2008) used the geological data models to constrain the Simpat geological modeling method. Wei et al. (2008b) combined the multipoint geostatistics and facies-controlled modeling to make geological-constrained modeling for the V reservoir in D Oilfield, Peru. Zhou Jinying et al. (2010) applied multipoint geostatistics to the geological modeling of coastal reservoirs, demonstrating that this method is obviously superior to traditional methods in reproducing the spatial structural characteristics of reservoirs. Jiahua and Wei (2011) conducted seismic-constrained geostatistical modeling, which is verified to increase the interwell deterministic information of the model, reduce the interwell uncertainty caused by mathematical interpolation and simulation, and ensure the model is more loyal to the underground conditions. Yanshu et al. (2011) summarized the progress of multipoint geostatistics and stated that its application has evolved from fluvial facies modeling to fan modeling, from reservoir structure modeling to physical property distribution simulation, from macroscopic geologic body prediction to microscopic pore-throat distribution, from geologic study to geostatistical inversion. Shi Shuyuan et al. (2011) established the multipoint geostatistic modeling method based on random walk modeling. Zhang et al. (2016) divided the deep-water turbidite complex channel sand bodies in the study area into four types of architectural elements: primary channel, channel edge, natural levee, and lateral accretion mud, defined the architecture of the deep-water turbidite complex channel sand bodies through stochastic simulation with the coordinated constraints of multipoint geostatistics and soft probability attributes, and then simulated the physical properties of the reservoir under the control of architectural elements. Dongping et al. (2012) determined the planar morphology and structural characteristics of underwater distributary channel and estuary sand bar microfacies through the study of modern sediments in the Poyang Lake Delta, made the quantitative training image based on the two microfacies widths from data statistics of dense well pattern area in Yong'anzhen Oilfield, and used the multipoint geostatistics to establish a geological model of the fan delta front. Hui et al. (2013) compared two improved multipoint geostatistical methods, namely, the multipoint geostatistical modeling method based on reservoir matrix and the multipoint geostatistical modeling method based on a random walk, and they indicated that the former is more applicable to the areas with small well density, while the latter to the areas with large well density. Xiaojun et al. (2015) used the multipoint geostatistics to build the geological model of the architecture of the middle fan of the alluvial fan of the Lower Karamay Formation in a block of Karamay Oilfield, producing the results that effectively illustrate the distribution morphology and contact relationship of different types of architectures, horizontally and laterally. Wenbiao et al. (2015) obtained the representative, quantitative 3D training images through high-precision inversion

**Table 3.20** Multipoint geostatistical modeling in China and abroad.

| Research local | Strength | Weakness |
|---|---|---|
| Abroad | (1) The geostatistics methods such as CoKriging are improved<br>(2) The theory of disturbance is combined with other multipoint geostatistics techniques<br>(3) With seismic constraints, rock geophysics is combined with multipoint geostatistics<br>(4) Multipoint geostatistics is applied to carbonate reservoirs, marine sandstone reservoirs, and the likes<br>(5) Multipoint geostatistics is applied to the modeling of reservoir pore structure | (1) Log data of dense well pattern are seldom used in multipoint geostatistical modeling<br>(2) Multipoint geostatistics has not been effectively applied in reservoir architecture modeling |
| China | (1) The Sensim multipoint geostatistics and other algorithms are improved<br>(2) Multipoint geostatistics is combined with random walk<br>(3) Multipoint geostatistics is applied to the sedimentary reservoirs such as fluvial facies and alluvial fan<br>(4) Multipoint geostatistics is introduced into reservoir architecture modeling<br>(5) Log data of dense well pattern are used in multipoint geostatistical modeling | (1) Seismic data have not been effectively used in modeling<br>(2) Field outcrop and modern sediment data have not been effectively used in multipoint geostatistical modeling<br>(3) Multipoint geostatistics has not been extended to the geological modeling of carbonate reservoirs and marine sandstone reservoirs |

of shallow seismic data to establish a deep-water sedimentary microfacies model. Siyu et al. (2016) proposed a multipoint geostatistics algorithm based on pattern scale-down cluster, which greatly improves the computational efficiency of the pattern-based multipoint geostatistics algorithm while ensuring the simulation quality and also saves the memory space. Table 3.20 compares the multipoint geostatistical modeling in China and abroad.

### *3.8.2.2 Principle*

In traditional two-point geostatistical modeling, it is assumed that the spatial variables are subject to the multivariate Gaussian distribution, and the mean and variance of the conditional distribution in each grid cell must be determined; thus a series of Kriging equations are required. In multipoint geostatistical modeling, local conditional distribution is built directly by scanning the training image. Jiahua and Hiamao (2012) illustrate the process. The training image can be seen as a priori structural model that is both quantitative and visual, needs not to be loyal to the well data at the corresponding location, but only shows how the patterns are connected in space. Interpreted images, remotely sensed data, or hand-drawn sketches can all be used as training images or as sources for building training images. For the sedimentary facies model, the training image represents the sedimentary pattern and only needs to reflect the geometry, contact relationship, plane distribution, and vertical distribution of different facies types. Before the simulation begins, a search template is determined to scan the training image. It is assumed that the position in the grid is being simulated (Fig. 3.74). In this template (red circle), there are four data values: two as sandstones (black pixels) and two as mudstones (white pixels). All four values, together with their geometry, are called data events. Then, this data event is used to scan the channel training

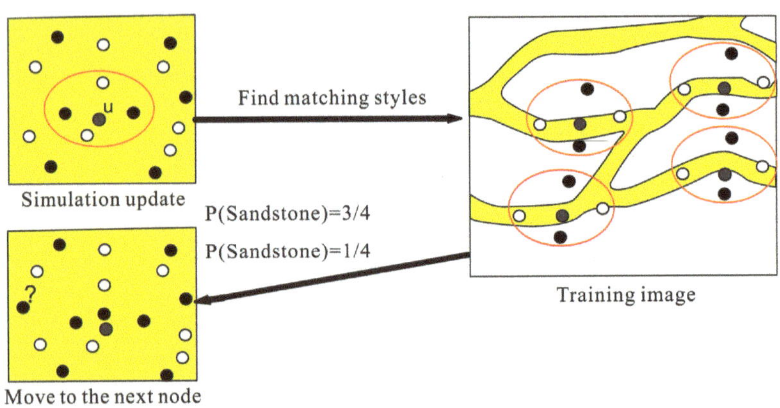


**Figure 3.74** Schematic illustration of data events and training images (Jiahua and Hiamao, 2012).

image to infer the sandstone probability at the data point. It is assumed that there are four repeated data events found in the training image: 3 sandstone data events and one mudstone data event. Thus the sandstone probability at the grid node is 3/4 and sampled to define one analog value. Either sandstone or mudstone may be sampled, but sandstone is more likely since sandstone exhibits a higher probability. Imagine that sandstone is sampled. The sandstone simulation values are then added to the conditional data to constrain the simulation of other grid nodes. Next, another grid position is simulated. This process is repeated until all nodes are accessed. Finally, a complete multipoint simulation of the channel is produced. The multipoint geostatistical stochastic modeling methods (e.g., SNESIM) are essentially different from traditional geostatistical stochastic modeling methods in the approach to determine the conditional probability distribution functions at the nonsampled points. The former applies a multipoint data template to scan the training image to construct a search tree and then obtains the conditional probability distribution function from the search tree. The latter uses a variogram analysis and the Kriging method to obtain the conditional probability distribution function. It is this difference that makes the multipoint geostatistics methods overcome the inadequacy of traditional two-point statistics methods, which cannot express the complex spatial structures and reproduce the target geometry (Jiahua and Hiamao, 2012).

#### *3.8.2.3 Comparison with traditional geological modeling methods*

Multipoint geostatistics is a concept relative to traditional two-point geostatistics. Geostatistics was founded in 1962 by G. Matheron of MINES ParisTech in Paris, France. Traditional geostatistics is mainly used in reservoir modeling in two aspects of model building: (1) the Kriging methods are used to establish deterministic models; and (2) the stochastic modeling methods are used to build optional and equiprobable geological models. All these methods are based upon variogram tools (Shenghe and Kewen, 2005). Li Shaohua et al. (2009) compared multipoint geostatistical methods with traditional geological modeling methods for simulating the physical property distribution of channel sand bodies. Traditional stochastic geostatistical modeling based on variogram is a common method for reservoir heterogeneity simulation and uncertainty evaluation. However, this method can only consider the correlation between two points in space, and thus it cannot reproduce a reservoir structure with complex facies associations. Multipoint geostatistics uses training images as a basic tool to emphasize the correlation between multiple points in space, avoiding the shortcomings of traditional geostatistics in describing geologic bodies with complex spatial geometry (Jiahua and Tao, 2013). Yin et al. (2012) compared the geological modeling methods for meandering-river reservoirs and revealed that the multipoint geostatistical modeling method is superior to the indicator Kriging and sequential indicator modeling methods and can effectively reflect the continuity of the sand body, represent the distribution characteristics of the river, and avoid the smoothing effect. Ding Fang et al. (2017) conducted geostatistical modeling with the best training images selected according to the sedimentary facies contact and proportion, mud content, and other parameters, and they concluded that this method is apparently superior to traditional methods in the modeling of sand body facies in complex stacking. Compared with traditional geological modeling methods, the multipoint geostatistical modeling method adds geological mode control during operation and replaces traditional two-point correlation analysis based on variogram with multipoint correlation based on training images so that the accuracy of interwell prediction of geological models is effectively improved. Besides, it takes the sequential algorithm, with the pixel node as the simulation unit, thereby ensuring a better loyalty to the discrete hard data of well. The biggest advantage of the training image is that it fully reflects the quantitative spatial distribution pattern of different sedimentary microfacies types, including the spatial stacking pattern, scale, and positions. Thus the researchers are allowed to have a deeper understanding of the three-dimensional distribution of sedimentary microfacies instead of the plane distribution of sedimentary microfacies by way of two-point geostatistics. The variogram in the traditional geological modeling method is special for its loyalty to the well data. In contrast, the training image in the multipoint geostatistical modeling is not completely loyal to well data and, instead, it is only the embodiment of the statistical law of well data. The researcher's geological thinking can be fully incorporated into the building of the training image, which, however, is basically impossible in the variogram analysis.

## 3.8.3 Examples

### *3.8.3.1 Foundation*

To establish a reservoir model using multipoint geostatistics, we should fully first understand the structural, stratigraphic, and sedimentary characteristics of the target reservoir in the work area. The structural study mainly relies on the well–seismic data together with fine seismic

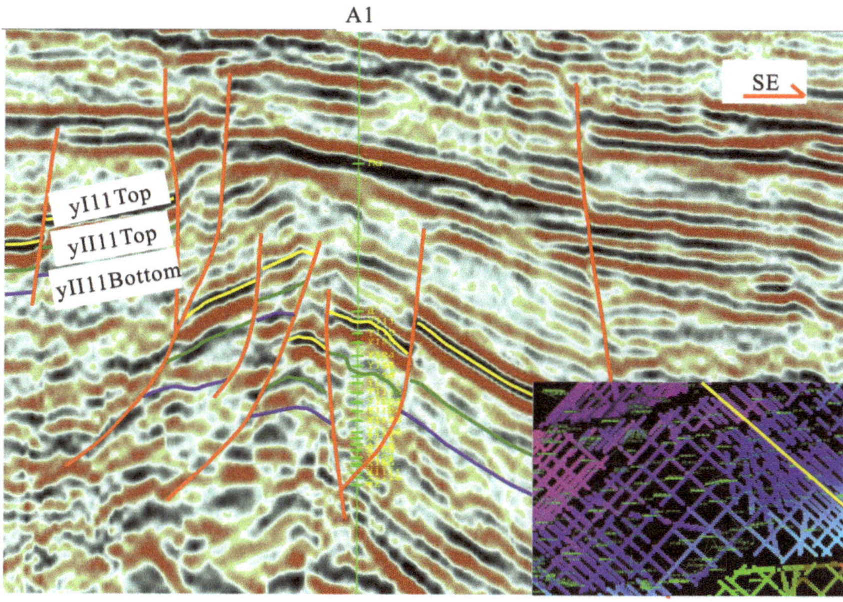


**Figure 3.75** Fault development in the Yulou reservoir in an area of Western Sag, the Liaohe Basin.

interpretation (Fig. 3.75). By synthesizing seismic records, the seismic data are calibrated by the core data to realize time-depth conversion. Coupled with conducting the regional geological setting, the sectional interpretation and plane combination of fault systems are conducted to achieve the structural analysis. Depending on the production requirements in the study area, the most popular classification system, by long-term base-level cycle, medium-term base-level cycle, and short-term base-level cycle, is adopted. The Yulou reservoir, the target in the study, is divided into 29 beds ($yI_1{}^{1a}$, $yI_1{}^{1b}$, $yI_1{}^{2a}$, $yI_1{}^{2b}$, $yI_1{}^{2a}$, $yI_2{}^{3a}$, $yI_2{}^{3b}$, $yI_2{}^{3c}$, $yI_2{}^{4a}$, $yI_2{}^{4b}$, $yI_2{}^{4c}$, $yI_3{}^{5a}$, $yI_3{}^{5b}$, $yI_3{}^{5c}$, $yI_3{}^{6a}$, $yI_3{}^{6b}$, $yI_3{}^{6c}$, $yII_1{}^{1a}$, $yII_1{}^{1b}$, $yII_1{}^{2a}$, $yII_1{}^{2b}$, $yII_1{}^{3a}$, $yII_1{}^{3b}$, $yII_2{}^{4a}$, $yII_2{}^{4b}$, $yII_3{}^{5a}$, $yII_3{}^{5b}$, $yII_3{}^{6a}$ and $yII_3{}^{6b}$), corresponding to 29 short-term base-level cycles (Fig. 3.76) (Huanqing et al., 2014a,b). Based on the core data of coring wells, the log data of noncoring wells and the well test data, together with rock type, sedimentary genesis, sedimentary structure, and logging facies, the Yulou fan delta front subfacies is divided into five types of sedimentary microfacies: underwater distributary channel, estuary sand bar, front sheet sand, underwater distributary interchannel sand, and underwater distributary interchannel mud (Fig. 3.76). Among them, the underwater distributary channel and estuary sand bar

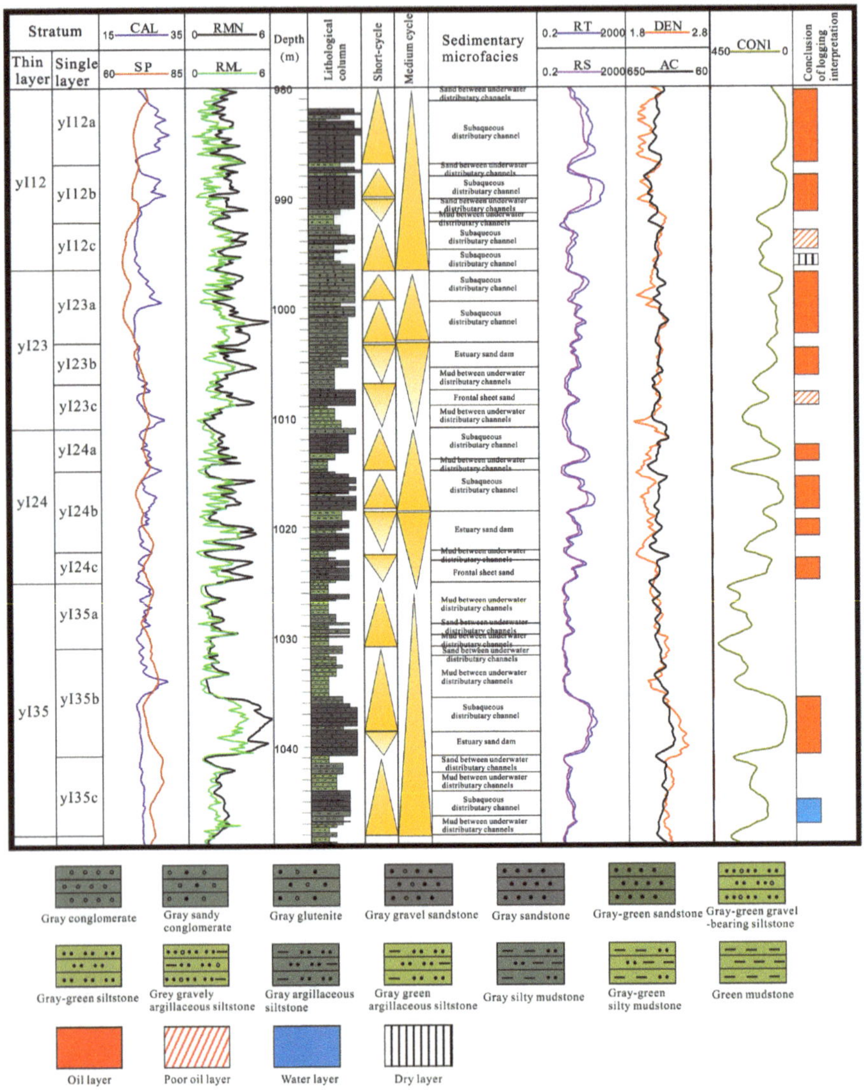


**Figure 3.76** Lithoelectric characteristics of sedimentary microfacies in the Yulou reservoir in an area of Western Sag, the Liaohe Basin.

are dominant. The statistical results of physical property data of coring wells indicate that the Yulou reservoir has the porosity of 25%–40%, with an average of 31.25%, and the permeability varied in a wide range from 1 to 5000 mD, with an average of 1829.3 mD, suggesting that it's a high-porosity high-permeability reservoir (Huanqing et al., 2016a,b,c,d).

#### 3.8.3.2 Determination of training image

As mentioned above, in the multipoint geostatistical modeling, the training image should be first determined, and then the sedimentary facies models of different horizons are built under the control of the training image. Based on the structural, stratigraphic, and sedimentary facies analysis, the plane maps of sedimentary microfacies in different beds are plotted and used as the basis for making training images. In this study, we analyzed, quantitatively and statistically, the development characteristics of different types of sedimentary microfacies in different beds (Fig. 3.77), and we identified the sedimentary pattern of the target reservoir, thus preparing for the making of training images in multipoint geostatistical modeling. Fig. 3.75 shows the quantitative statistical data of spatial development of the underwater distributary channel microfacies. Determination of training images is the most critical step in multipoint geostatistical

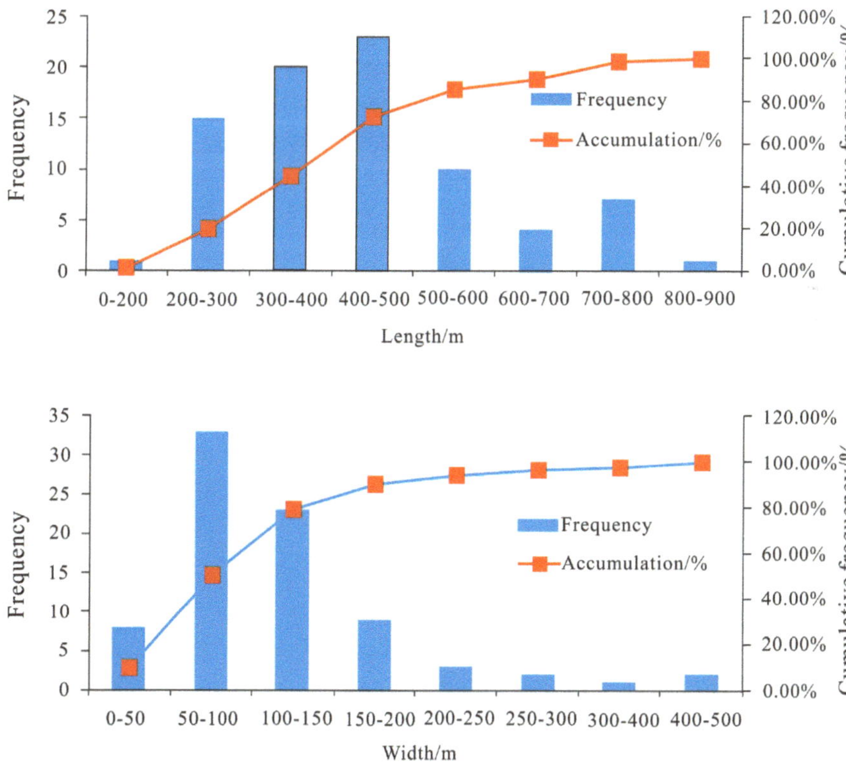


**Figure 3.77** Statistical data of the length and width of the underwater distributary channel in the Yulou reservoir in an area of Western Sag, the Liaohe Basin.

modeling, and it is directly related to the success of modeling. The training image itself is a facies model. In this study, based on the geometrical parameters of microfacies derived from the abovementioned statistical analysis and the plane map of sedimentary microfacies, the deterministic modeling method was adopted to create the training image, as shown in Fig. 3.76. This is a training image shared by layers of the same type and represents a priori and generalized sedimentary microfacies model. Each microfacies has an aspect ratio meeting the statistical range and includes all possible types and forms of sedimentary microfacies. The training images can be 2D or 3D (Fig. 3.78). The 3D training images can fully reflect the lateral migration and vertical accretion of different sedimentary microfacies and, thus, can better reflect the spatial structure of sediments. Therefore the 3D training image is mainly adopted in practical applications. Generally, training images can be obtained from isopach map of sand bodies, geological understanding, artificial facies division, geological knowledge base, physical simulation experiments, and marked point process. In this study, the 3D training image was acquired by using the quantitative statistical data of different sedimentary microfacies types in different horizons, together with the depositional mode. In the process, the spatial stacking of different sedimentary microfacies types was fully considered.

#### *3.8.3.3 Building of geological model*

We completed the geological modeling of the Yulou reservoir in an area of Western Sag, the Liaohe Basin, by using traditional methods and multipoint geostatistical method (Fig. 3.79). The results show that, for the fan delta sedimentary reservoir, the multipoint geostatistical method produces a geological model apparently superior to the sedimentary microfacies model built by the traditional sequential indicator method.

### 3.8.4 Problems and trend

#### *3.8.4.1 Problems*

The benefits of the multipoint geostatistical modeling method put it far above traditional methods. Nevertheless, it has some problems due to technical restrictions. Shenghe and Kewen (2005) summarized the problems in three technical aspects of modeling.

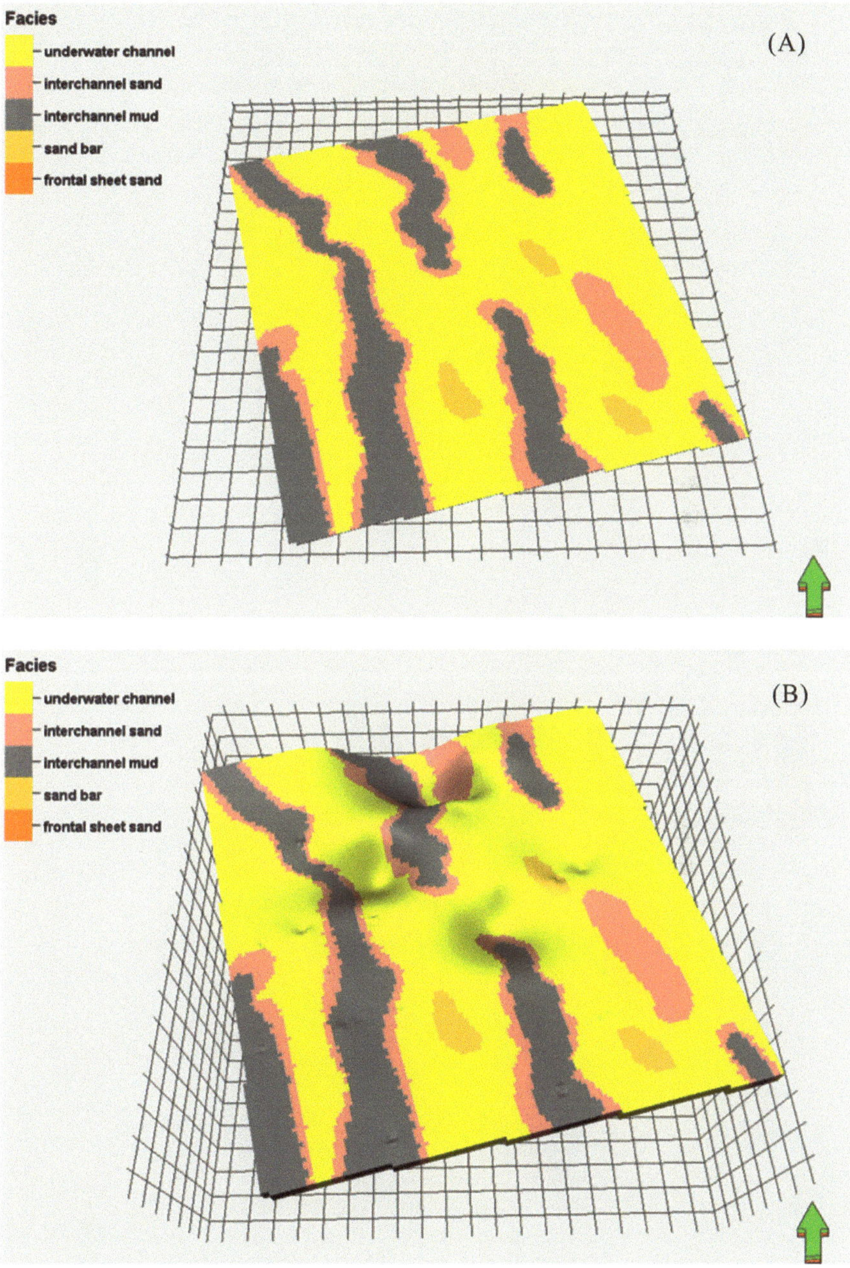


**Figure 3.78** Training images for multipoint geostatistical modeling for an area of Western Sag, the Liaohe Basin. (A) Features of plane training image; (B) Features of 3D training image.

(A)

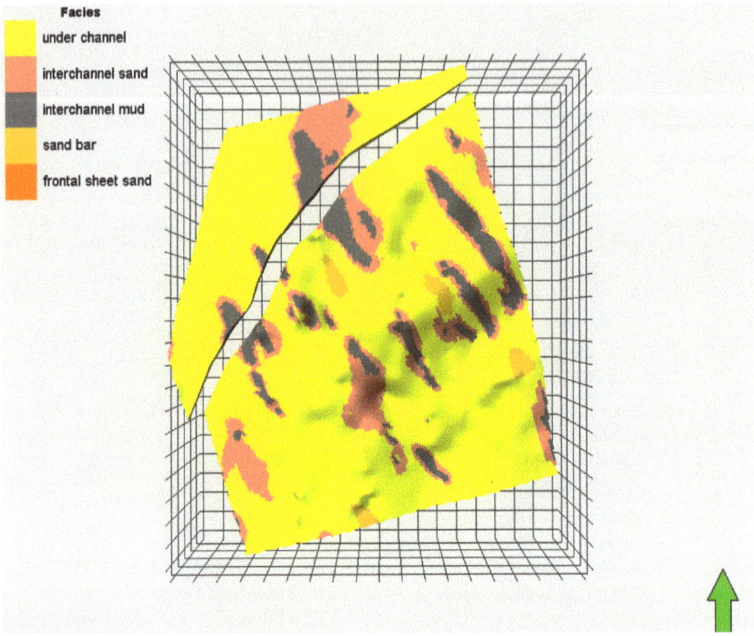


(B)

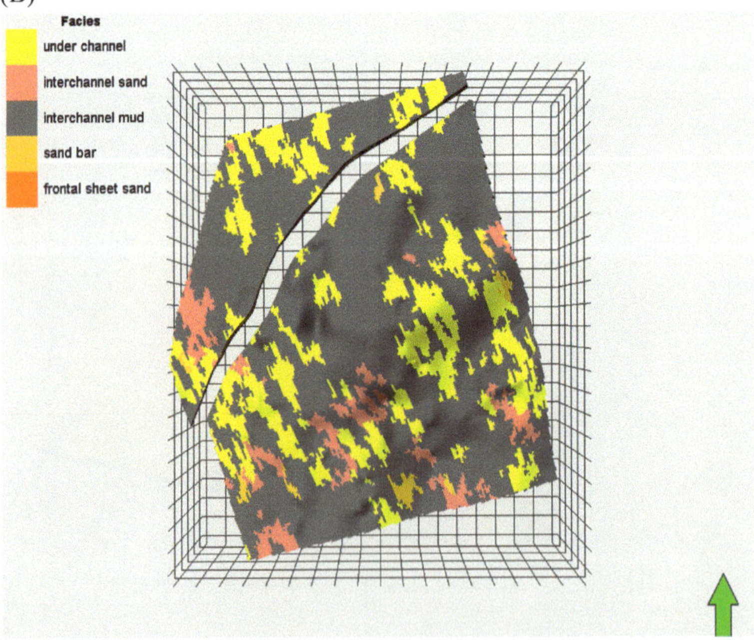


**Figure 3.79** Results of sedimentary microfacies modeling by multipoint geostatistical method and sequential indicator method for an area of Western Sag, the Liaohe Basin. (A) $yl_1^{2a}$ multipoint simulation results; (B) $yl1_2^{a}$ sequential indicator simulation results; (C) $yl_2^{4b}$ multipoint simulation results; (D) $yl_2^{4b}$ sequential indicator simulation results.

(C)

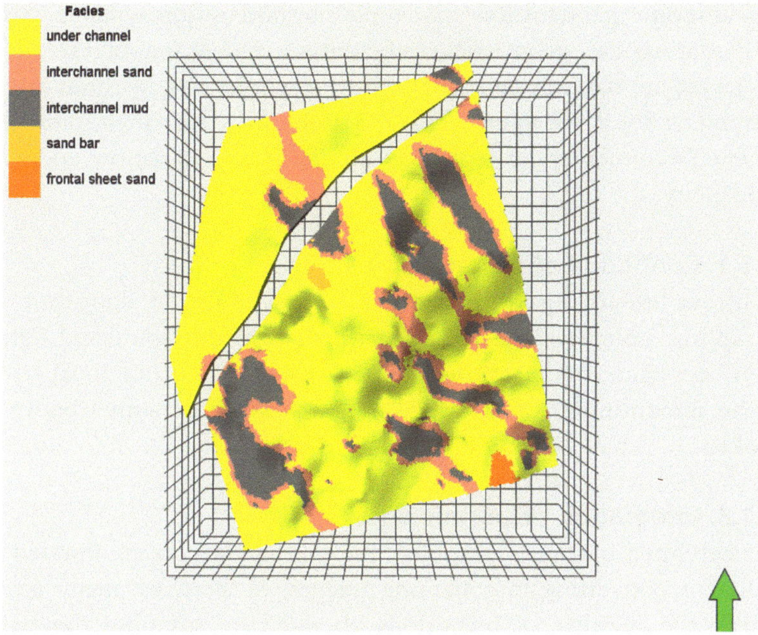


(D)

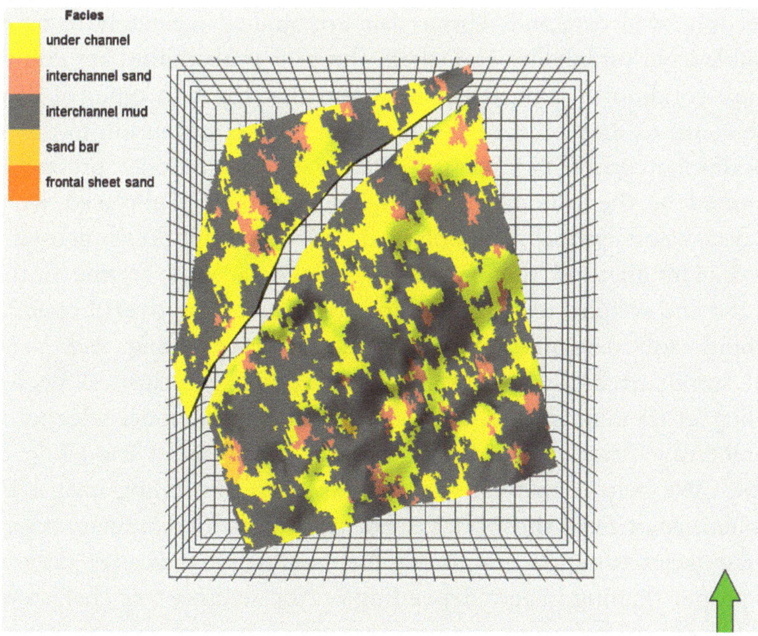


**Figure 3.79** (*Continued*)

#### 3.8.4.1.1 Stability of training image

The multipoint geostatistical modeling method requires stable training images; that is, the geometric configuration and shape of the training image target are basically unchanged in the whole area, without any evident trend or local variability. This requirement is extremely difficult for continental sedimentary reservoirs with fast facies variation and strong heterogeneity.

#### 3.8.4.1.2 Continuity of target

The current Snesim algorithm is a sequential simulation algorithm. Each nonsampling point is accessed only once, and the simulated value is "frozen" to hard data. Although this method is fast and loyal to hard data, the discontinuity of the target (e.g., simulated channel body) may be broken.

#### 3.8.4.1.3 Integration of seismic data

This is exhibited in three cases. First, the seismic data are interpreted geologically for converting into training images. If there are many types of soft data, the number of repetitions obtained by scanning the training images is too small, thus, affecting the derivation of conditional probability. Second, well data and seismic data are applied separately to calculate the conditional probability and, then, the two probabilities are combined into one conditional probability, provided that the two types of data are independent, or they must be weighted for their contribution to the overall conditional probability. Third, the overall conditional probability is determined by the collocated CoKriging method and replaced with the hard data-based probability obtained by the multipoint geostatistical method. This method requires less seismic data (same as the simulated grid), and the weights of hard data and soft data to the overall conditional probability still depend on the Kriging variance. Yang and Yanchao (2008) summarized the problems in multipoint geostatistical geological modeling as training images, target continuity, data model selection, and comprehensive seismic data by taking the fluvial facies modeling as an example. We believe that the representativeness of training images is also a problem, apart from those above. For continental sedimentary reservoirs with fast facies variation and strong heterogeneity, it is very difficult to select proper training images depending on the sedimentary characteristics. A feasible option is to create and select suitable training images depending on horizons and blocks.

**Figure 3.80** Modern sediments or field outcrop of different sedimentary and tectonic processes. (A) Meandering-river deposit, Ribbon River, Hulunbeir, Inner Mongolia; (B) Meandering-river deposit, Yarlung Zangbo River, Tibet (Nangaba Bawa Peak section); (C) Alluvial fan deposit, Huanlegu Desert, Dunhuang, Gansu; (D) Fluvial deposit, Heishanzuizi, Dunhuang, Gansu; (E) Braided-river deposit, Qianjiadian area, Yanqing, Beijing; (F) Fault in Yanchang Formation, Yanhe section, Ordos Basin, Shaanxi.

#### 3.8.4.2 Trend

We believe that the multipoint geostatistical modeling will further evolve in the following aspects of efficient model building:

1. Training image is acquired based on multiinformation analysis of geological genesis (Fig. 3.80). Many scholars have worked on the determination of training images. Zhang (2008) integrated the reservoir geological conceptual model into the making of training images for multipoint geostatistical geological modeling, with more accurate training images obtained. Yin Yanshu et al. (2014) proposed a multipoint geostatistical method based on the sedimentary model, which is reproduced by ways of overall replacement, structured random path, and multigrid through correlating the reservoir characteristics with the deposition location via the distance function. In practice, based on the geological genesis analysis, multiple information such as field outcrops, modern sediments, seismic data, and log data should be combined to create training images closer to actual underground geology by blocks and horizons.
2. The multipoint geostatistical algorithm is improved to secure higher modeling accuracy. Yin Yanshu et al. (2011) integrated the random walk process and multipoint geostatistics to construct the geological model of fluvial sedimentary reservoir, where the random walk process is used to generate the main channel line to constrain the multipoint statistical prediction, thereby avoiding the channel discontinuity caused by random sampling in multipoint geostatistical modeling. Zhang Li et al. (2012) attempted to build the three-dimensional pore structure model of the core using multipoint geostatistics and demonstrated that the pore structure model built is very close to the three-dimensional pore distribution of the real core, with similar homogeneity and pore connectivity. Zhongshan et al. (2013) calculated the reserves by using the results of multipoint geostatistical modeling. Shen's results showed that the sedimentary microfacies map and the porosity, permeability, and oil saturation models generated by multipoint geostatistics are most significant geologically, revealing clear shape and good continuity of channels, and the reserves determined by the multipoint geostatistical modeling method exhibit the smallest variance and uncertainty. Shen Zhongshan et al. also combined the multipoint geostatistics with other methods to improve the accuracy of interwell prediction based on multipoint geostatistics, and they discussed the application of

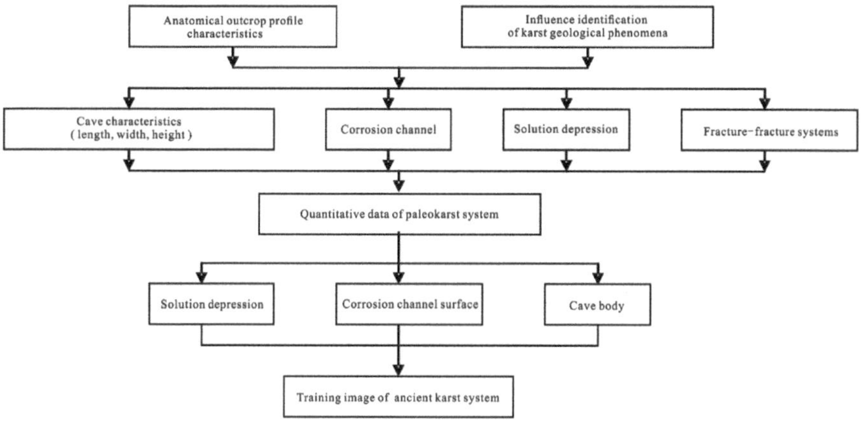


**Figure 3.81** Workflow of training image making for ancient karst. *From Shuyuan, S., Hui, Q., Xuan. Z., et al., 2014. A method making paleokarst training image based on GoogleEarth. Complex Hydrocarbon Reservoir 7(4):6–10.*

multipoint geostatistics in the geological modeling of carbonate reservoirs. Shuyuan et al. (2014) proposed an approach for making training images of paleo-karst systems based on GoogleEarth image data (Fig. 3.81). Hongkai and Zhijiang (2015) took the S80 fractured-vuggy unit of Tahe Oilfield as an example to make classification modeling of the dissolved pores in fractured-vuggy carbonate reservoirs. In Li's study, the dissolved vugs induced by the collapse of caves were modeled by the multipoint geostatistical method, the dissolved vugs around the faults were modeled by the fault-constrained partition modeling method, and the layered dissolved vugs on the weathering crust surface were modeled by the karst facies-controlled modeling method, thereby, revealing the spatial distribution characteristics of dissolved vugs with different genesis and improving the accuracy of geological modeling of dissolved pores.

3. The multipoint geostatistical modeling method is applied in wider coverage, with a higher level and accuracy. In addition to the 3D sedimentary facies modeling, the reservoir pore structure modeling is more focused on providing quantitative data basis for the prediction of remaining oil distribution between wells. Lihui et al. (2015) discussed the application of direct-sampling multipoint geostatistics (DS-MPS) in geological modeling, which saves the computer memory since it doesn't need to build the search tree in advance and achieves higher target continuity.

Zitao et al. (2017) conducted a sensitivity analysis of key input parameters in the Snesim algorithm during multipoint geostatistical modeling, and they showed that the closer the target ratio to the marginal facies probability of the training image, the better the simulation results. Wen Zitao et al. also extended the multipoint geostatistics to the geological modeling of unconventional reservoirs such as tight oil and shale gas. Bin et al. (2014) applied the multipoint geostatistics to the geological modeling of tight sandstone gas reservoirs in Block s48−17−64, and they proposed the idea of "well−seismic-geostatistics" integrated stochastic geological modeling, which realizes a higher accuracy. Some scholars have applied multipoint geostatistics to the modeling of reservoir architecture, flow unit, and fracture, making the multipoint geostatistical modeling increasingly capable for the interwell prediction of different reservoir properties in fine reservoir description. Chuangang (2015) used multipoint geostatistics to randomly simulate the underwater distributary channel in the area with closer-well-spacing to determine the channel width and drilling probability. Keke et al. (2016) successfully applied multipoint geostatistics to the geological modeling of the internal architecture of the point bar. It is recommended to organically integrate the data of dense well pattern and the quantitative statistical data of development scale of sedimentary facies/microfacies into the multipoint geostatistical modeling, making the modeling increasingly quantitative and accurate. Multipoint geostatistics is one of the core reservoir modeling methods. This method featured with high accuracy of interwell prediction will certainly be more and more attractive for the researchers.

## 3.9 Remaining oil characterization

The remaining oil, in a broad sense, refers to the crude oil preserved or retained in the reservoirs under different geological environments after the reservoirs are developed by different modes or in different stages (Shouyu, 2005). One of the most important goals of fine reservoir description is to quantify the spatial distribution of the remaining oil. There are diverse methods to characterize the remaining oil, including the geological method, the reservoir engineering method, well testing and numerical simulation method, and the indoor laboratory techniques (Ping et al., 2004). The researchers may select different characterization methods depending on the research objectives and data availability. Essentially, the remaining oil characterization is to determine the remaining oil saturation in beds of a single well. Sealed-coring wells, especially in the observation wells in the middle–late development stage, are the most direct means for observing and detecting the remaining oil distribution in beds, and the saturation logging technique is the key tool to determine the remaining oil saturation in the site (Yang, 2011). Han Dakuang (2010) pointed out that China's oilfields are basically continental reservoirs, which are characterized by strong heterogeneity, high crude oil viscosity, and low recovery by water flooding, so they have great potential for recovery enhancement. In the high water-cut stage, the remaining oil shows a generally highly scattered and locally relatively enriched distribution pattern, so the remaining oil characterization is very difficult. Here, we introduce the remaining oil characterization from the aspects of current status, contents, methods, problems, and trends.

## 3.9.1 Current status

The remaining oil characterization is always a focus of research in the middle–late development stage. In China, the strata are generally continental, with strong heterogeneity of reservoir. With the progress of oilfield development, the underground oil–water relationship is increasingly complicated, and the distribution of remaining oil is also complicated, making the remaining oil characterization more challenging. Many scholars in China and abroad have studied remaining oil. Chaochen and Wenxiang (1995) introduced in detail the methods for determining remaining oil saturation, including the development geology method, seismic techniques, logging techniques, tracer testing methods, and core analysis method. Qitai (1997) derived four types of remaining oil by the volume scale, that is, microscale, small-scale, large-scale, and macro-scale. Chengyan (2000) summarized the theories, methods, and techniques for the formation and distribution of remaining oil with a case study of the Guantao Formation fluvial sand sedimentary reservoirs in the Central No.1 Zone of Gudao Oilfield and the No.7 Zone of Gudong Oilfield. Jianmin and Shouyu (2003) studied the sedimentary model of fluvial reservoir and its control on remaining oil distribution by taking the Guantao Formation reservoirs in the Gudao Oilfield, eastern Zhanhua Sag of Jiyang Depression, and in Gudong Oilfield as examples, and concluded that the sedimentary model is the most important factor controlling the formation and distribution of remaining oil. Youguo et al. (2003) discussed the application of log-based comprehensive reservoir evaluation in remaining oil characterization. Wu (2004) discussed the geological methods for characterizing the remaining oil distribution in fault-block reservoirs. Bonnett et al. (2006) analyzed the application of wellbore trajectory design to the recovery of the potential remaining oil in the heavy oil development zone in the North Sea's Captain field. Babadagli (2007) discussed the development of mature oil fields and pointed out that the remaining oil characterization is the basis for the research on EOR in oilfields. Levin Barrios Vera et al. (2010) assessed fluid migration and quantified remaining oil saturation in a mature carbonate reservoir with an example from Dukhan Arab D, and the results showed that the differences in oilfield development and production lead to the migration of fluids underground after passing through the sedimentary boundary, which increases the uncertainty of remaining oil distribution underground, and given the classical material balance and volumetric detection techniques can be used to monitor the change of reservoir during fluid migration and to characterize the location of remaining oil, thereby, making it possible for enhancing oil recovery and reducing the research limitations and uncertainties. Zhizhang et al. (2010) carried out phased numerical simulation and remaining oil distribution prediction for complex fractured reservoirs, with a case study of H14 in Huoshaoshan Oilfield. Congjun et al. (2012) predicted the remaining oil distribution based on single-factor and multifactor coupling analysis in the typical lake-basin sedimentary reservoir in Quan 4 Member, Fuyu Oilfield, which provides a new approach for tapping the potential remaining oil in mature oilfields in the high water-cut stage. Yan et al. (2013) discussed the application of high-resolution sequence stratigraphic classification in the characterization of remaining oil in continental reservoirs using the Karamay Formation in the I-East Zone of Karamay Oilfield as an example. Currently, the remaining oil characterization is mainly completed on the basis of the results of microstructure analysis, a fine description of the single sand body, and logging interpretation of swept layers. Yusheng et al. (2015) studied the flooded model for evaluating remaining oil distribution in the braided-river delta front by taking the evaluation of water flooded layers in the Wenxisankuai block. Zhaolun et al. (2016) defined the remaining oil distribution pattern in Kukmol South Oilfield, the South Turgai Basin, Kazakhstan, based on the architecture characterization of sand bodies. Through literature review, it is found that the research on remaining oil is increasingly quantitative—from the viewpoint of the remaining oil classification initially (Table 3.21) (Naiju et al., 1999)—leading to the description of the remaining oil distribution and the identification of remaining

**Table 3.21** Distribution of remaining oil in the high water-cut stage of Lasaxing Oilfield, Daqing, 1994 (after Naiju et al., 1999).

| Area percentage of effective thickness to total effective thickness, % Type | West part of central zone | Eastern part of north II Zone | Lamadian north block |
|---|---|---|---|
| Uncontrolled by well pattern | 3.1 | 8.4 | 2.0 |
| Low-permeability zone/belt | 4.3 | 11.6 | 1.8 |
| Imperfect injection-production system | 20.1 | 31.4 | 26.7 |
| Water flooding response at the second line | 11.0 | 4.4 | 5.9 |
| Single response of production well | 8.1 | 5.0 | 7.3 |
| Reservoir bed not recovered | 5.4 | 12.0 | 4.0 |
| Interstrata interference | 5.6 | 13.6 | 0.5 |
| No water flooding inside strata | 30.0 | 0.0 | 40.1 |
| Sheltering by barrier | 12.3 | 13.6 | 11.0 |

oil distribution patterns, and leading to the quantitative prediction of spatial remaining oil development currently. Also, the methods turn from the development geology method, logging interpretation method, and numerical simulation method to the analysis and testing, physical simulation, and mathematical-statistical analysis of the sealed-coring wells. The object also gradually changes from the description of the remaining oil after water flooding to the characterization of remaining oil after polymer flooding and gas flooding. More and more efforts are devoted to microscopic remaining oil description (Fig. 3.82), in stark contrast to macroscopic remaining oil description. Moreover, high-resolution sequence stratigraphy, reservoir heterogeneity analysis, and flow unit characterization have also been applied to the research on remaining oil, making the work improve in both level and accuracy. The focus has also shifted from macroscopic remaining oil distribution to intensive study of microscopic remaining oil distribution.

### 3.9.2 Contents

The remaining oil characterization is mainly conducted in the middle–late development stage. It involves diverse matters, including both the reservoir factors and the impact of the oilfield development process.

(A)

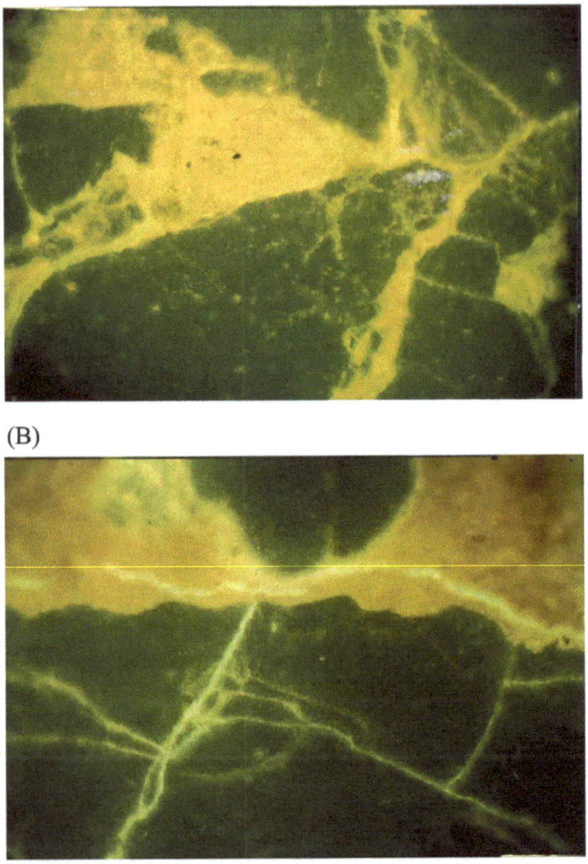

Figure 3.82 Fluorescent thin-section showing the characteristics of the glutenite reservoir of an alluvial fan in the Lower Karamay Formation in an area at the northwestern margin of the Junggar Basin. (A) Well T16, brown, oil immersed, glutenite, 1069.23 m; (B) Well T16, medium-bright blue fluorescent display in microfractures, gray, oil immersed, glutenite, 1074.18 m.

In this study, we summarized key contents of the remaining oil characterization as six aspects of research methodology.

1. Characterization of types and distribution patterns of remaining oil reservoir

   Dong et al. (1999) studied the types and distribution of remaining oil in fluvial reservoirs. Songjiang and Jiaxi (2003) studied the distribution of remaining oil in complex fault-block reservoirs and supporting recovery measures with a case study of the Guangdong development

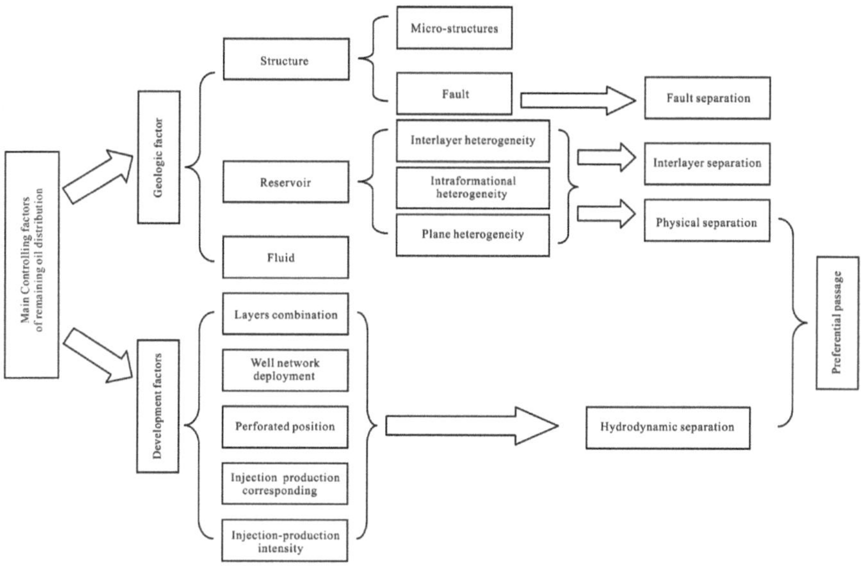


**Figure 3.83** Block diagram of factors controlling remaining oil formation. *From Yang, L., 2011. Characterization of Remaining Oil Enrichment Region of Terrestrial Reservoir with Water Drive. Beijing: Petroleum Industry Press.*

zone of Dagang Oilfield. The remaining oil include macroscopic remaining oil (which refers to the distribution of remaining oil in the reservoir scale) and microscopic remaining oil (which refers to the distribution of remaining oil in pore structure).

2. Characterization of remaining oil formation and distribution pattern and analysis of controlling factors (Fig. 3.83)

   Zhigao et al. (2004) took the Dalinghe reservoir in Shu 2 zone of Liaohe Oilfield as an example to analyze the formation and distribution pattern of heavy oil and controlling factors. In this regard, geological and development characteristics are combined to classify, describe, and predict the remaining oil depending on the genesis and distribution of the remaining oil.

3. Description of remaining oil by geology, logging interpretation, seismic monitoring, numerical simulation, well testing, and other methods

   Ruili et al. (2004) discussed the application of through-casing resistivity logging technique in the evaluation of remaining oil saturation in Daqing Oilfield, and they showed that the measurement of resistivity outside the casing could help evaluate the oil saturation of the reservoir and monitor the dynamic change of fluid, and to provide the interpretation results with similar accuracy to core analysis. Taiju et al. (2006) presented the remaining oil prediction by the comprehensive geological method, which consists of three parts: establishing a geological knowledge base, building a reservoir geological model, analyzing the development characteristics of the sand body, and predicting the remaining oil distribution with the genetic sand body as a unit.

4. Application of sequence stratigraphy division, fine structure interpretation, reservoir architecture characterization, reservoir heterogeneity analysis, and flow unit classification in remaining oil characterization

   Yining et al. (2015) studied the application of a high-precision structural model in reservoir prediction and remaining oil recovery in the case of the dense well pattern. Wangshui et al. (2012) analyzed the macroscopic heterogeneity of reservoirs and its influence on the distribution of remaining oil with a case study of the Tenggeer Formation in Xilinhaolai Baiyinchagan Sag. Cheng et al. (2012) took Block S17−19 of Fuyu Oilfield, Jilin, as an example to study the distribution pattern of dominant flowpath inside point bar and its control on the remaining oil distribution.

5. Prediction of remaining oil distribution characteristics

   Taiju et al. (2004) used Machang Oilfield as an example to predict the distribution of remaining oil in complex fault blocks in the high water-cut stage, and they indicated that the remaining oil prediction involves the remaining oil both in and between wells.

6. Description of remaining oil distribution after tertiary oil recovery

   Kaoping et al. (2004) analyzed the influencing factors of microscopic distribution of remaining oil in case of polymer flooding, and the results showed that: (1) the polymer solution reduced the mobility ratio and enlarged the swept volume macroscopically; (2) the viscoelasticity of the polymer solution increases the friction between the polymer and oil film, thereby, improving the microscopic oil displacement efficiency; (3) different degrees of sweep produce the remaining oil with different characteristics, and the blind end remaining oil is most affected by the polymer flooding; and (4) the polymer flooded remaining oil distribution is affected by inaccessible pore volume multiplier, and it mainly exists in clusters.

### 3.9.3 Methods

The remaining oil characterization methods are introduced as follows.

#### *3.9.3.1 Development geology methods*

##### 3.9.3.1.1 Remaining oil characterization method is based on the macroscopic and microscopic description of reservoirs

The reservoir development characteristics are the material basis for determining the distribution of the remaining oil. From the macroscopic point of view, the sedimentary facies type of sand body, the spatial scale and connectivity of sand body, and the development characteristics of reservoir intercalation have great controls on the distribution of remaining oil. From a microscopic point of view, the size, type, and connectivity of reservoir pore structure, the type of clay mineral, and the likes can also have important effects on the distribution of remaining oil. All of these can be addressed with the development geology methods. Lijun and Xinjun (2003) studied the effect of reservoir heterogeneity on remaining oil distribution. Huanqing et al. (2015a,b,c,d), in their study on the architecture of the Lower Karamay Formation alluvial fan reservoir in an area at the northwestern margin of the Junggar Basin, statistically analyzed the scale features (e.g., length and width) of different fourth-order architectures like trough flow gravel bodies. Based on these quantitative data, it is possible to judge the position of the remaining oil distribution in the conglomerate reservoir. The remaining oil is mostly distributed at the edges of sand bodies of different sedimentary genesis or the position where two sets of sand bodies are in contact. Depending on the remaining oil distribution positions, development adjustment wells can be deployed to recover the remaining oil. This also proves the importance of development geology methods for remaining oil characterization.

##### 3.9.3.1.2 Remaining oil characterization method based on structural analysis

In addition to the reservoir, the structure can also have important impacts on the distribution of the remaining oil. Du et al. (1998) studied the relationship between microstructural morphology and remaining oil distribution by taking No.7 and 8 sand groups in Sha 2 Member of Shengtuo Oilfield as examples. Dezhi et al. (2003) analyzed the hydrodynamic reasons for the formation of remaining oil in reservoir microstructures. We believe that, from the perspective of structure interpretation, the remaining oil can be studied in respect of: (1) the influence of third- and higher-order faults on remaining oil distribution, and (2) the influence of microstructure on remaining oil distribution. Previously, all wells were drilled bypass faults; as a result, oil and gas close to and sheltered by the faults could not be recovered but formed remaining oil. With the deepening aims and methods of studies on oilfield development, the current oilfield development has turned to fault searching. Well placement near the fault has become one of the most important directions for recovering the remaining oil. In this regard, Daqing Oilfield has worked successfully, and other oilfields are taking active actions. Microstructures include small nose structures, small depressions, and also lower-order (e.g., fourth-order and fifth-order) faults. Shengli Oilfield, Liaohe Oilfield, and Changqing Oilfield have achieved evident results in the identification of microstructures for tapping the remaining oil. In general, the development geology methods are most fundamental for the remaining oil characterization by enabling an intensive understanding of the remaining oil genesis and distribution pattern. However, the methods are somewhat less quantitative, so they often need to be combined with other methods for the remaining oil description.

### 3.9.3.2 Core observation description and test methods

A core is the most intuitive tool for understanding underground reservoirs. Through the observation and description of cores, the difference in oil content of reservoir can be qualitatively judged (Fig. 3.84). The oil-bearing level of the core is mainly determined by the oil-bearing occurrence, oil saturation, and oil-bearing area, and it can be divided into four levels: oil sand, oil containing, oil immersion, and oil patch (Yingchang and Zaixing, 2003). According to this division, the oil-bearing property can be preliminarily identified. Testing of the core is also optional, which includes the description of physical properties and oil-bearing properties of the reservoir. By comparing the relationship between oil-bearing properties and physical properties, the boundaries of the effective reservoir can be determined; then, the reserves are estimated, and the remaining oil is characterized. Fig. 3.85 shows the relationship between reservoir physical properties and oil-bearing properties for the Yulou reservoir in an area of Western Sag, the Liaohe Basin, suggesting that the physical property boundary of the effective reservoir can be determined to provide quantitative data for understanding the oil-bearing properties of the reservoir. Jie and Yanyi (2004) took Pu I Formation of Daqing Oilfield as an example to study the microscopic distribution of remaining oil after polymer flooding through fluorescence analysis of core thin-section. Li Jie et al. concluded that the proportions of different types of microscopic remaining oil distribution (mainly four types: cluster, membrane, blind end, and corner)

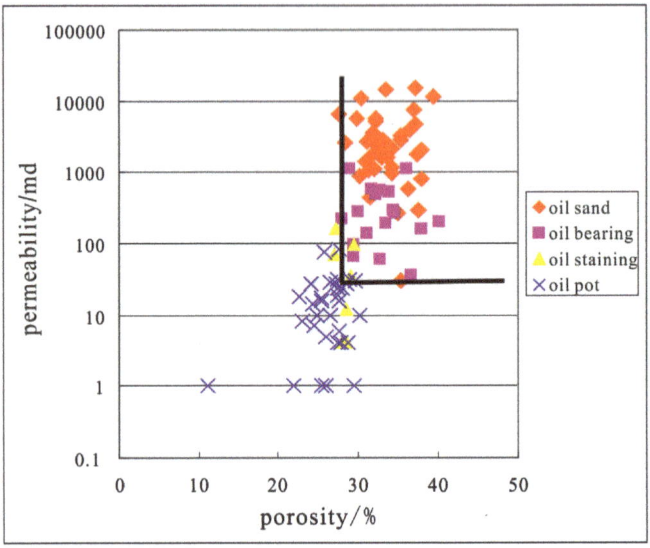


**Figure 3.85** Relationship between oil-bearing property and physical properties of the Yulou reservoir in an area of Western Sag, the Liaohe Basin.

measured quantitatively during water flooding reduced to different extents after polymer flooding, and they summarized the distribution law of remaining oil at positions with high, medium, and low displacement intensities. Data of core, especially the core from sealed-coring well, have become one of the most important information for remaining oil description in the middle–late development stage and have been proved effective in practice. However, not all target formations can provide ideal core data, and human or instrumental errors in core sampling and testing may result in mistakes in results, which limit the application and promotion of these methods. Besides, due to the strong heterogeneity of continental sedimentary reservoirs, the core samples may not be representative adequately, which affects the application of these methods.

### *3.9.3.3 Swept layer logging interpretation methods*

The results of swept layer logging interpretation can be combined with regional geology and production data to predict the distribution of the remaining oil between wells and highlight the remaining oil enrichment zones and variation, thereby providing a reliable basis for adjusting oilfield development plan, controlling water output while stabilizing oil production, and improving the oilfield development results (Peihua, 2003a).

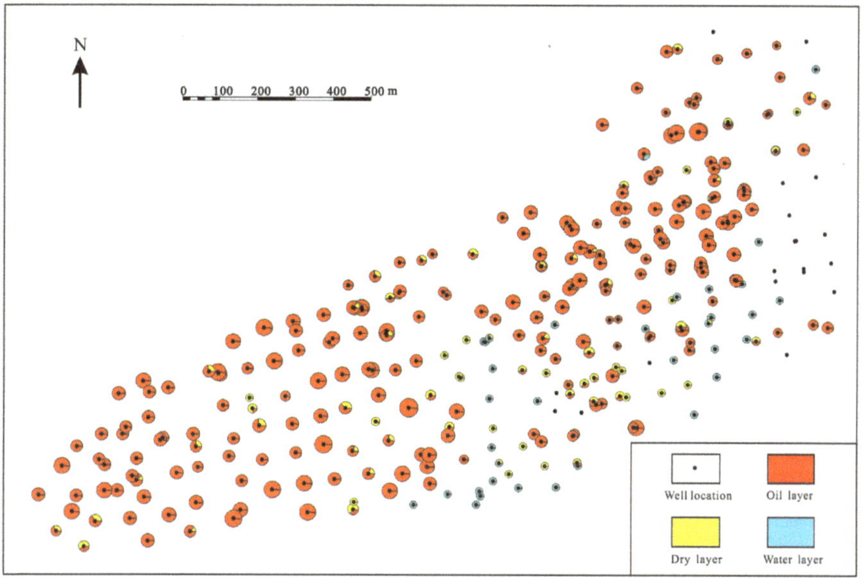


**Figure 3.86** Plane map of oil–water distribution of $yl_3^{5a}$ in the Yulou reservoir in an area of Western Sag, the Liaohe Basin.

Jun and Jinliang (2004) took the oil reservoir developed by water injection in the Dong 2 Member in the H31 fault block in Haiwaihe Oilfield to characterize the remaining oil distribution on the basis of analysis of swept layer. Shao Weizhi et al. (2004) analyzed the NMR logging evaluation method. The fine logging interpretation of swept layer can quantify the remaining oil. In the study on remaining oil in an area of Western Sag, the Liaohe Basin, we plotted the plane map of oil–water distribution in different beds for various wells (Fig. 3.86) and compared the variation of oil–water distribution characteristics in different beds. Since log data are readily available, this method is currently widely used. However, the accuracy of the logging interpretation model is unqualified for the production practice in some oilfields. For example, for the glutenite reservoir of an alluvial fan in Xinjiang Oilfield, the accuracy of swept layer logging interpretation is less than 70%, which somewhat limits the application of the swept layer logging interpretation method in remaining oil characterization.

#### *3.9.3.4 4D seismic method*

The 4D seismic method, in a broad sense, is a time-lapse geophysical method; namely, seismic data are repeatedly observed in a time-lapse manner so as to monitor the movement of gas or liquid during reservoir production. Wang Yuncheng et al. (2003) introduced the application of 4D seismic method in remaining oil characterization for the middle–late development stage by virtue of several examples. Yumei et al. (2003) studied the feasibility of seismic method to monitor the remaining oil distribution in water flooded thin interbed reservoirs. Lideng et al. (2010), taking an area of Jidong Oilfield as an example, introduced the approach and principle of using the 4D seismic method to characterize the remaining oil. The 4D seismic method has been satisfactorily adopted broadly for remaining oil characterization abroad, such as the North Sea Oilfield. In China, this method has been tested in the Liaohe Oilfield and the Jidong Oilfield but without outstanding effects on the remaining oil recovery, mainly due to technology and costs. This method can intuitively obtain quantitative data of the spatial distribution of remaining oil, but its accuracy is far behind the requirement of highly heterogeneous continental reservoirs in China. Moreover, the high cost makes the method unacceptable under the climate of low oil prices now.

#### *3.9.3.5 Mathematical statistics methods*

The mathematical statistics methods have been widely used in the remaining oil characterization because they can generate quantitative results. Shengli et al. (2003) took the HB reservoir of Xijiang 30–2 Oilfield as an example to characterize the remaining oil distribution by the gray relational analysis (GRA) method, and they obtained the results in good agreement with the actual production. Guomin et al. (2004) quantitatively evaluated the potential of remaining oil in complex and heterogeneous reservoirs by way of multilevel fuzzy analysis, based on 11 static and production dynamic indicators (i.e., remaining oil saturation, reserves abundance, sand body type, sand body position, sand body location, connectivity, microstructure morphology, water injection distance, perforation perfectness, injection-production perfectness, and variation coefficient of permeability) selected after the analysis of the formation conditions, distribution laws, and controlling factors of the remaining oil in clastic reservoirs in the high water-cut stage. The mathematical statistics methods are diverse and unique individually. Generally, their application in remaining oil characterization is one of the relevant orientations. It should be noted, however, that the geophysical significance of different parameters should be clarified.

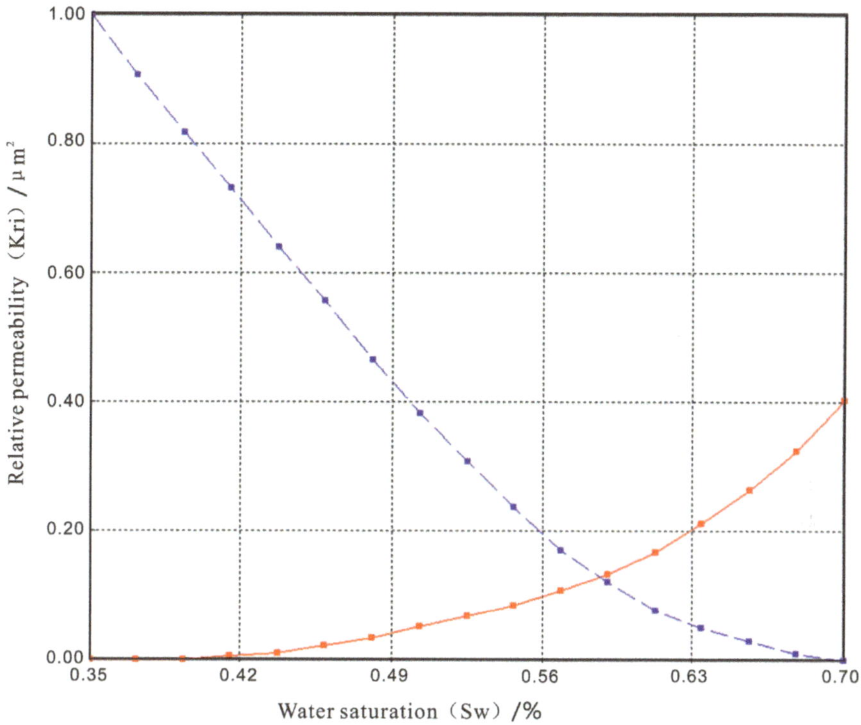


**Figure 3.87** Oil–water relative permeability curve of the Yulou reservoir in an area of Western Sag, the Liaohe Basin.

### *3.9.3.6 Numerical simulation methods*

Reservoir numerical simulation is one of the most widely used methods for remaining oil characterization. Mingli (2003), taking the complex fault-block oilfield as an example, used the numerical simulation method to study the distribution characteristics of the remaining oil, and the research results were satisfactorily applied in the development of the Wenzhong Old III block. Yanhui et al. (2011) carried out the numerical simulation of tertiary-mainlayers polymer flooding in Daqing Oilfield with the Sabei development zone as an example, and they showed that tertiary-mainlayers polymer flooding realized the best performance in EOR and water-cut reduction by perforating the reservoir with effective thickness less than 0.5 m in the first stage and re-perforating the reservoir with effective thickness more than 0.5 m. In this study, we made the numerical simulation of the remaining oil distribution in an area of Western Sag, the Liaohe Basin (Figs. 3.87 and 3.88). The distribution

(A)

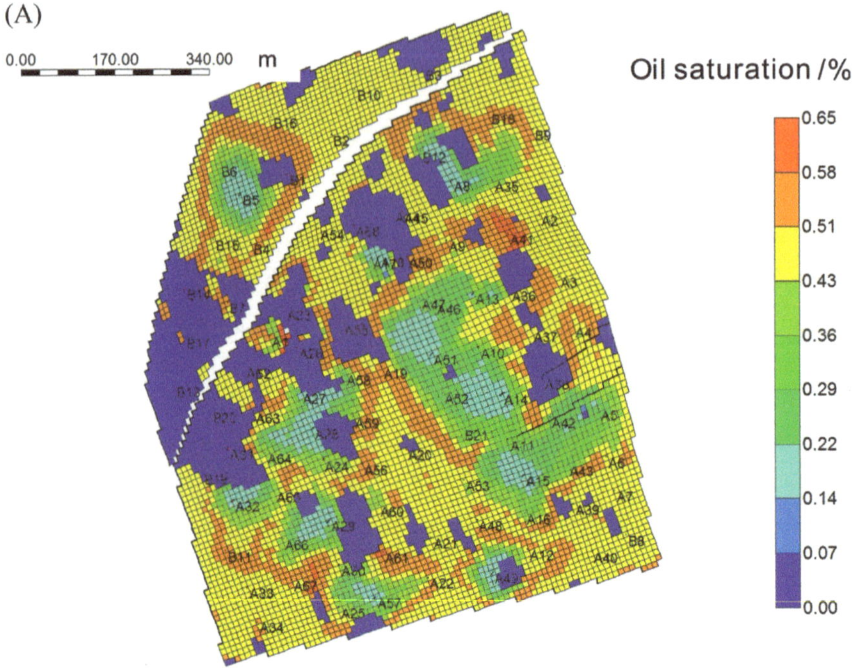


(B)

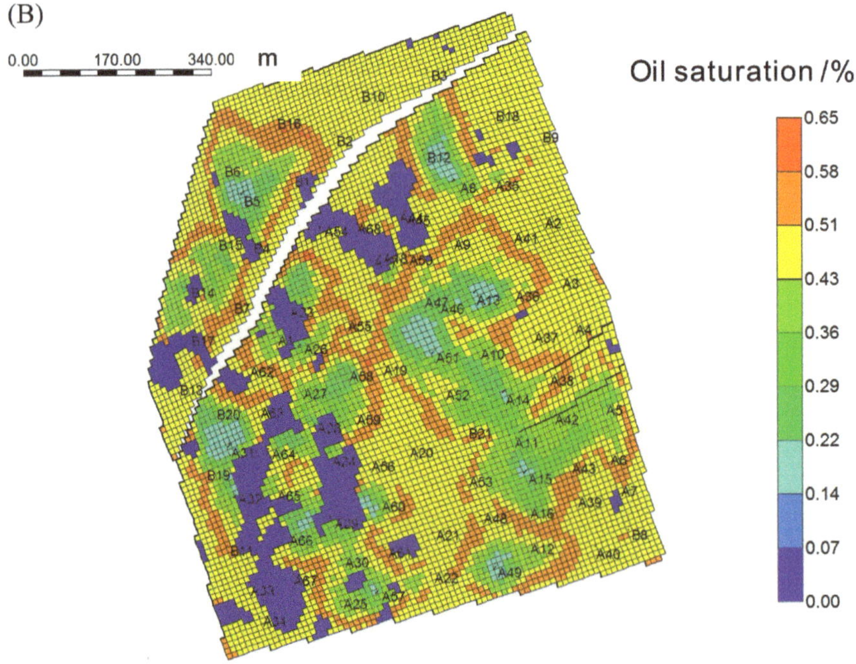


**Figure 3.88** Distribution of remaining oil in beds of Yulou reservoir in an area of Western Sag, the Liaohe Basin. (A) Single layer of $yl1_1^a$; (B) Single layer of $yl1_1^b$.

characteristics of the remaining oil in the target reservoir at different periods were obtained, which provided the basis for the conversion of development mode from steam stimulation to steam flooding. The numerical simulation method is the most widely used method for the remaining oil study from the perspective of reservoirs. Uniquely, this method can quantitatively predict the distribution of the remaining oil. However, the precision and accuracy are limited by the data and software platform and are also closely related to the experience and competence of the researchers.

#### *3.9.3.7 Dynamic monitoring analysis method*

A large amount of production performance data and dynamic monitoring data are generally acquired from reservoirs in the middle–late development stage. These data can be used to characterize the remaining oil distribution. Chengwen et al. (2004) explored the monitoring and evaluation techniques for the remaining oil saturation in ultra-deep ultra-high temperature reservoirs in the ultra-deep high-temperature strata of Donghetang Oilfield, the Tarim Basin. Baoxi et al. (2005) took the water injection cluster in the No.2 fault block in the Guangxi No.3 Zone of Dagang Oilfield as an example to explain the technique of determining remaining oil saturation in water flooding reservoir by monitoring the interwell tracer. Hujun et al. (2007), taking the M reservoir of Laojunmiao Oilfield, Yumen, as an example, used the isotope tracing technique to study the distribution of the remaining oil, which provides a simple and intuitive method for the remaining oil characterization for the late stage of water flooding in mature oilfields. Mingjun and Jiagen (2009) conducted numerical simulation with a streamlined model based on the results of interwell monitoring of two clusters (Yang 13–32 and Yang 13–14) in Yangsanmu Oilfield, Dagang Oilfield, and obtained the remaining oil distribution of Guan II3, Guan II4, and Guan $II_{5+6}$. The dynamic monitoring analysis method can provide rich and intuitive data for characterizing the remaining oil distribution. However, the results are multiple, so the necessary screening and quality control of the data are very important. Moreover, due to the limitations of the method and the production practice, some monitoring data won't necessarily have the accuracy to meet the requirements for reservoirs in the middle–late development stage at the scale of beds. For example, when studying the remaining oil in the Yulou reservoir in an area of Western Sag, the Liaohe Basin, we used tracer data relational analysis, and we found that the data segments corresponding to

**Table 3.22** Adaptation and characteristics of reservoir engineering methods.

| Method | Application | Advantage and disadvantage |
|---|---|---|
| Relative permeability curve method | Strata or block with complete data of relative permeability curve | It produces better results for zones around the well from which the relative permeability curve is acquired but poor results for zones far from the well |
| Water flooding characteristic curve method | Wells with high and extra-high water-cut | The predicted single well swept reserves and remaining recoverable reserves may be too great or too small |
| Dimensionless injection-production method | Oilfields with regular well pattern and relatively fixed injection-production mode | It is simple and accurate and requires less parameters, but it involves the production splitting |
| Material balance method | Reservoirs with regular well pattern and relatively fixed injection-production mode but with formation pressure higher than saturation pressure | The calculation formula is relatively simple, but it involves longitudinal and lateral production splitting. It is not suitable for reservoirs with formation pressure lower than saturation pressure and serious degassing |
| Front advance velocity method | Different homogeneous sections in the target layer | It analyzes the interreservoir or intrareservoir sweep by using the data of coring well, C/O, isotope logging, and others |

*Source*: From Yang, L., 2011. Characterization of Remaining Oil Enrichment Region of Terrestrial Reservoir with Water Drive. Beijing: Petroleum Industry Press.

the injection well and production well can only reach the requirements at the layer level, and below that at the bed level; as a result, other relevant analysis must be conducted to meet the research objectives.

#### 3.9.3.8 *Reservoir engineering methods*

The reservoir engineering method is a relatively straightforward and simple method for quantitatively describing the remaining oil distribution. By using oilfield development data, one or more reservoir engineering methods can be used to quickly identify the underground situations (Ping et al., 2004). Xiarong et al. (2009) quantitatively calculated the remaining oil saturation with the reservoir engineering method, and they demonstrated that the method could quickly and simply evaluate the remaining oil and has a certain guiding effect on the history matching in the numerical simulation process. Yang (2011) summarized the reservoir engineering methods for quantitatively studying the remaining oil saturation of wells (Table 3.22). This method is advantageous in that the development performance characteristics are actually the comprehensive reflection of reservoir characteristics and flow patterns of fluid. Table 3.22 shows the advantages and disadvantages of this category of methods.

#### 3.9.3.9 *Well test interpretation methods*

Well testing is the testing of oil, gas, or water wells (Nengqiang, 2008). Li Xing et al. (2008) determined the distribution of remaining oil saturation by defining the oil–water relatively permeability curve based on the data of pressure drawndown test of water injection well, with consideration to the relationship between the derivative of water-cut at the front of water injection or pseudo-skin factor caused by mobility ratio in oil–water zone and the oil–water relative permeability. Minglu et al. (2009) studied the distribution of the remaining oil in polymer flooded reservoirs by streamlining the numerical well test method for the No.437 fault block in Shuanghe Oilfield, and they demonstrated that the method is suitable for polymer flooded reservoirs with medium–high water-cut. The well test interpretation method can obtain the radial flow parameters of the formation to calculate the quantitative data of oil saturation. However, the accuracy of data measured cannot be ensured, and the precision of calculation results needs to be improved.

### 3.9.4 Problems and trends

#### 3.9.4.1 *Problems*

The remaining oil characterization is one of the most important parts of fine reservoir description and covers a wide range of aspects. It is beset with the following problems.

1. There are many controlling factors for the remaining oil genesis, and the main factors are difficult to identify and classify. The distribution of the remaining oil is affected by both geological factors and development processes. Identification of the major controlling factors and division of the remaining oil distribution types are always critical means to guide the recovery of the remaining oil. However, the problems in structural interpretation, reservoir characterization, analysis of reservoir variation during development, and description of flow paths cause many problems in the analysis of remaining oil genesis. Jing et al. (2016) discussed the application of fine microstructure characterization in the remaining oil prediction.
2. Each method of the remaining oil characterization has its advantages and disadvantages. It is difficult to select one or more suitable methods depending on the research objectives and the available database. For example, in the core analysis method, how to select the sample points that can fully reflect the remaining oil saturation data of the target layer in the study area, and at the same time reduce the cost of analysis and testing as much as possible, which is the great challenge that researchers need to face. Also, it is extremely difficult to closely combine the development geology methods with the mathematical statistics methods to reflect, both qualitatively and quantitatively, the remaining oil characterization.
3. The accuracy of the remaining oil description by relevant methods needs to be improved. Taking the swept layer logging interpretation method as an example, establishing a more sophisticated logging interpretation model to provide accurate results for special lithologic reservoirs such as glutenite was difficult.
4. The depiction of the remaining oil in the well can be accurately achieved by numerical simulation and well testing methods, but the prediction of remaining oil between wells is very difficult. Taiju et al. (2004) discussed the prediction of the remaining oil distribution in the high water-cut stage in complex fault-block areas with a case study using the Machang Oilfield.
5. Application of study results in some fast-growing disciplines, such as high-resolution sequence stratigraphy, reservoir architecture, reservoir heterogeneity, flow unit, and geological modeling, into the remaining oil characterization is still in its infancy. Zhijie et al. (2008) used the N reservoir in the offshore M Oilfield as an example to introduce how a 4D model for facies-controlled remaining oil distribution is established.

Wenqi et al. (2016) studied the recovery of the remaining oil in the ultra-low-permeability reservoirs based on the division of stress intercalation by taking the Yanchang Formation reservoir in Baiyushan area, the Ordos Basin, as an example. Rongqiang et al. (2016) studied the distribution of the remaining oil based on the division of flow units under the control of point-bar architecture.

6. The remaining oil characterization for water flooding is mature, while that for tertiary oil recovery techniques such as polymer flooding and gas flooding is still in its infancy. Fenglan et al. (2004) analyzed the oil displacement effect and the remaining oil characteristics of sealed-coring wells before and after polymer flooding in the Sazhong area, and they revealed that the polymer flooding adjusted the interlayer and intralayer profiles to improve the thickness of recovery and that the water-washing degree can be increased to enhance the oil displacement efficiency, but it is very difficult to reduce the remaining oil after water flooding. In case of polymer flooding, little remaining oil is left, and it distributes dispersedly, thereby making it difficult to recover; improvement of oil displacement efficiency should be the focus.
7. The remaining oil characterization in China remains to target clastic rock reservoirs, while the remaining oil characterization of complex lithologic reservoirs such as carbonate rock, volcanic rock, and metamorphic rock is unsatisfactory due to limited databases and technologies.
8. The remaining oil characterization in China is more qualitative and semiquantitative rather than quantitative.
9. Macroscopic remaining oil characterization is popular among researchers, but microscopic remaining oil characterization is weak.

### 3.9.4.2 Trend

The remaining oil characterization is believed to develop in the following aspects of research methodology.

1. Genesis analysis and development pattern identification are fundamental for the remaining oil characterization, and both geology and development factors should be considered. Wenjie et al. (2015a,b) studied the turbulent channels of alluvial fan reservoirs and the remaining oil distribution patterns controlled by turbulent channels, with an example from the Lower Karamay Formation in the Central I zone Karamay Oilfield.
2. Existing methods are improved, and new methods are developed. Haibo et al. (2016) performed an NMR analysis on the remaining

movable oil in low-permeability reservoirs, and they quantified the parameters such as current remaining oil saturation, relative oil production, movable oil saturation, and oil displacement efficiency. Zhongsheng et al. (2016) evaluated the remaining oil and flooding degree by using the C/H logging technique, and they showed that the method could effectively evaluate the flooding degree, oil saturation, and oil—water contact, providing valuable data for detecting missing oil layers and determining water plugging measure, and it can be widely applied to sand-mud profiles regardless of sandstone, conglomerate, or low-resistivity reservoirs. Qiaolin et al. (2017) divided the pore structure of the JIV reservoir in the Lower Jurassic Yangxia Formation in Lunnan 2 Oilfield, the Tarim Basin, based on the analysis of thin-sections, conventional physical properties, SEM, and mercury intrusion test, and they selected four types of pore structures for microscopic water flooding experiments of sandstone cores to identify the microscopic remaining oil distribution characteristics.

3. For the mathematical statistics methods, the physical meaning of parameters should be fully recognized, and the application effect of the results in production practice should be improved.
4. The data processing and calculation under various quantitative methods should be optimized to improve the precision and accuracy of the remaining oil prediction between wells. Youxi and Caijing (2004) used fractal condition simulation and a streamlined model to predict the remaining oil saturation distribution, and they revealed that the method could calculate fast while ensuring accuracy, so it is prospective for promotion. Wenlin (2010) explored the petrophysical basis and method for predicting remaining oil distribution in water flooded oilfields. The time-lapse seismic method, which is used to obtain the amplitude difference between two 3D seismic data before and after water flooding, is based on the seismic response differences of fluids, more effective to predict the remaining oil distribution than conventional oil and gas seismic detection techniques. Its feasibility has also been proved by models and examples.
5. High-resolution sequence stratigraphy, reservoir architecture characterization, dominant water flowpath description, flow unit characterization, geological modeling, and other related disciplines are considered for the remaining oil characterization. Yefei et al. (2003) studied the distribution of the remaining oil in high water-cut oilfields by using the flow unit method, with a case study of the oil reservoir in the

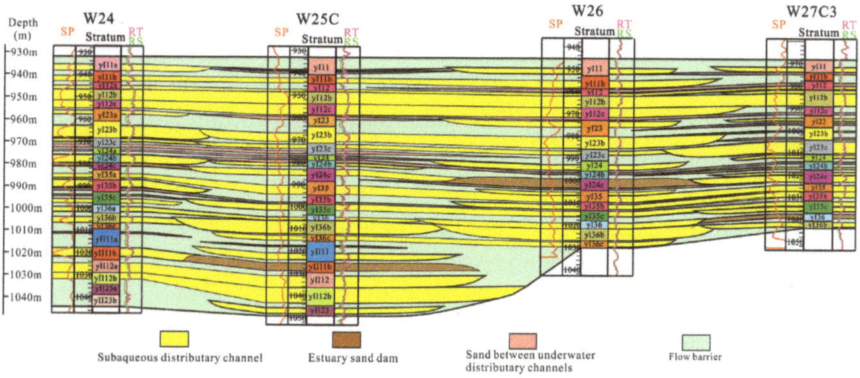


**Figure 3.89** Flow barrier of the Yulou reservoir in an area of Western Sag, the Liaohe Basin. *From Huanqing, C., Chao, D, Yijing, D, Jue, W., 2015a. Advances of reservoir evaluation researches. Geological Science and Technology Information 34(5), 66–74. Huanqing, C., Chengfang, S., Chen, C., 2015b. Advances in the study of fine reservoir description. Geological Review 61(5), 1135–1146. Huanqing, C., Shuxian, L., Peng, J., et al., 2015c. Flow barrier characteristics of thermal rescovery of heavy oil reservoir: taking Yulou oil-bearing sets in some experimental area in the south end of westward slope of West depression in Liaohe Basin as an example. Acta Sedimentologica Sinica 33(3), 616–624. Huanqing, C., Shuxian, L., Zhirui, S., et al., 2015d. Caracteristics of conglomerate reservoir architecture of alluvial fan and its controlling effects to reservoir development: taking alluvial fan reservoir in some area of Northwest margin of Junggar Basin as an example. Journal of Jilin University (Earth science edition) 45(1), 13–24.*

middle Sha 3 Member of the North Zone, Liangjialou Oilfield, Dongying Sag. Dali et al. (2013) conducted a physical simulation of the remaining oil distribution controlled by internal architecture of point bar of composite meandering point-bar sand body. Huanqing et al. (2015a,b,c,d) characterized the distribution of the remaining oil at the end of steam stimulation of the Yulou reservoir in an area of Western Sag, the Liaohe Basin, on the basis of flow barrier analysis. In Chen's study, the flow barrier of the target layer was first finely described; then, the corresponding relationship between the flow barrier and the remaining oil distribution was analyzed; finally, the remaining oil distribution was characterized from the geological point of view (Fig. 3.87). Ningning et al. (2016) took the Hei46 fault block as an example to study the heterogeneity of low-permeability reservoir and its effects on remaining oil distribution (Fig. 3.89).

6. According to the implementation of tertiary oil recovery techniques such as polymer flooding and gas flooding, relevant remaining oil characterization methods can be improved by adding appropriate

contents and simulation experiments so as to establish a series of remaining oil characterization methods and techniques in the tertiary oil recovery stage. Guo (2010) used Eclipse software to study the application of residual polymer after polymer flooding to the recovery of remaining oil in Daqing Oilfield.

7. More efforts should be made for the remaining oil characterization for unconventional reservoirs such as carbonate rocks and volcanic rocks and to analyze the differences between unconventional reservoirs and clastic reservoirs in geological characteristics and development measures. Shiwen et al. (2015) carried out research on multidisciplinary integrated remaining oil recovery in the later stage of offshore oilfield development by taking the H4C thin oil reservoir in X3 Oilfield, the Pearl River Mouth Basin, as an example. Zeyu et al. (2016) carried out a physical visualization experiment for the fractured-vuggy carbonate reservoir in Tahe Oilfield, which visually displayed the distribution of the remaining oil after gas flooding, and they discussed relevant influencing factors.
8. The application of logging interpretation, 4D seismic, mathematical statistics, geological modeling, and well test interpretation in remaining oil characterization should be strengthened to ensure an increasingly quantitative remaining oil characterization by virtue of the geology and development combination and the geology and engineering combination. Shaoying et al. (2017), taking the YM32 dolomite reservoir in the Tarim Basin as an example, analyzed the practical effects of geology-engineering integration in the efficient recovery of remaining oil.
9. The remaining oil characterization methods such as microscopic simulation experiments and tests are considered to continuously improve the level and accuracy of microscopic remaining oil characterization. Zhenquan et al. (2005) established an oil and gas two-phase 3D network simulation model to analyze the influence of microscopic reservoir parameters on remaining oil distribution through microscopic simulation.

CHAPTER 4

# Challenges and solutions of fine reservoir characterization

Large-scale fine reservoir characterization was proposed by PetroChina in 2003: when they said reservoir characterization must be conducted before an oilfield is put into production. In addition, fine reservoir characterization should be updated for the developed high water-cut oilfields. Through decades of development, this new technology has been applied widely to achieve significant economic and social benefits in production (Zerong et al., 1993; Yinan and Ziqi, 1996; Jie, 1996; Zhizhang et al., 1999a,b; Longxin et al., 2006; Ailin and Lihua, 2012; Huanqing et al., 2015a,b,c,d). However, international oil prices fell since the second half of 2014 and settled near $60 per barrel despite the slight rebound in the recent 2 years (Fig. 4.1) (Ling, 2014; Kun et al., 2015; Huanqing, 2016). The continuous fall in world oil price has had a significant impact on the world petroleum industry, including the geophysical industry (Tao et al., 2015), deep-water oil & gas exploration (Xiaoxia et al., 2015), and tight oil development (Huanzhi et al., 2015). Fine reservoir characterization study is facing unprecedented challenges.

## 4.1 Challenges of fine reservoir characterization

The oil price in China tends to drop as a result of international oil price fall (Keyao et al., 2015). International oil prices dropped below $30 per barrel on January 16, 2016, for the first time in the last 12 years. Many petroleum corporations made the strategy of "cost decreasing and benefit increasing" in response to the sharp profit decline, and attempts to deal with the severe challenge posed by the fall in oil prices were made by cutting their investments in exploration and production. As a result of this effect, oilfield development study work is facing unprecedented challenges. It is well known that more than 50% of oil and gas consumption needs to rely on imports. The CNPC (China National Petroleum Corporation LTD) Research Institute Economics & Technology (CNPC ETRI) report titled "Report on Oil & Gas Industry Development 2015"

*Fine Reservoir Description*
DOI: https://doi.org/10.1016/B978-0-323-95401-3.00004-4

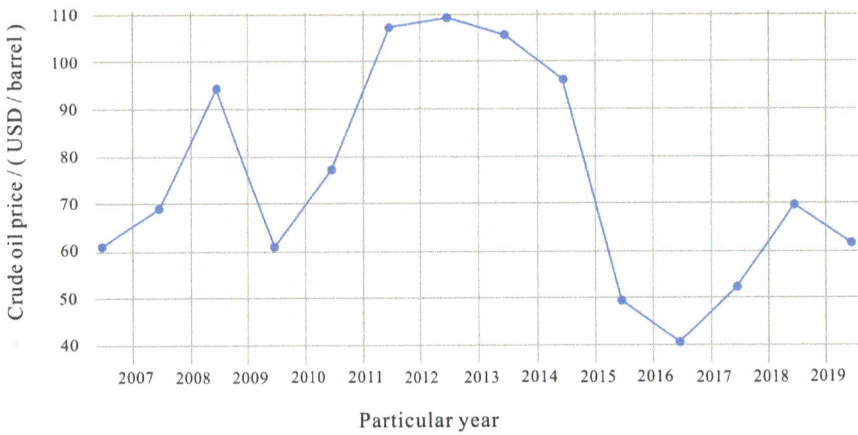


**Figure 4.1** Trend of average crude oil price. *Data from OPEC, as of March 11, 2019.*

published on January 26, 2016, discloses that the net oil import for China in 2015 was 2.8 billion tons, increasing by 6.4%, which was 0.6% faster than imports in 2014. Oil consumption in China during the same period continues to grow at a medium-low rate resuscitated by the external dependence reaching 60.6%, which exceeded 60% for the first time (Huanqing et al., 2016a,b,c,d). Given the need to ensure national energy safety, oil and gas production in China would be unlikely to decrease as oil price falls but remains stable at a certain level. However, a continuous fall in the international oil price would bring huge losses to oil companies for maintaining the current production. To solve this issue, major international oil companies have taken measures such as layoffs and investment cuttings. Three major oil companies in China also propose the strategy of "cost decreasing and benefit increasing." In this context, CNPC, for example, decided to greatly cut the projects and investments for fine reservoir characterization study. However, the responsibility of fine reservoir characterization study, which is deemed to be the fundamental core work for ensuring the rational, high-efficiency development of oilfields, grows all the more. Enhancing and stabilizing crude oil production in the context of significant cutting in investments are critical challenges facing fine reservoir characterization engineers. Fine reservoir characterization in the backdrop of low oil price is facing a series of challenges. (1) The significant cutting in projects and investments behind fine reservoir description efforts causes the quantity of data recorded to fall, study direction to deflect away from fine reservoir description work, and scope of work for study to be compressed, limiting the breadth and comprehensiveness of fine reservoir characterization studies, such as the abundance of the basic data recorded and the exploring of new methods/techniques undertaken. Fine reservoir characterization can be divided into two parts: that is, the structural study and the reservoir study. The reservoir study, for example, can be further divided into the fine stratigraphic classification and correlation, sedimentary microfacies and reservoir architecture study, reservoir heterogeneity study, quantitative reservoir evaluation, and so forth. Scientific design and planning are required to ensure that, under the significant cutting in investments, the data recorded shall meet the demand of the study and the major subject of reservoir characterization study should be retained. (2) Fine reservoir characterization for new and mature oilfields becomes significantly difficult. In China, fine reservoir characterization is conducted mostly for mature oilfields that were put into production over 30 years ago. Given the highly heterogeneous continental sedimentary reservoirs, limitation in oilfield development technology, and high water-cut, the subsurface oil-water relation can be very complex and the remaining oil dispersed over a broad area and concentrated in local areas. Thus, it is difficult to precisely capture the static and dynamic reservoir features, especially under the condition of investment cutting. As for newly discovered oilfields, it is also difficult to carry out the fine reservoir characterization that covers the whole block and captures the overall objective understanding of the static and dynamic reservoir features under the condition of limited investment, which requires scientific planning of scope of work for the study. (3) The decreased manpower and resources have an influence on the development of fine reservoir characterization studies. Fine reservoir characterization, in general, relies on the statistic and analytical calculation of a large quantity of data, which requires teamwork using advanced computer technologies. The decreased manpower and equipment cost will inevitably affect the progress of the study and is unfavorable for improving the study's prowess.

## 4.2 Solutions of fine reservoir characterization

Considering the above challenges and the requirements for cost reduction and benefit improvement, we propose the following solutions.

### 4.2.1 Planning the basic data acquisition scientifically

As the total investment decreases, the budget for data acquisition is also significantly reduced. It is suggested in this book that a key point should be emphasized while ensuring the comprehensiveness of the study so as to ensure the investment is well spent. It is well known that the availability of abundant static and dynamic data provides a basis for conducting the fine reservoir characterization study and to properly understand the fundamental geological condition of the reservoir. The dramatic reduction in investment has limited the increase in data richness. To solve this issue, researchers shall review the data and information recorded during the exploration and the early development stages, and they should properly design the data acquisition program under the current situation and study object of fine reservoir characterization study for the purpose of decreasing the proportion of data acquisition in total study budget while avoiding losing data potency. The principle is to obtain the richest data with a minimum budget, thereby laying a solid foundation for realizing the study object. While dealing with the conventional basic data, it is important to review and dig out the information from existing data. However, the budget for recording data required for solving critical bottleneck problems must be ensured to provide a solid data basis for the study. Chen Huanqing et al. (2013a,b,c,d), when analyzing the pore structure of the Ying-1 Member volcanic reservoir in the Xudong area, the Songliao Basin, has utilized limited data and selected rock samples from the typical well for CT scanning to derive the direct data of reservoir pore structure. Contrastive analysis of these data leads to good results in understanding the microscopic pore structure of volcanic reservoirs.

### 4.2.2 Setting the study project properly

Fine reservoir characterization study projects can be divided into two types, depending on the study phase. One is the first fine reservoir characterization study, which involves the fine stratigraphic classification and correlation, fault system interpretation, sedimentary microfacies study, comprehensive reservoir evaluation, geological modeling, and flow unit characterization, all based on the basic condition of the reservoir. The other is the second or finer reservoir characterization, which, in comparison with the former study, incorporates fracture study, reservoir architecture characterization, secondary fine well-log interpretation, interlayer study, and percolation barrier study. Thus, it covers more tasks and goes deeper than the first fine reservoir characterization study. When conducting the fine reservoir characterization for the first time, it is required that a variety of related data should be fully and accurately recorded for carrying out the comprehensive analysis and characterization of the reservoir and providing support to designing and implementation of the development plan. Since it is so important, the study must be accomplished with nice quality and quantity, even in the low oil price context. As for the second or subsequent fine reservoir characterization, full consideration should be given to the challenges that would be faced in the production practice of oilfields when designing the project. The principle is to emphasize the key point of work and make the breakthrough at selected points. For example, CNPC suggests holding a joint study for oilfields having the same or similar challenges in order to realize the complementary advantages and benefits of a collaborative study by brainstorming and thereby avoid the repeatability in project designing on the company-level such that the study cost can be reduced to the extent needed. Meanwhile, the study results can be shared over oilfields. The author has designed the specific study project of reservoir architecture characterization when conducting the fine characterization of the Lower Karamay Formation glutenite reservoir in an area at the northwestern margin of the Junggar Basin, which places emphasis on reservoir architecture characterization and provides a solid basis for deeply understanding the distribution law of glutenite reservoir through classifying the reservoir architecture by origin, characterizing the spatial development law of reservoir architecture, and defining the spatial scale of different types of reservoir architectures (Fig. 4.2).

### 4.2.3 Stressing on key contents of the study

As for planning the scope of work for study, the cut in investment would significantly reduce the number of project aims and corresponding budget. However, a fine reservoir characterization study is necessarily required since the exploitation of old fields should be expanded and the newly discovered fields need to be put into production. As a result, researchers shall find the solution to decrease the cost and increase the benefit. For example, the quantity of data for recording can be optimized and the exploration data should be reviewed to dig more information. It is noteworthy that necessary works must be done with assured quality, although we are going to decrease the cost and increase the benefit. The purpose of a fine reservoir characterization study is to provide a solid basis for optimizing the design of development adjustment plans of oilfields and enhancing the oil recovery. Meanwhile, fine reservoir characterization is a dynamic study process throughout the entire oilfield development process. The majority of oilfields in China are old fields, especially those located in the central

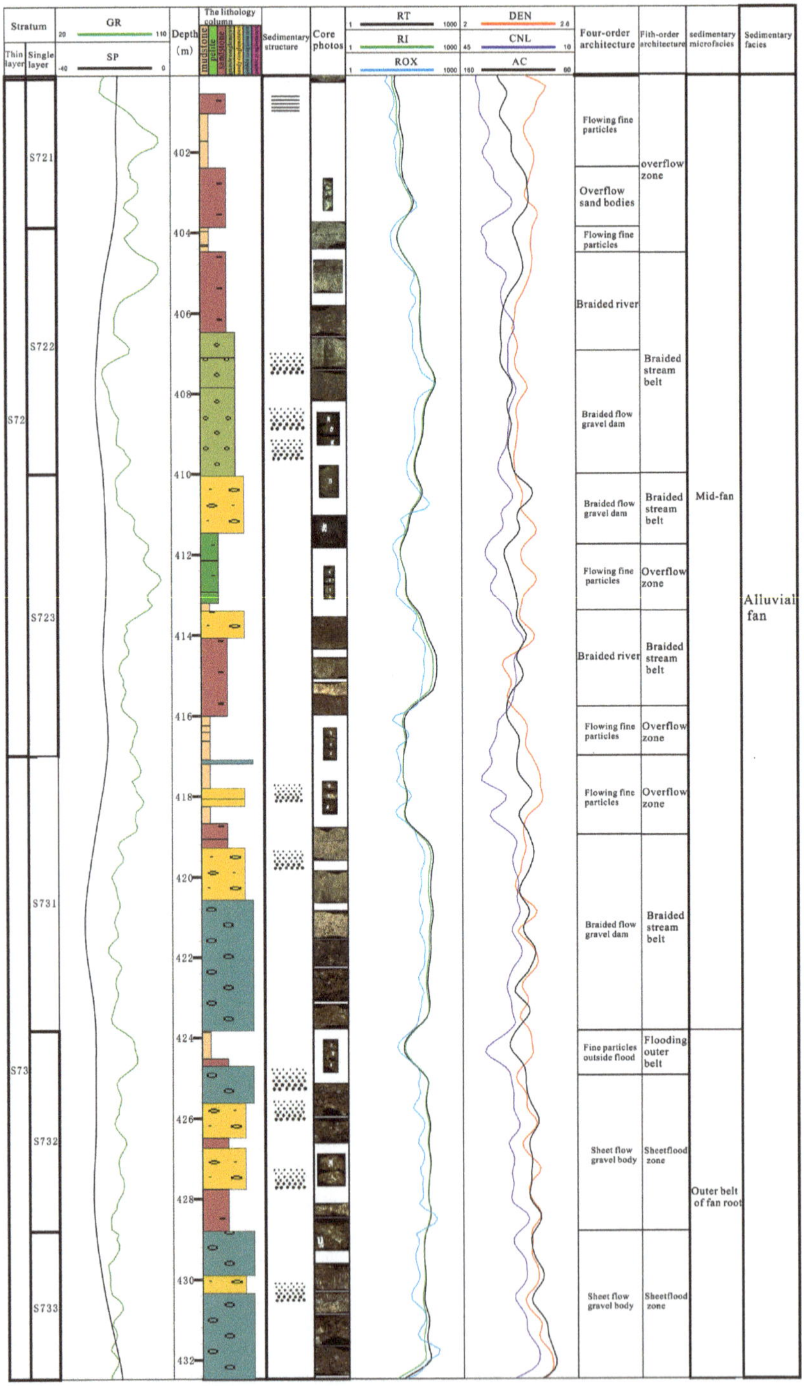


**Figure 4.2** Division of architecture for Well J8: the Lower Karamay Formation alluvial reservoir in an area at the northwestern margin of the Junggar Basin.

and western areas that have been produced for at least 30 years. For these fields, the second or third fine reservoir characterization should be conducted to continuously deepen the understanding of the reservoir, exploit the remaining oil, and enhance the oil recovery. Thus, it is necessary to carry out the fine reservoir characterization repeatedly. It is recommended that emphasis should be placed on key points while conducting this type of reservoir characterization study. This requires an analysis and determination of key bottleneck issues that may exist in different development periods of oilfields. In this way, the challenges to be solved urgently in production can be solved by designing a more practical project, and the fixed assets and human resource costs can be reduced by avoiding repetitive jobs. Meanwhile, the results of previous fine reservoir characterization studies can be partially utilized to shorten the study period and reduce the cost. For example, most of the CNPC's oilfields are fault block-type, where the distribution of oil and gas is closely related to the development characteristics of the fault system, and the distribution law of a large quantity of remaining oil resources are dependent upon the fault system. However, fine characterization of the fault system, especially the fourth- and fifth-order fault system, remains challenging due to the limitation of various factors, such as the complexity in geological conditions, availability of data, and technical methodologies. Fine characterization of fault systems has become one of the challenges that may influence the remaining oil exploitation in the high water-cut old fields. For such types of oilfields, it is recommended to place emphasis on logging–seismic fine characterization of fault system when conducting the second or third fine reservoir characterization. Stress field analysis and logging–seismic fine characterization of the low-order fault and fractured reservoirs should be conducted for the purpose of making a breakthrough in key bottleneck issues. If conditions allow, it is a good idea to organize several oilfields facing the same challenge to carry out a collaborative study. In this way, the technical and personal advantages of different fields can be combined by teamwork and result-sharing mode to significantly reduce the investment for project and study, thereby reaching the target of cost to decrease and benefits to increase. Reservoir percolation barrier study is determined to be a key point of study by Chen Huanqing et al. (2015a,b,c,d) in his fine reservoir characterization study of the Yulou reservoir in an area of Western Sag, the Liaohe Basin. His study has successfully defined the spatial distribution law of the percolation barrier, which supports the transition in the development mode for thermal recovery of heavy oil reservoir from overall cyclic steam stimulation to steam driving (Fig. 4.3).

**Figure 4.3** Diagenetic percolation barrier feature of the Yulou reservoir in a block located in the Western Sag, the Liaohe Basin. (A) Kaolinite transferring into illite forms the diagenetic percolation barrier, depth: 997.02 m, × 2400. (B) Level I of overgrowth of feldspar, diagenetic percolation barrier, depth: 990.18 m, × 6000. (C) Grains in point-line contact, compaction, diagenetic percolation barrier, depth: 954.04 m, × 100. (D) Moderate sandstone, siliceous cement, diagenetic percolation barrier, Well J93, depth: 1033.84–1033.94 m. *Data from Huanqing, C., Chao, D., Yijing, D., Jue, W., 2015a. Advances of reservoir evaluation researches. Geological Science and Technology Information 34 (5), 66–74. Huanqing, C., Chengfang, S., Chen, C., 2015b. Advances in the study of fine reservoir description. Geological Review 61 (5), 1135–1146. Huanqing, C., Shuxian, L., Peng, J., et al., 2015c. Flow barrier characteristics of thermal rescovery of heavy oil reservoir: taking Yulou oil-bearing sets in some experimental area in the south end of westward slope of West depression in Liaohe Basin as an example. Acta Sedimentologica Sinica 33 (3), 616–624. Huanqing, C., Shuxian, L., Zhirui, S., et al., 2015d. Characteristics of conglomerate reservoir architecture of alluvial fan and its controlling effects to reservoir development: taking alluvial fan reservoir in some area of Northwest margin of Junggar Basin as an example. Journal of Jilin University (Earth science edition) 45 (1), 13–24.*

### 4.2.4 Paying attention to the practical application of study results

The ultimate goal of fine reservoir characterization is to deeply understand the static and dynamic reservoir features so as to provide a basis for designing the development plan or development adjustment plan, thereby exploiting the oilfield in a high-efficiency way and enhancing the oil recovery. Thus, it should always be noted as to what kind of problem in production practice can be solved by fine reservoir characterization, how to apply the study results in practice, and what kind of effect can be achieved when a fine reservoir characterization study is being conducted. For example, CNPC has proposed a variety of measures to enhance the development degree and effect of oilfields, such as the secondary development of old wells or the significant development test and large-scale application of horizontal drilling. Application of fine reservoir characterization study resulting in these jobs has achieved good effects, thereby providing technical support to the implementation of related measures. Currently, some applications are in progress. Huanqing et al. (2015a,b,c,d) made a statistic of the quantitative information of the development scale for different types of reservoir architectures in the space while conducting the architecture characterization of the Lower Karamay glutenite reservoir in an area at the northwestern margin of the Junggar Basin, Xinjiang Oilfield, which provides a solid basis for infilling the development well pattern for the high water-cut glutenite oil reservoir at the late development stage.

### 4.2.5 Enhancing the level of informatization by constructing the digital reservoir

Informatization is the inevitable request for social development and also the oilfield development. With the continual progress in computer technology, the informatization level of oilfield development increases. Some oilfields even put forward a slogan "Digital Oilfield." In contrast to petroleum exploration and evaluation, oilfield development study requires huge data volume. No doubt, the improvement in informatization (digitization) would significantly enhance the efficiency of oilfield development. The Daqing Oilfield has introduced the concept of Digital Oilfield by year-end 1999. This concept of the Digital Oilfield and the target to build it into oilfield development were firmly confirmed in June 2000. Since then, it has become a strategic target for the development of enterprise, as the fundamental idea was generally recognized in the industry (Zhiliang, 2011). Digital oilfield is essentially the "digitized oilfield" and a systems engineering (Fig. 4.4). Digital oilfield provides support to exploration, development, production, and surface gathering and transferring of the whole field on the basis of IT and spatial data infrastructure construction. Informatization should be ensured over the whole processes of production and operation of oilfields, such as petroleum exploration, development and production management, drilling, well logging, oil recovery engineering, downhole operation, oilfield chemistry, reservoir geology, reservoir engineering, law and regulation, economic evaluation, land, HSE, surface gathering & transferring, oilfield management, enterprise strategy, and operation. Scope of work includes (1) digital surface, digital subsurface (digital structure, digital trap, digital reservoir, and digital borehole), and digital surface facility (digital pipeline network, digital gathering and transferring, and digital oil depot); (2) digitalization during the petroleum exploration and development study; and (3) digitalization during the oilfield operating management (e.g., ERP) (Jianfeng et al., 2006).

With the improvement in computer technology and software and hardware, digital reservoir construction has been gradually conducted in China's

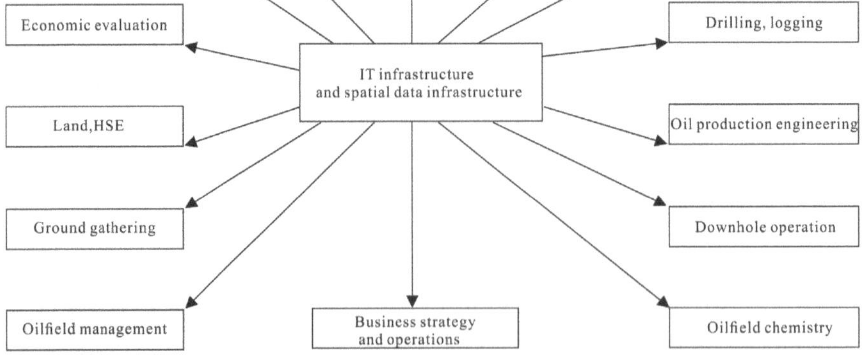


**Figure 4.4** Composition of Digital Oilfield. *From Jianfeng, L., Shuzhong, L., Zhilin, Z., 2006. Digital Oilfield. Chemical Industry Press, Beijing, pp. 11–27.*

oilfields. At present, the first digital reservoir platform has been constructed in the Xinjiang Oilfield. In addition, significant progress has been made in CNPC's Dagang and Daqing oilfields regarding digital reservoir construction. Other CNPC's fields are either carrying out or planning this work. In addition, other large oil companies in China, such as Sinopec and China National Offshore Oil Corporation, are also conducting the study of the digital reservoir. Constructing the digital reservoir provides a broad development platform for improving the informatization level of fine reservoir characterization. In CNPC's Xinjiang Oilfield, for example, the static geological data, dynamic production data, and oilfield production monitoring data that are required for fine reservoir characterization study can be stored onto one platform by constructing the digital reservoir. In this way, all the related researchers can share the data and also the study results. Furthermore, the software applications in fine reservoir characterization, such as Petrel, Discovery, Forward, Eclipse, and Geomap, can be shared on the platform, such that the software resources can be fully utilized to reduce the per head cost for purchasing the software and enhancing the efficiency of utilizing the software. Meanwhile, researchers from different departments can collaborate with one another through the platform, thereby saving the manpower and material costs and reaching the target of cost decreasing and benefit increasing.

### 4.2.6 Enhancing the quality control of study process and result

Quality management and control are key links to enterprise management. Its function is to define, implement, and realize the quality policy, quality responsibility, and quality target (Juran et al., 1979; Zhizheng, 1984; Guohong et al., 2005). Quality control has been applied more broadly to machinery, aircraft, and electronic information industries than to the petroleum industry. The author believes that the application of quality control to fine reservoir characterization study should be emphasized, particularly under the current background of low oil prices. First, the data quality should be controlled while recording it to eliminate the abnormal value, thereby providing the study with real data. Second, the quality control workflow should be set up according to the characteristics of different study contents to determine the quality control points. Lastly, quality control should be conducted when delivering the final result to ensure the precision and standardability of study results. The author who set up the workflow of quality control for the fine stratigraphic classification and correlation of the Yulou reservoir in an area of Western Sag, the Liaohe

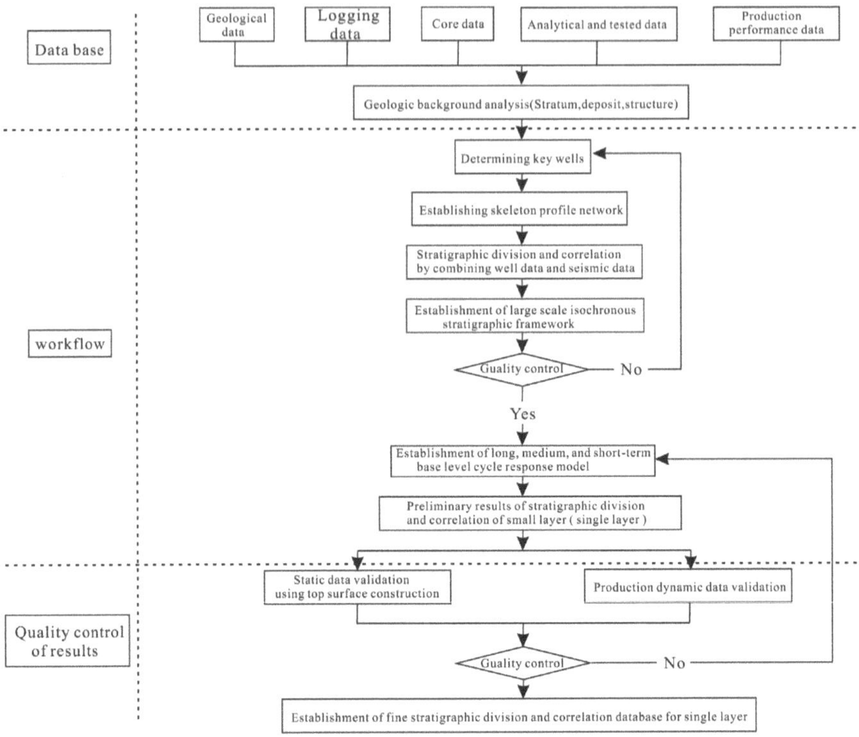


**Figure 4.5** Workflow of the quality control system for fine stratigraphic classification and correlation of the Yulou reservoir in an area of Western Sag, the Liaohe Basin.

Basin, also defined the quality control points (Fig. 4.5). Accordingly, the time cost can be saved by avoiding detours in the study and the precision of study results can be ensured. At present, the quality control method is being applied broadly by major international oil companies. In China, however, scant attention has yet been paid to this study. Researchers for fine reservoir characterization shall explore the application of quality control systems to the study to control the key points of their job roles, such that the standardability and level of study can be improved continuously.

### 4.2.7 Reducing the cost and improving the efficiency with new methods/techniques

Fine reservoir characterization study is a systematic engineering analysis that covers all aspects of reservoir geological study, as well as reservoir engineering. Accomplishing the studies mentioned above requires multidisciplinary collaborative studies, such as mathematics, chemistry, and computer technology. With the progress in these methods/techniques, fine reservoir characterization study has developed continuously. Developing new methods/techniques enables us to solve the key urgent bottlenecks in the production practice of oilfields. Meanwhile, the manpower cost can be reduced by updating the technology, and the study cost can be decreased by shortening the study cycle. Accordingly, it is possible to achieve the target of costs to decrease and benefits to increase. It is recommended to further explore the following new methods/techniques: that is, the application of ground-penetrating radar to reservoir architecture study, analysis of reservoir microscopic pore structure with constant-rate mercury intrusion test and CT scanning technique, and 4D seismic technique for monitoring the variation in the remaining oil. Attempts should be made to decrease the cost and increase the benefit by making progress in technologies and methodologies, thereby facilitating the development and progress of fine reservoir characterization study. For example, we shall strive to make a breakthrough in studying the 4D seismic monitoring technique for the purpose of providing an effective tool for studying the variation law of the remaining oil during the development of oilfields (Lideng et al., 2010). Huanqing et al. (2011a,b,c) combined the traditional coherence analysis with the ant-tracking technique to successfully characterize the Ying-1 Member volcanic reservoir in the Xudong area, the Songliao Basin. It provides an effective tool for studying the fine fracture characterization in the case where this is now well (or very few wells) drilled. The ant-tracking technique yields more precise characterization of the fracture system than the traditional coherence analysis with a low error rate. Taking $CO_2$ displacing oil and $CO_2$ sequestrated research as an example, $CO_2$ displacing oil had got good economic benefits outside China and was carried out in China in oilfield experiments today. $CO_2$ sequestrated research was in the exploring stage all over the world. The key problems in $CO_2$ displacing oil and $CO_2$ sequestrated research contained five parts, such as enlarging the sweep volume of EOR, the carrier and medium choice of $CO_2$ sequestrated research, the formation damage in the process of $CO_2$ displacing oil, air source, industrial coordination, and overall planning. The application of $CO_2$ displacing oil and $CO_2$ sequestrated research in fine reservoir description mainly uses $CO_2$ flooding to enhance oil recovery and realize partial $CO_2$ storage at the same time. This is not only of great practical value for efficient development of oil and gas reservoirs but also of great significance for realizing carbon neutralization and carbon peak, environmental protection, energy conservation, and emission reduction.

### 4.2.8 Considering the economic effectiveness carefully

Economic effectiveness shall also be considered in a fine reservoir characterization project. In the case of relatively high oil prices, the importance of economic effectiveness seems to be of less concern to researchers. As the oil price continuously drops, the evaluation and analysis of economic effectiveness become significantly important. It is well known that one of the major goals of fine reservoir characterization is to design the development plan or development adjustment plan of the oilfield in which the economic effectiveness of the development plan is a critical factor. The development plan that was feasible may become infeasible in a low oil price context since the implementation may not yield economic benefits. It is recommended to pay full attention to evaluating the economic effectiveness of the development plan when conducting a fine reservoir characterization study, and to give full consideration to the factors that may reduce the cost while selecting the methods/techniques, such that the development plan or development adjustment plan designed based on a fine reservoir characterization study remains implementable and economically feasible even in a relatively low oil price level. The current low oil price is a challenge to fine reservoir characterization study and also provides an opportunity for transition and development. As discussed above, fine reservoir characterization study is a systematic engineering project that requires multidiscipline collaborative studies, such as geology and reservoir engineering. The author believes that we shall always adhere to the principle of "cost decreasing and benefit increasing" in a low oil price context, reject the previous robust management and development mode in project setup and management, and make the planning and design in a scientific manner. The scope of work for study should be fully covered while performing the fine reservoir characterization study for a new oilfield, and the study should emphasize finding the solution to the bottlenecks that limit the production practice while performing the fine reservoir characterization for a mature oilfield. It is required to strive to reach the target of cost decreasing and benefit increasing by technological progress. Only by doing this can we realize the scientific and effective development of an oilfield in a low oil price context and meet the challenge from a low oil price situation.

# Bibliography


Adeniran A, Elshafei M, Hamada G. 2010. Artificial Intelligence Techniques in Reservoir Characterization-Functional Networks Softsensor for Formation Porosity and Water Saturation in Oil Wells. VDM Verlag Dr. Muller: USA.

Aggoun, R.C., Tiab, D., Owayed, J.F., 2006. Characterization of flow units in shaly sand reservoirs – Hassi R'mel Oil Rim, Algeria. Journal of Petroleum Science and Engineering 50, 211–226.

Aghighi, M.A., Rahman, S.S., 2010. Horizontal permeability anisotropy: effect upon the evaluation and design of primary and secondary hydraulic fracture treatments in tight gas reservoirs. Journal of Petroleum Science and Engineering 74, 4–13.

Aguilera, R., 2006. Sandstone vs. carbonate petroleum reservoirs: a global perspective on porosity-depth and porosity-permeability relationships: discussion. AAPG Bulletin 90 (5), 807–810.

Aguilera R. 2014. Flow Units: From Conventional to Tight-Gas to Shale-Gas to Tight-Oil to Shale-Oil Reservoirs. SPE165360, 17 (02): 190–208.

Ailin, J., 2010. Fine Oil Reservoir Description and Geological Modeling Techniques. Petroleum Industry Press, Beijing.

Ailin, J., Lihua, C., 2012. Fine Reservoir Description Program Method. Petroleum Industry Press, Beijing.

Aiping, F., Juan, Z., Renchao, Y., et al., 2011. Pore structure of reservoir rocks in Shanl and He8 members, the East II Block of Sulige Gas Field. Natural Gas Geoscience 22 (3), 482–487.

Al-Ghamdi, A., Chen, B., Behmanesh, H., et al., 2011. An improved triple-porosity model for evaluation of naturally fractured reservoirs. SPE132879 14 (04), 377–384.

Ali, F., Hamed, A., Nawi, D.M., et al., 2012. Inpact of reservoir heterogeneity on steam assisted gravity drainage in heavy oil fractured reservoirs. Energy Exploration & Exploition 30 (4), 553–566.

Allen, J.P., Fielding, C.R., 2007. Sedimentology and stratigraphic architecture of the Late Permian Betts Creek Beds, Queensland, Australia. Sedimentary Geology 202, 5–34.

Amro, E., Michel, D., 2001. A markov chain model for subsurface haracterization: theory and applications. Mathematical Geology 33 (5), 569–589.

Anna, X., Yuexia, D., Dakuang, H., et al., 2009. Integrated description and evaluation of reservoirs based on seismic, logging and geological data: taking Dongying formation member 1 oil reservoir of No. 1 structure, Nanpu Sag as an example. Petroleum Exploration and Development 36 (5), 541–551.

Artun, E., Mohaghegh, S., 2011. Intelligent seismic inversion workflow for high-resolution reservoir characterization. Computers & Geosciences 37, 143–157.

Baas, J.H., Van Kesteren, W., Postma, G., 2004. Deposits of depletive high-density turbidity currents: a flume analogue of bed geometry, structure and texture. Sedimentology 51, 1053–1088.

Babadagli, T., 2007. Development of mature oil fields – a review. Journal of Petroleum Science and Engineering 57, 221–246.

Babak, O., Deutsch, C.V., 2009. Improved spatial modeling by merging multiple secondary data for intrinsic collocated cokriging. Journal of Petroleum Science and Engineering 69, 93–99.

Backert, N., Ford, M., Malartre, F., 2010. Architecture and sedimentology of the Kerinitis Gilbert-type fan delta, Corinth Rift, Greece. Sedimentology 57, 543–586.

Baofeng, Z., Xiaojiang, Z., Naiqin, H., et al., 2009. Reservoir evaluation technology combining well test curve with geophysical data. Oil and Gas Well Testing 18 (5), 20–22.

Baoxi, G., Xiuqin, W., Ximei, W., 2005. Application of inter-well tracer test in determining residual oil saturation of the water flood reservoir. Well Logging Technology 29 (3), 240–243.

Bastia, R., Das, S., Radhakrishna, M., 2010. Pre- and post-collisional depositional history in the upper and middle Bengal fan and evaluation of deepwater reservoir potential along the northeast Continental Margin of India. Marine and Petroleum Geology 27, 2051–2061.

Benjamin, B., Benoît, V., Christophe, D., et al., 2014. Characterization and origin of permeabilityeporosity heterogeneity in shallow-marine carbonates: from core scale to 3D reservoir dimension (Middle Jurassic, Paris Basin, France). Marine and Petroleum Geology 57, 631–651.

Benoît, I., Simon, F., Yann, L.G., et al., 2013. Modelling of $CO_2$ injection in fluvial sedimentary heterogeneous reservoirs to assess the impact of geological heterogeneities on $CO_2$ storage capacity and performance. Energy Procedia 37, 5181–5190.

Berger, A., Gier, S., Krois, P., 2009. Porosity-preserving chlorite cements in shallow-marine volcaniclastic sandstones: evidence from cretaceous sandstones of the Sawan gas field, Pakistan. AAPG Bulletin 93 (5), 595–615.

Bhattacharya, S., Nikolaou, M., 2016. Comprehensive optimization methodology for stimulation design of low-permeability unconventional gas reservoirs. SPE Journal 6, 947–964.

Bhattacharya, S., Nikolaou, M., Economides, M.J., 2012. Unified fracture design for very low permeability reservoirs. Journal of Natural Gas Science and Engineering 9, 184–195.

Bin, W., Jianwen, C., Junmao, Z., et al., 2000. Utilization of reservoir flow unit to study remaining oil distribution in high water containing oilfield. Earth Science Frontiers (China University of Geosciences, Beijing) 7 (4), 403–410.

Bin, Y., Xiaodong, L., Guosheng, X., et al., 2010. Logging reservoir evaluation of Carboniferous gas reservoirs in Shaguanping, eastern Sichuan. Geophysical and Geochemical Exploration Calculation Technology 32 (1), 35–40.

Bin, F., Linhui, S., Lei, J., et al., 2014. Application of multiple-point geostatics method in reservoir modeling of tight sandstone gas reservoir: taking Block s48-17-64 as an example. Fault-Block Oil & Gas Field 21 (6), 726–729.

Bintao, W., Yanfang, G., Yuxiu, F., et al., 2014. Remaining oil logging evaluation methods for Lamadian Oilfield. Journal of Oil and Gas Technology 36 (7), 84–88.

Bloch, S., Lander, R.H., Bonnell, L., 2002. Anomalously high porosity and permeability in deeply buried sandstone reservoirs: origin and predictability. AAPG Bulletin 86 (2), 301–328.

Bo, G., Jianchun, G., 2013. Fracture parameter optimization of large hydraulic fracturing based on the fine geological model. Oil & Gas Geology 34 (6), 809–815.

Bonnett, N., Dalton, J., Van Dijk, P., 2006. Infill drillng on captain targets remaining oil pockets. Offshore 52–54.

Brenda, B.B., Brigette, A.M., Marjorie, A.C., et al., 2007. Reflectance spectroscopic mapping of diagenetic heterogeneities and fluid-flow pathways in the Jurassic Navajo Sandstone. AAPG Bulletin 91 (2), 173–190.

Caineng, Z., Shizhen, T., Lianhua, H., et al., 2011. Unconventional Oil and Gas Geology. Geology Press, Beijing.

Camp, W.K., 2011. Pore-throat sizes in sandstones, tight sandstones, and shales: discussion. AAPG Bulletin 95 (8), 1443–1447.

Cancan, Z., Tangyan, L., Zaitian, M., et al., 2006. Evaluation of pore structure using sphere-cylinder model. Acta Petrolei Sinica 27 (1), 92–96.

Caumon, G., 2010. Towards stochastic time-varying geological modeling. Mathematical Geosciences 42, 555–569.

Caumon, G., Collon-Drouaillet, P., Le Carlier de Veslud, C., et al., 2009. Surface-based 3D modeling of geological structures. Mathematical Geosciences 41, 927–945.
Cerepi, A., Durand, C., Brosse, E., 2002. Pore microgeometry analysis in low-resistivity sandstone reservoirs. Journal of Petroleum Science and Engineering 35, 205–232.
Chaika, C., Dvorkin, J., 2000. Porosity reduction during diagenesis of diatomaceous rocks. AAPG Bulletin 84 (8), 1173–1184.
Chalmers, G.R.L., Marc Bustin, R., 2012. Geological evaluation of halfwaye–Doige–Montney hybrid gas shale–tight gas reservoir, northeastern British Columbia. Marine and Petroleum Geology 38, 53–72.
Chao, Y., Xiaoming, X., Mei, Q., et al., 2015. Injector-producer connectivity identification and water injection optimization of mature oil fields in high water cut period. Journal of Central South University (Science and Technology) 46 (12), 4592–4601.
Chao, L., Shuiliang, L., Liwei, D., et al., 2016. Research on flow units based on high resolution stratigraphic sequence. Journal of China University of Petroleum 40 (6), 22–32.
Chaochen, Z., Wenxiang, W., 1995. Determination of Residual Oil Distribution Technology. Information Institute of CNPC.
Chaoli, L., Jun, W., Weimin, Z., et al., 2001. Architectural element analysis for alluvial deposits: principle and applicability. Geological Science and Technology Information 20 (2), 37–40.
Chaoli, L., Jinxiu, W., Minghui, Y., et al., 2008. A preliminary view about indexes of low permeable volcanic reservoir evaluation. Petroleum Geology and Recovery Efficiency 15 (6), 32–34.
Chaomo, Z., Zhenbiao, C., Zhansong, Z., et al., 2007. Researches on reservoir pore fractal structure which based on (NMR) T2 data. Journal of Oil and Gas Technology 29 (4), 80–86.
Chekani, M., Kharrat, R., 2012. An integrated reservoir characterization analysis in a carbonate reservoir: a case study. Petroleum Science and Technology 30, 1468–1485.
Cheng, C., Xinmin, S., Jun, L., 2012. Dominant flow channels of point-bar reservoirs and their control on the distribution of remaining oils. Acta Petrolei Sinica 33 (2), 257–263.
Chengbo, L., Jianhua, Z., Yufang, Y., et al., 2010. Physical properties and microscopic pore structure characteristics of low porosity and low permeability sandstone reservoir in Q4 Member of Longxi area, Songliao Basin. Geological Science 45 (4), 1170–1179.
Chenglin, Z., Menghui, L., Aimei, H., et al., 1997. Special Oil and Gas Reservoir. Petroleum Industry Press, Beijing.
Chengwen, X., Shijun, Y., Gang, W., et al., 2004. Monitoring and evaluation for remaining oil saturation in the superdeep and high temperature reservoir. Well Logging Technology 28 (1), 68–70.
Chengyan, L., 2000. Residual Oil Formation and Distribution. University of Petroleum Press, Shandong Province Dongying.
Chengyan, L., Tingbin, S., Chunmei, D., et al., 2013. Fine characterization of remaining oil based on a single sand body in the high water cut period. Acta Petrolei Sinica 34 (6), 1131–1136.
Chiyin, Y., 2004. Controlling factor of fluid heterogeneity of condensate gas reservoir of Qianmiqiao main buried hill. Natural Gas Industry 24 (11), 34–37.
Chuang, Z., Wei, S.X.W., 2011. Reservoir microscopic characteristics and displacement eficiency of $E_1f_3$ reservoir in Shanian Oilfield, Subei Basin. Geoscience 25 (1), 70–77.
Chuangang, X., 2015. Determination of underwater distributary channel width and drilled probability by multiple-point geostatistics. Fault-block Oil & Gas field 22 (2), 164–167.

Chuanqing, C., Ke, Y., Shaochun, Y., et al., 2008. Sedimentary microfacies and heterogeneity of the turbidite sand bodies: a case study in the Middle 3 Member of the Shahejie Formation in the T74 District of Shengtuo Oilfield, Dongying. Geological Journal of China Universities 14 (3), 419–425.

Chunhua, X., Jiagen, H., Xiyuan, Z., et al., 2008. The application of cased hole Formation resistivity logging in Karamay Low permeable reservoir. Journal of Southwest Petroleum University (Science & Technology Edition) 30 (4), 55–59.

Chunjie, D., Xuguang, W., Xiaodong, L., et al., 2013. Formation sensitivity of the deep section high pressure and low-permeability sandstone reservoir. Geological Science and Technology Information 32 (3), 94–99.

Chunjun, Z., Yanbin, W., 2011. Resevoir genesis and evaluation for the low porosity and permeability sandstone in the Daniudi Gasfield. Journal of Southwest Petroleum University (Science & Technology Edition) 33 (1), 49–56.

Chunlin, L., Li, L., Li, W., 2004. The mechanism of volcanic rock fracture formation of Eastern Sag of Liaohe Depression. Journal of Jilin University (Earth Science Edition) 34 (5), 46–50.

Chunning, G., Pingcang, W., Junxiang, N., et al., 2011. Scale compositions of water injection formations in ultra-low permeability reservoir and it's effect. Oilfield Chemistry 28 (1), 28–31.

Chunying, W., Huiping, H., Jihui, J., et al., 2008. Application of fuzzy math method to reservoir evaluation-taking Chang 4 + 5 Reservior of Yanchang Formation in Baiyushan Area in Ordos Basin as an example. Journal of Earth Sciences and Environment 30 (2), 156–160.

Clarkson, C.R., Qanbari, F., Williams-Kovacs, J.D., 2014. Innovative use of rate-transient analysis methods to obtain hydraulic-fracture properties for low-permeability reservoirs exhibiting multiphase flow. The Leading Edge 10, 1108–1122.

Clarkson, C.R., Qanbari, F., 2016. A semi-analytical method for forecasting wells completed in low permeability, undersaturated CBM reservoirs. Journal of Natural Gas Science and Engineering 30, 19–27.

Comunian, A., Renard, P., Straubhaar, J., 2011. 3D multiple-point statistics simulation using 2D training images. Computer & Geoscience 1–17.

Congjiao, X., Zhenliang, G., Jiangshan, 2005. Reservoir pore structure model based on microcosmic network stochastic simulation. Geological Science and Technology Information 24 (2), 97–100.

Congjun, F., Zhidong, B., Bingchun, C., 2012. Application of the single factor analysis and multifactor coupling method to he remaining oil prediction in Fuyu oilfield. Acta Petrolei Sinica 33 (3), 465–471.

Congjun, F., Qitong, S., Weicheng, S., et al., 2013. Reservoirs heterogeneity and its control on remaining il distribution of $K_1q^4$, Fuyu Oilfield. Journal of China University of Petroleum (Natural Science Edition) 37 (1), 1–7.

Crawford, B.L., Betts, P.G., Aillères, L., 2010. An aeromagnetic approach to revealing buried basement structures and their role in the Proterozoic evolution of the Wernecke Inlier, Yukon Territory, Canada. Tectonophysics 490, 28–46.

Dakuang, H., 2010. Discussions on concepts countermeasures and technical routes for the redevelopment of high water-cut oilfields. Petroleum Exploration and Development 37 (5), 583–591.

Daley, T.M., Schoenberg, M.A., Rutqvist, J., et al., 2006. Fractured reservoirs: an analysis of coupled elastodynamic and permeability changes from pore-pressure variation. Geophysics 71 (5), 033–041.

Dali, Y., Shenghe, W., Chengyan, L., et al., 2005. Reservoir heterogeneity and remaining oil distribution pattern of reef limestone reservoir in Liuhua 11-1 oilfield. Geological Science and Technology Information 24 (2), 90–96.

Dali, Y., Shenghe, W., Jianmin, L., 2007. An accurate method for anatomizing architecture of subsurface reservoir in point bar of meandering river. Acta Petrolei Sinica 28 (4), 99–103.

Dali, Y., Shenghe, W., Lifeng, W., et al., 2013. Physical simulation of remaining oil distribution controlled by reservoir architecture within point-bar of meandering river. China Sciencepaper 8 (5), 473–476.

Daniela, E.B., Claiton, M.S.S., 2010. Facies architecture and heterogeneity of the fluvial-aeolian reservoirs of the Sergi formation (Upper Jurassic), Reconcavo Basin, NE Brazil. Marine and Petroleum Geology 27, 1885–1897.

Daopin, L., 1999. Development of Low Permeability Oil Fields. Petroleum Industry Press, Beijing.

Daopin, L., 1997. Development of Low Permeability Sandstone Oilfield. Petroleum Industry Press, Beijing.

Daopin, L., 2003. Low Permeability Oilfield Efficient Development Decision Theory. Petroleum Industry Press, Beijing.

Daqing, Y., Hong, C., Yong, H., 2013. Fine structure characterization, adjustment and potential tapping for Lamadian Oilfield. Petroleum Geology and Oilfield Development in Daqing 32 (4), 48–52.

Darabi, H., Kavousi, A., Moraveji, M., et al., 2010. 3D fracture modeling in Parsi oil field using artificial intelligence tools. Journal of Petroleum Science and Engineering 71, 67–76.

Deptuck, M.E., Steffens, G.S., Barton, M., et al., 2003. Architecture and evolution of upper fan channel-belts on the Niger Deltaslope and in the Arabian Sea. Marine and Petroleum Geology 20, 649–676.

Deschamps, R., Guy, N., Preux, C., et al., 2012. Analysis of heavy oil recovery by thermal EOR in a meander belt: from geological to reservoir modeling. Oil & Gas Science and Technology 67 (6), 999–1018.

Dezhi, G., Huaimin, W., Cuiling, L., 2003. Hydrodynamic cause for residual oil stagnated in the reservior micro-structure. Petroleum Geology and Development in Daqing 22 (1), 32–34.

d'Huteau, E., Repsol, Y.P.F., Breda E., et al. 2001. Stimulation with hydraulic fracture of an Upper Cretaceous fissured tuff system in the San Jorge Basin, Argentina. In: Proceedings of the SPE Latin American and Caribbean Petroleum Engineering Conference, Buenos Aires, Argentina, pp. 1-11. Paper Number: SPE69586.

Dianxing, R., Changbing, T., et al., 2012. Multi - Condition Constrained Reservoir Geological Modeling. Petroleum Industry Press, Beijing.

Dianxing, R., Jiuning, Z., Changbing, T., et al., 2015. A solution to geological modeling and numerical simulation upscaling in complex reverse fault-block reservoirs: a case study of Yingdong-1 reservoir. China Petroleum Exploration 20 (6), 29–38.

Dong, D., Jie, C., Mingwen, Q., 1999. Types and distribution of remaining oil in fluvial reservoirs. Oil and Gas Recovery Technology 6 (3), 39–46.

Dong, S., Juan, C., Qing, Z., 2004. The implementation process of oil and gas reservoir evaluation using GIS technology. Journal of Jianghan Petroleum Institute 26 (1), 23–24.

Dong, S., Chunsheng, Z., Jing, X., 2009. Study on grey comprehensive GIS evaluation of reservoir heterogeneity. Computing Techniques for Geophysical and Geochemical Exploration 31 (1), 48–52.

Donghui, W., Zhanyang, Z., Jun, L., et al., 2014. Application of multi-point geostatistics method in lithofacies simulation of Dongsheng gas field. Petroleum Geology and Engineering 28 (3), 27–30.

Dongping, D., Jiagen, H., Yuming, L., et al., 2012. Application of multi-point geostatistics in delta front microfacies simulation. Journal of China University of Petroleum 36 (2), 22–26.

Donselaar, M.E., Geel, C.R., 2007. Facies architecture of heterolithic tidal deposits: the Holocene Holland Tidal Basin. Netherlands Journal of Geosciences - Geologie en Mijnbouw 86 (4), 389–402.

Dubost, F.X., Zheng, S.Y., Corbett, P.W.M., 2004. Analysis and numerical modelling of wireline pressure tests in thin-bedded turbidites. Journal of Petroleum Science and Engineering 45, 247–261.

Ehrenberg, S.N., Nadeau, P.H., 2005. Sandstone vs. carbonate petroleum reservoirs: a global perspective on porosity-depth and porosity-permeability relationships. AAPG Bulletin 89 (4), 435–445.

Ehrenberg, S.N., Eberli, G.P., Keramati, M., et al., 2006. Porosity-permeability relationships in interlayered limestone-dolostone reservoirs. AAPG Bulletin 90 (1), 91–114.

Ehrenberg, S.N., Nadeau, P.H., Steen, Ø., 2009. Petroleum reservoir porosity versus depth: influence of geological age. AAPG Bulletin 93 (10), 1281–1296.

Ehrenberg, S.N., Walderhaug, O., Bjørlykke, K., 2012. Carbonate porosity creation by mesogenetic dissolution: reality or illusion? AAPG Bulletin 96 (2), 217–233.

Enayati-Bidgoli, A.H., Rahimpour-Bonab, H., Mehrabi, H., 2014. Flow unit characterisation in the Permian-Triassic carbonate reservoir succession at South Pars gasfield, offshore Iran. Journal of Petroleum Geology 37 (3), 205–230.

Enayatie Bidgoli, A.H., Rahimpoure Bonab, H., 2016. A geological based reservoir zonation scheme in a sequence stratigraphic framework: a case study from the PermoeTriassic gas reservoirs, Offshore Iran. Marine and Petroleum Geology 73, 36–58.

Ercilla, G., García-Gil, S., Estrada, F., et al., 2008. High-resolution seismic stratigraphy of the Galicia Bank Region and neighbouring abyssal plains (NW Iberian continental margin). Marine Geology 249, 108–127.

Escuder-Viruete, J., Carbonell, R., Prez-Soba, C., et al., 2004. Geological, geophysical and geochemical structure of a fault zone developed in granitic rocks: implications for fault zone modeling in 3-D. Int J Earth Sci (Geol Rundsch) 93, 172–188.

Fabricius, I.L., Gommesen, L., Krogsbøll, A., et al., 2008. Chalk porosity and sonic velocity versus burial depth: influence of fluid pressure, hydrocarbons, and mineralogy. AAPG Bulletin 92 (2), 201–223.

Fan, Y., Yushan, S., Yinmin, S., et al., 2006. Using diagenetic facies to analyze heterogeneous reservoir of Donghe sandstone of east slope of Lunnan, Tarim Basin. Petroleum Exploration and Development 33 (2), 136–140.

Fang, D., Yan, L., Dongping, D., et al., 2017. Complex superimposition type sandbody characterized by multiple-point geostatisties modeling method. Complex Hydrocarbon Reservoirs 10 (1), 34–38.

Fanhao, G., Zhidong, B., Hancheng, J., et al., 2011. Study on the reservoir diagenesis heterogeneity of Chang 6 Member of Upper Triassic in Jiyuan Area of Ordos Basin. Journal of Jilin University (Earth Science Edition) 41 (3), 639–646.

Fanhui, Z., Jianchun, G., 2011. Damage and acidizing of the low permeability formation in Donghe Oilfield $C_{III}$. Oilfield Chemistry 28 (3), 296–299, 304.

Feng, Z., Xinguang, W., 2009. Numerical simulation for influence factor of pulsed neutron-neutron logging. Journal of China University of Petroleum 33 (6), 46–51.

Feng, W., Liqiang, S., Hongming, Y., et al., 2014. Research on NMR $T_2C$ cutoff of complex lithology in the West Qaidam Basin. Well Logging Technology 38 (2), 144–149.

Fengjuan, D., Xuefei, L., Huijiao, J., et al., 2012. Division of flow units of the low permeability sandstone reservoir based on Entropy-Weighting TOPSIS Method. Geological Science and Technology Information 31 (6), 124–128.

Fenglan, W., Tianzhi, W., Lijuan, L., 2004. Oil displacement effect of closed core wells before and after polymer flooding and residual oil analysis in Sazhong Area. Petroleum Geology and Oilfield Development in Daqing 23 (2), 59–60.

Fenglan, W., Zhenqiang, B., Wei, Z., 2011. Study on geological 3D reservoir architecture modeling and distribution of remaining oil of different development stage in meandering reservoir. Acta Sedimentologica Sinica 29 (3), 512–519.

Fenglian, Z., Guoyin, C., Yuqing, L., et al., 2007. Seismic property analysis in the application of the lava crack research at the area of Xudong in Songbei. Journal of Daqing Petroleum Institute 31 (2), 12–14.

Fengqi, Z., Jishui, S., Zhongyue, S., et al., 2007. A study on fracture orientation and characteristic of remnant paleomagnetization of deep-burial volcanic rocks, north of the Songliao Basin. Chinese Journal of Geophysics 50 (4), 1167–1173.

Fengxiang, Y., Shisheng, Y., Min, Z., et al., 2002. Application of laser scanning confocal microscope to the measurement of pore texture in reservoir. Acta Sedimentologica Sinica 20 (1), 75–79.

Fic, J., Pedersen, P.K., 2013. Reservoir characterization of a "tight" oil reservoir, the middle Jurassic Upper Shaunavon Member in the Whitemud and Eastbrook pools, SW Saskatchewan. Marine and Petroleum Geology 44, 41–59.

Fuqing, Y., 2005. Numerical simulation study on the influence of heterogeneity on polymer flooding effect. Journal of Oil and Gas Technology (J.JPI) 27 (2), 239–241.

Galloway, W.E., Hobday, D.K., 1983. Terrigenous Clastic Depositional Systems-Applications to Petroleum, Coal, and Uranium Exploration. Spring-Verlag, New York, Inc., New York.

Gang, X., Zhenping, H., Li, L., et al., 2007. Logging identification of complicated lithology with constrained least square algorithm. Well Logging Technology 31 (4), 354–356.

Gang, D., Chuanping, L., Shujun, Y., 2014. Prediction of the productivity for the complex lithology reservoirs in hailar Basin. Petroleum Geology and Oilfield Development in Daqing 33 (2), 158–160.

Gaofeng, W., Xiongjie, Z., Yu, Z., et al., 2015. A new screening method of low permeability reservoirs suitable for $CO_2$ flooding. Petroleum Exploration and Development 42 (3), 358–363.

Ghiasi-Freez, J., Kadkhodaie-Ilkhchi, A., Ziaii, M., 2012. Improving the accuracy of flow units prediction through two committee machine models: an example from the South Pars Gas Field, Persian Gulf Basin, Iran. Computers & Geosciences 46, 10–23.

Ghosh, A., Singh, R.K., Nangia, V., et al. 2012. Multi-dimensional NMR technique for wireline formation tester sampling optimization and perforation strategy determination in a Brown Field, India. Proceedings of the SPE Oil and Gas India Conference and Exhibition, Mumbai, India, March 2012, pp. 28-30, Paper Number: SPE-153657-MS.

González, E.F., Mukerji, T., Mavko, G., 2008. Seismic inversion combining rock physics and multiple-point geostatistics. Geophysics 73 (1), 11–21.

Gouw, M.J.P., 2007. Alluvial architecture of fluvio-deltaic successions: a review with special reference to Holocene settings. Netherlands Journal of Geosciences 83 (3), 211–227.

Grana, D., Mukerji, T., 2015. Bayesian inversion of time-lapse seismic data for the estimation of static reservoir properties and dynamic property changes. Geophysical Prospecting 63, 637–655.

Guangqing, Y., Zhongxian, C., 2005. Principles and Methods of Petroleum Reservoir Geology. China University of Geosciences Press, Wuhan, pp. 109–146.

Guo, L., 2010. Research on tapping remaining oil with residual polymer after polymer flooding. Oilfield Chemistry 27 (2), 192–195.

Guohong, S., Lichun, Y., Wenwei, G., 2005. Quality Control and Reliability Engineering Foundation. Chemical Industry Press, Beijing.

Guomin, F., Lining, M., Xinzhong, Q., 2004. Quantitative evaluation for the potentiality of remaining oil by multilevel fuzzy judgement method. Journal of Earth Sciences and Environment 26 (2), 39–41.

Guoping, L., Zhenliang, G., Zhenfeng, G., 2010. High-reality 3D digital core modeling and digital image analyzing method. Petroleum Geology and Oilfield Development in Daqing 29 (3), 101–106.

Gupta, S.D., Chatterjee, R., Farooqui, M.Y., 2012. Rock physics template (RPT) analysis of well logs and seismic data for lithology and fluid classification in Cambay Basin. International Journal of Earth Sciences (Geol Rundsch) 101, 1407–1426.

Haibo, L., Hekun, G., Shangwen, Z., et al., 2016. NMR analysis of movable remaining oil of low permeability reservoir. Journal of Southwest Petroleum University (Science & Technology Edition) 38 (1), 119–127.

Haicheng, S., Zhuhong, T., Zhebin, M., 2008. The application of NMR technique in the reconstruction of complex lithologic reservoirs. Journal of Yangtze University (Natural Sciences Edit) 5 (3), 48–50.

Haifa, T., Shimi, P., Yanchao, Z., 2006. Characteristics of pore structure and reservoir evaluation in He2 + 3 tight gas reservoir, Daniudi Gas Field. Journal of Mineralogical and Petrological Sciences 26 (3), 107–113.

Haifa, T., Shimi, P., Yanyao, S., et al., 2009. A novel method for discriminating flow units of thick proluvial fan conglomerate reservoir-Taking Badaowan Formation in Karamay oilfield as an example. Petroleum Geology & Experiment 31 (3), 307–322.

Haiping, H., Gates, I., Latter, S., 2010. Application of fluid heterogeneity in heavy oil reservoir appraisal and production. Sino-Global Energy 15 (10), 34–42.

Haiyan, L., Dali, Y., Xiujuan, Z., 2012. Characteristics of pore structure and reservoir evaluation of low permeability reservoir in Sulige gas field. Earth Science Frontiers 19 (2), 133–140.

Hampson, G.J., Sixsmith, P.J., Johnson, H.D., 2004. A sedimentological approach to refining reservoir architecture in a mature hydrocarbon province:the Brent Province, UK North Sea. Marine and Petroleum Geology 21, 457–484.

Hanlin, L., 1998. In: Yongjun, Z. (Ed.), Petroleum Mathematical Geology. Petroleum University Press, Dongying, Shandong.

Hanqing, Z., 2001. Understanding and suggestions to flow unit research. Petroleum Geology and Oilfield Development in Daqing 20 (3), 8–10.

Hanqing, Z., 2002. Research ideas of reservoir heterogeneity system, sand body internal architecture and flow unit. Petroleum Geology and Oilfield Development in Daqing 21 (6), 16–18, 43.

Hanqing, Z., 2005. High resolution sequence stratigraphy contrast compared with small layers carved up of our country. Petroleum Geology and Oilfield Development in Daqing 24 (1), 5–9, 12.

Hao, S., Weihong, W., Hong, L., et al., 2005. Well-logging evaluation for threefold fracture and pore construction of complex reservoir and its application. Journal of the University of Petroleum, China 29 (6), 23–26.

Hao, W., Dehua, L., Hongru, L., et al., 2015. Quantitative evaluation of remaining oil distribution at the high water cut stage-Taking second single layer of forth stratum of the Third section of Hetaoyuan Formation in Anpeng Area in Zhaoao Oilfield as an example. Journal of Yangtze University (Natural Science Edition) 12 (29), 66–70, 86.

Hatampour, A., Schaffie, M., Jafari, S., 2015. Hydraulic flow units, depositional facies and pore type of Kangan and Dalan Formations, South Pars Gas Field, Iran. Journal of Natural Gas Science and Engineering 23, 171–183.

He, W., et al., 1990. Development of Low Permeability Oil and Gas Fields – Volume I (Reservoir Description). Petroleum Industry Press, Beijing.

Heath, J.E., Kobos, P.H., Roach, J.D., et al., 2012. Geologic heterogeneity and economic uncertainty of subsurface carbon dioxide storage. SPE Economics & Management 4, 32–41.

Heqing, P., Yan, Z., Chengchuan, L., et al., 2017. Characteristics and controlling factors of micro-pore structure of Xu 5 reservoir in western Sichuan depression. China Petroleum Exploration 22 (4), 48–60.

Hesselinka, A.W., Weertsb, H.J.T., Berendsen, H.J.A., 2003. Alluvial architecture of the human-influenced river Rhine, The Netherlands. Sedimentary Geology 161, 229–248.

Heydari, E., 2000. Porosity loss, fluid flow, and mass transfer in limestone reservoirs: application to the upper Jurassic Smackover Formation, Mississippi. AAPG Bulletin 84 (1), 100–118.

Higgs, K.E., Funnell, R.H., Reyes, A.G., 2013. Changes in reservoir heterogeneity and quality as a response to high partial pressures of $CO_2$ in a gas reservoir, New Zealand. Marine and Petroleum Geology 48, 293–322.

Hill, E.J., Griffiths, C.M., 2008. Formal description of sedimentary architecture of analogue models for use in 2D reservoir simulation. Marine and Petroleum Geology 25, 131–141.

Hisafumi, A., Katsuaki, K., Toru, Y., Shinichi, T., 2006. Magnetotelluric resistivity modeling for 3D characterization of geothermal reservoirs in the Western side of Mt. Aso, SW Japan. Journal of Applied Geophysics 58, 296–312.

Hoffman, B.T., Kovscek, A.R., 2004. Efficiency and oil recovery mechanisms of steam injection into low permeability, hydraulically fractured reservoirs. Petroleum Science and Technology 22 (5), 537–564.

Hollis, C., Vahrenkamp, V., Tull, S., et al., 2010. Pore system characterisation in heterogeneous carbonates: an alternative approach to widely-used rock-typing methodologies. Marine and Petroleum Geology 27, 772–793.

Honghua, L., Mingda, R., Jincheng, L., et al., 2006. Application of Q-model principal factor analysis and clustering method in evaluation on sandstone reservoirs of the Neogene in Huatugou Oilfield, Qaidam Basin. Acta Scientiarum Naturalium Universitatis Pekinensis 42 (6), 740–745.

Hongkai, L., Zhijiang, K., 2015. Differential modeling of dissolved vugs in carbonate fracture and vug reservoirs. Special Oil and Gas Reservoirs 22 (5), 50–54.

Hongming, T., Jingsong, Z., Zhong, C., et al., 2000. Effects of steam displacing oil to reservoir pore structure and mineral components. Journal of Southwest Petroleum Institute 22 (2), 11–14.

Hongwen, D., Hongliang, W., Xiaomeng, L., 1997. Application of high-resolution sequence stratigraphic correlation to fluvial facies. Oil &Gas Geology 18 (2), 90–95,114.

Hongwen, D., Hongliang, W., Yongjun, Z., et al., 2002. High Resolution Sequence Stratigraphy - Principles and Applications. Geology Press, Beijing.

Hongxian, L., Changfu, X., Jianhua, Q., et al., 2010. Pore structure and displacing oil rate of conglomerate accumulation. Journal of Oil and Gas Technology 32 (4), 189–191.

Hu, W., Wei, S., Le, Q., et al., 2011. Ticro-pore structure characteristics and its influence on gas well production performance in the northern part of sandstone reservoir in Jingbian Gasfield. Geological Science and Technology Information 30 (2), 85–90.

Huaiqiang, R., Jinyang, L., Shaochun, Y., et al., 2008. Sandstone reservoir microscopic characteristics of braid delta of Hongtai area in Tuha Basin. Journal of China University of Petroleum 32 (5), 12–17.

Huaiyou, J., Pingping, S., Cunzhang, Y., et al., 2008. Present situation and prospect of $CO_2$ storage in world oil industrycompiled Foreign Oil Field Engineering 24 (7), 50–54.

Huaiyou, J., Binshan, J., Liangji, J., et al., 2011. Current situation and prospect of global igneous reservoir exploration and development. Special Oil and Gas Reservoirs 18 (2), 1–6.

Huanqing, C. 2006. Fine reservoir description of Chang2 reservoir of Yanchang formation in the first part of Dalugou Area of Jing'an Oilfield. Xi'an in Shann'xi province, Department of geology, Northwestern University.

Huanqing, C., 2016. Reflections on oilfield exploitation and research in the context of low oil prices. Journal of Southwest Petroleum University (Social Sciences Edition) 18 (3), 19–26.

Huanqing, C., 2019. Flow unit characteristics of fan delta front deposits and its influence on reservoir development - Taking yulou oil bearing sets in some experimental area

in west depression in Liaohe Basin in China as an example. Journal of Petroleum Science and Engineering 179, 44–55.
Huanqing, C., Xiaomin, Z., 2008. Microfacies modeling in fine reservoir description. Geological Science and Technology Information 27 (2), 73–79.
Huanqing, C., Yushuang, Z., Qingyin, L., et al., 2006. Sedimentary microfacies in Chang 6 oil-bearing formation of Yanchang formation of Xinghe district in Ansai Oilfield. Journal of Northwest University (Natural Science Edition) 36 (2), 295–300.
Huanqing, C., Xiaomin, Z., Yanlei, D., et al., 2009. Sequence stratigraphy analysis and lithologic-stratigraphic reservoir predication in deepwater faulted basins- an example from the Paleogene T Formation of the deepwater C Basin in South China Sea. Oil & Gas Geology 30 (5), 626–634.
Huanqing, C., Yongle, H., Lin, Y., et al., 2010. Advances in the study of reservoir flow unit. Acta Geoscientica Sinica 31 (6), 875–884.
Huanqing, C., Yongle, H., Jiuqiang, J., et al., 2011a. Multiple sets of information synthesized to describe fractures of volcanic reservoir: taking volcanic reservoir of the Member1 of Yingcheng Formation in Xudong Area of Xushen Gas Field as an example. Earth Science Frontiers 18 (2), 294–303.
Huanqing, C., Yongle, H., Jiuqiang, J., et al., 2011b. Researches on flow unit of the volcanic reservoir in 1st member of lower Cretaceous Yingcheng Formation, Xudong area, Songliao Basin. Geology in China 38 (6), 1430–1439.
Huanqing, C., Yongle, H., Jiuqiang, J., et al., 2011c. Microcosmic heterogeneity characteristics of volcanic gas reservoir of the first member of Yingcheng Formation in Xudong Area. Fault-Block Oil & Gas Field 18 (1), 9–13.
Huanqing, C., Yongle, H., Changbing, T., 2012a. Advances in $CO_2$ displacing oil and $CO_2$ sequestrated researches. Oilfield Chemistry 29 (1), 116–121.
Huanqing, C., Yongle, H., Qiquan, R., Jing, Z., 2012b. Characteristics of the accumulation space (as) in No. 1 Member volcanic reservoir of Yingcheng Formation in Xudong Area of Songliao Basin. Petroleum Geology and Oilfield Development in Daqing 31 (3), 7–12.
Huanqing, C., Yongle, H., Yingcheng, Z., et al., 2012c. Research advances in geology of volcanic reservoir. Fault-block Oil & Gas Field 19 (1), 75–79.
Huanqing, C., Chen, C., Shuxian, L., Jiahong, L., 2013a. Research advances on reservoir pores. Natural Gas Geoscience 24 (2), 227–237.
Huanqing, C., Chunyan, L., Jing, Z., et al., 2013b. Research advances of reservoir diageneses. Petroleum Geology and Oilfield Development in Daqing 32 (2), 1–9.
Huanqing, C., Ping, J., Danfeng, Z., Jiahong, L., 2013c. Pore structure of volcanic reservoir classified and distribution characteristics evaluated: Taking Member 1 of Yingcheng Formation in Xudong Area in Songliao Basin as an example. Journal of Central South University (Science and Technology) 44 (4), 1453–1463.
Huanqing, C., Yingcheng, Z., Zhirui, S., et al., 2013d. Research advances of reservoir architecture. Special Oil and Gas Reservoirs 20 (5), 7–13.
Huanqing, C., Yingcheng, Z., Shuqing, L., et al., 2014a. Significance of lithology analyses on tje researches of conglomerate reservoir architecture of alluvial fan sediments: taking alluvial fan reservoir in some area of Northwest Margin of Junggar Basin as an example. Natural Gas Geoscience 25 (5), 721–731.
Huanqing, C., Yingcheng, Z., Xingjun, G., et al., 2014b. High-resolution sequence stratigraphy and its application in the fine-scale stratigraphic correlation of the Yulou reservoir in the West Depression of the Liaohe Basin. Journal of Stratigraphy 38 (3), 317–323.
Huanqing, C., Chao, D., Yijing, D., Jue, W., 2015a. Advances of reservoir evaluation researches. Geological Science and Technology Information 34 (5), 66–74.
Huanqing, C., Chengfang, S., Chen, C., 2015b. Advances in the study of fine reservoir description. Geological Review 61 (5), 1135–1146.

Huanqing, C., Shuxian, L., Peng, J., et al., 2015c. Flow barrier characteristics of thermal rescovery of heavy oil reservoir: taking Yulou oil-bearing sets in some experimental area in the south end of westward slope of West depression in Liaohe Basin as an example. Acta Sedimentologica Sinica 33 (3), 616–624.
Huanqing, C., Shuxian, L., Zhirui, S., et al., 2015d. Caracteristics of conglomerate reservoir architecture of alluvial fan and its controlling effects to reservoir development: taking alluvial fan reservoir in some area of Northwest margin of Junggar Basin as an example. Journal of Jilin University (Earth science edition) 45 (1), 13–24.
Huanqing, C., Chengfang, S., Chen, C., 2016a. Problems with the fine description of reservoirs. Petroleum Geology & Experiment 38 (5), 569–576.
Huanqing, C., Yijing, D., Jue, W., 2016b. Pore structure characteristics and its influence to reservoir development: taking Yulou Oil Bearing Sets in some experimental area in South End of Westward Slope of West Depression in Liaohe Basin as an example. Science Technology and Engineering 32 (16), 28–35.
Huanqing, C., Yongle, H., Chengfang, S., 2016c. The Single Element Characterization and Quantitative Evaluation of Volcanic Rocks in Songliao Basin. Petroleum Industry Press, Beijing.
Huanqing, C., Yongle, H., Lin, Y., et al., 2016d. Comprehensive quantitative evaluation of Yingcheng volcanic reservoirs in Xudong. Special Oil and Gas Reservoir 23 (1), 21–24.
Huanqing, C., Chengfang, S., Hiayan, H., et al., 2017a. Reflection and countermeasure of fine reservoir description in low oil price. Geological Science and Technology Information 36 (5), 85–91.
Huanqing, C., Jue, W., Liping, Y., et al., 2017b. Heterogeneity and its effects to oilfield development studied of thermal recovery of heavy oil reservoir: taking Yulou oil bearing sets in some experimental area in south end of westward slope of west depression in Liaohe Basin as an example. Chinese Journal of Geology 52 (3), 998–1009.
Huanzhi, Z., Yanqing, H., Maoxin, Q., et al., 2015. Impact of low oil price on tight oil development. Petroleum Science and Technology Forum 3, 68–71.
Hui, Q., Zhizhang, W., Haiming, L., et al., 2013. Comparative research of two modified muhiple point geostatistics methods. Complex Hydrocarbon Reservoirs 6 (3), 10–14.
Huili, X., Liqun, W., Yangquan, J., et al., 2016. The quantitative evaluation index system for uranium reservoir heterogeneity in Hantaimiao Region, Ordos Basin. Earth Science 41 (2), 279–292.
Hujun, Z., Yajun, L., Hui, Y., et al., 2007. Application of Isotopic Tracer to Study of Residual oil Distribution – an exampie of M pool in Yumen Laojunmiao oilfield. Xinjiang Petroleum Geology 28 (1), 97–100.
Hynes, G.F., Dixon, J.M., 2005. Geological mapping and analogue modeling of the Liard, Kotaneelee and Tlogotsho ranges, Northwest Territories. Bulletin of Candian Petroleum Geology 53 (1), 67–83.
Isha, S., Roland, N.H., 2005. Multiresolution wavelet analysis for improved reservoir description. SPE Reservoir Evaluation & Engineering 8, 53–69.
Jackson, M.D., Valvatne, P.H., Blunt, M.J., 2003. Prediction of wettability variation and its impact on flow using pore- to reservoir-scale simulations. Journal of Petroleum Science and Engineering 39, 231–246.
Jamaloei, B.Y., Kharrat, R., Asghari, K., et al., 2011. The influence of pore wettability on the microstructure of residual oil in surfactant-enhanced water flooding in heavy oil reservoirs: implications for pore-scale flow characterization. Journal of Petroleum Science and Engineering 77, 121–134.
Jazi, R.S., Eckstein, Y., 2015. Simulation of groundwater flow system in alluvium and fractured weathered bedrock zone: sand Lick watershed, Boone County, West Virginia, USA. Environmental Earth Sciences 74, 2247–2258.

Jeff, C., Chen, J., Cheng, G., 2014. translated by Sirina Petroleum Geology Statistics. Petroleum Industry Press, Beijing.
Jeff, C. indigenous, Cheng, G., Li, X., et al., 2016. translation Uncertainty Modeling in Earth Science. Petroleum Industry Press, Beijing.
Jerry, L.F., 2007. Carbonate reservoir characterization-An integrated approach, second ed. Springer-Verlag Berlin Heidelberg, Berlin.
Jesús, O., Jeannette, W., Michael, H.G., 2013. Recognition criteria for distinguishing between hemipelagic and pelagic mudrocks in the characterization of deep-water reservoir heterogeneity. AAPG Bulletin 97 (10), 1785–1803.
Jia, L., Ross, C.M., Kovscek, A.R., 2007. A Pore-network-modeling approach to predict petrophysical properties of diatomaceous reservoir rock. SPE Reservoir Evaluation & Engineering 10, 597–608.
Jiafan, Z., Xiaohong, C., Qin, Z., 2003. Application of grey association analysis in reservoir evaluation. Progress in Exploration Geophysics 26 (4), 282–286.
Jiagen, H., Yuming, L., Fang, X., et al., 2008. Architecture of braided fluvial sandbody and origin for petroliferous diference of the Guantao Formation of Neogene in Kongdian Oilfield of Huanghua Depression. Journal of Palaeogeography 10 (5), 459–464.
Jiahua, W., Hiamao, Y., 2012. Research on multiple-point geostatistics modeling. Petrochemical Industry Application 31 (10), 72–74.
Jiahua, W., Jian, H., 2007. Research on multipoint geostatistics reservoir modeling method based on workflow technology. Journal of Xi'an Shiyou University (Natural Science Edition) 22 (6), 101–103, 109.
Jiahua, W., Tao, C., 2013. Research on reservoir facies modeling using multiple-point geostatistics. Petrochemical Industry Application 32 (8), 57–59.
Jiahua, W., Tuanfeng, Z., 2001. indigenous Stochastic Modeling of Oil and Gas Reservoirs. Petroleum Industry Press, Beijing.
Jiahua, W., Wei, Z., 2011. Study of geostatistical modeling method based on seismic constraint. Tuha Oil & Gas 16 (1), 1–5.
Jiajia, Z., Hongbing, L., Fengchang, Y., 2012. A Geophysical method for the identification and evaluation of oil shale. Acta Petrolei Sinica 33 (4), 625–632.
Jianchao, K., Guosheng, X., Yuncheng, W., et al., 2001. Single-well modeling for fracture and productivity prediction of tight clastic reservoir: a case study of Shaximiao Formation in XZ Gas Field. Mineral Rock 21 (2), 62–67.
Jiandong, W., Jiyu, L., Runtao, Y., et al., 2003. Applications of hierarchy analysis method in reservoir evaluation. Journal of Daqing Petroleum Institute 27 (3), 12–14.
Jiandong, S., Jinshan, H., Kuntai, Q., et al., 2013. To improve the effect of polymer flooding process control technology-Taking H3ⅣI-3 stratum series in North Block in Shuanghe Oilfield in Henan Oil Region for an example. Petroleum Geology and Recovery Efficiency 20 (2), 91–94, 98.
Jianfeng, L., Shuzhong, L., Zhilin, Z., 2006. Digital Oilfield. Chemical Industry Press, Beijing, pp. 11–27.
Jianfeng, W., Mingshun, Z., Zhongwen, H., et al., 2008. Interpretation and evaluation techniques of complex lithologic reservoirs in erlian basin. China Petroleum Exploration 71–73, 89. second phase.
Jiang, L., Guangju, L., Jiangyu, L., et al., 2013. High-watercut-late-stage interperetation difficulties and direction of water-flooded-layer loggings. Petroleum Geology and Oilfield Development in Daqing 32 (3), 126–130.
Jiang, Y., Yan, L., Yili, S., et al. 2014. Variation Characteristics of Reservoir Dominant Channel Before and after Water Flooding Shuanghe Oilfield VII Lower Series as an example. 23(14): 168-173.

Jianjiang, W., Jianguo, G., 2013. Application of boron (gadolinium) – Neutron lifetime logging technology in oilfield development. Xinjiang Petroleum Technology 2 (23), 17–21.
Jianmin, L., Shouyu, X., 2003. Reservoir sedimentary model of fluvial facies and it's control to remaining oil distribution. Acta Petrolei Sinica 24 (1), 58–62.
Jianqiong, C., Huifang, Y., Zhihong, Z., et al., 2006. Cases Well-Chosen on Statistical Analysis of SPSS. Tsinghua University Press, Beijing.
Jie, W., 1996. Reservoir Description Technology – Exploration Stage. Petroleum Industry Press, Beijing.
Jie, L., Yanyi, T., 2004. Research on microscopic remaining oil distribution after polymer flooding by fluorescence analysis of polished core section-taking Pu I member in Daqing oilfield as example. Petroleum Geology and Recovery Efficiency 11 (6), 64–66.
Jie, L., Ya, L., Congcong, Z., 2009. Effect of pores structure parameter on polymer flooding recovery-applying 3D porosity network model. Petroleum Geology and Oilfield Development in Daqing 28 (3), 110–115.
Jihong, L., Zhihao, Q., Qinghua, C., 2001. Effect of injecting water development on microstructure of reservoirs in Gudao Oilfield. Petroleum Geology & Experiment 23 (4), 424–428.
Jin, L., Guiwen, W., Yu, C., et al., 2014. Mechanism analysis and quantitative assessment of pore structure for tight sandstone reservoir: an example from Chang 8 Oil Layer in the Jiyuan Area of Ordos Basin. Acta Geologica Sinica 88 (11), 2119–2130.
Jin, L., Minghua, L., Libin, G., et al., 2016. Methodology of reservoir prediction in areas with complicated lithology in saline lake basins: a case study of the Qianjiang Sag. Jianghan Basin 38 (2), 273–277.
Jing, F., Shenghe, W., Jinhua, F., et al., 2013. Research on quantitative diagenetic facies of the Yanchang Formation in Longdong Area, Ordos Basin. Earth Science Frontiers 20 (2), 086–097.
Jing, Y., Wentao, C., Yunfeng, G., et al., 2016. Microstructure fine characterization and its application in remaining oil prediction. Science Technology and Engineering 16 (3), 143–148, 161.
Jingang, H., Yili, K., Lijun, Y., et al., 2013. Damage evaluation of fracturing fluids on sandstone reservoir with extra-low permeability. Oilfield Chemistry 30 (2), 173–178.
Jinglong, L., Jing, N., Pengju, L., et al., 2009. Micropore structure type identification based on BP neural network. Well Logging Technology 33 (4), 355–359.
Jingshan, C., Yan, Z., Jun, P., et al., 2007. Genetic type and intrafomational heterogeneity pattern of low-porosity and low-permeability sandbodies in Fuxian Exploration Area. Acta Sedimentologica Sinica 25 (1), 53–58.
Jingwei, C., Rukai, Z., Songtao, W., et al., 2013. Heterogeneity and lower oily limits for tight sandstones: a case study on Chang-7 oil layers of the Yanchang Formation, Ordos Basin. Acta Petrolei Sinica 34 (5), 877–882.
Jingyao, W., Jiaming, Z., 2011. Change laws of rock physical properties in $CO_2$ oil flooding. Petroleum Geology and Oilfield Development in Daqing 30 (4), 141–143.
Jinliang, Z., Jun, X., 2011. Petroleum Development Geology. Petroleum Industry Press, Beijing.
Jinying, Z., Biwen, G., Wen, L., 2010. Application of multipoint geostatistics in coastal reservoir modeling. Journal of Southwest Petroleum University (Science&Technology Edition) 32 (6), 70–74.
Jiyu, L., Jingbo, H., Wanquan, Y., et al., 1998. Studying method and its significance of flow units. Journal of Daqing Petroleum Institute 22 (1), 5–7.
Jiyu, L., Zhichun, P., Xiaobo, G., 2005. Applied of grey correlation analysis on reservoir evaluated-taking Beier Area in Saertu Oilfield in Daqing as an example. Petroleum Geology and Recovery Efficiency 12 (2), 13–15.

Jo, H.R., Chough, S.K., 2001. Architectural analysis of fluvial sequences in the northwestern part of Kyongsang Basin (Early Cretaceous), SE Korea. Sedimentary Geology 144, 307–334.

Jones, G.D., Xiao, Y., 2005. Dolomitization, anhydrite cementation, and porosity evolution in a reflux system: insights from reactive transport models. AAPG Bulletin 89 (5), 577–601.

Jun, X., Jinliang, Z., 2003. Description and Prediction of Remaining Oil. Petroleum Industry Press, Beijing.

Jun, X., Jinliang, Z., 2004. Characteristics of sand body architecture and distribution pattern of remaining oil: a case study of Kumkol South oilfield in South Turgay Basin, Kazakhstan. Journal of Shandong University of Science and Technology (Natural Science edition) 23 (1), 108–111.

Jun, T., Qi, W., Xiaofeng, M., et al., 2012. Application of Q-type cluster analysis and discriminatory analysis in reservoir evaluation-taking reservoir Chang 81 of Jiyuan Area in Ordos Basin as an example. Special Oil and Gas Reservoirs 19 (6), 28–31.

Jun, C., Hengrong, Z., Shaojun, Z., et al., 2016. Joint inversion method of electromagnetic wave resistivity logging while drilling and its application. Acta Petrolei Sinica 37 (3), 371–381.

Junhua, Z., 2012. was an editor Fault Block, Fractured Reservoir Seismic Fine Description Technology. China University of Petroleum Press, Dongying.

Junhua, L., Yunzhi, J., Shiyu, G., 2008. Application of the optimization method to complex lithology reservoir log interpretation. Well Logging Technology 2 (6), 542–561.

Junlong, Z., Juntao, C., Chuanhao, L., et al., 2013. A Review and perspective of logging evaluation technology of oil remaining in middle and high water cut stage. Progress in Geophysics 28 (2), 838–845.

Junsheng, D., Jianchun, X., Zhaoping, M., et al., 2003. Predication of volcanic rock fissure with finite deformation method. Journal of the Universitu of Petroleum, China (Edition of Natural Science) 27 (1), 1–3, 10.

Junwei, T., Ailin, J., Dongbo, H., et al., 2006. Development technologies for the Sulige gas field with low permeability and strong heterogeneity. Petroleum Exploration and Development 33 (1), 107–110.

Junwei, B., Juxing, Z., Songhua, N., et al., 2014. Heterogeneity of shale in Damuntun and its control on shale oil distribution. Special Oil and Gas Reservoirs 21 (5), 29–33.

Kadkhodaie-Ilkhchi, R., Rezaee, R., Moussavi-Harami, R., et al., 2013. Analysis of the reservoir electrofacies in the framework of hydraulic flow units in the Whicher Range Field, Perth Basin, Western Australia. Journal of Petroleum Science and Engineering 111, 106–120.

Kaixuan, N., Zhengyuan, L., Huaizhong, W., et al., 2007. Application of rock flexural strength in prediction of fractural vulcanite reservoir. Geophysical Petroleum Exploration 42 (2), 186–189.

Kaoping, S., Zhao, Y., Zhihua, S., et al., 2004. Influence factors of microscopic distribution of polymer flooding residual oil. Journal of Daqing Petroleum University 28 (2), 25–27.

Kato, A., Onozuka, S., Nakayama, T., 2008. Elastic property changes in a bitumen reservoir during steam injection. The Leading Edge 27, 1124–1131.

Keke, L., Jiagen, H., Yuming, L., et al., 2016. Application of multiple-point geostatistics in 3D internal architecture modeling of point bar. Oil & Gas Geology 37 (4), 577–583.

Keqi, L., Junjie, X., Xifeng, Y., 2005. Application of "weight method" for reservoir evaluation in s4of Weicheng81block in Dongpu Sag. Special Oil and Gas Reservoirs 12 (1), 46–48,55.

Kewen, W., Ning, L., 2009. Numerical simulation on effects of reservoir characteristics and saturation on T2 spectra of nuclear magnetic resonance. Acta Petrolei Sinica 30 (3), 422–426.

Keyao, J., Yan, H., Li, M., et al., 2015. The upstream effect of current low oil prices and suggestions. International Petroleum Economics 23 (4), 59–62.
Khidir, A., Catuneanu, O., 2010. Reservoir characterization of Scollard-age fluvial sandstones, Alberta foredeep. Marine and Petroleum Geology 27, 2037–2050.
Kun, X., Guangwen, H., Shihong, W., et al., 2015. Development strategy of China's petroleum enterprises under background of low oil price. Foreign Oil & Gas Information 4, 1–4.
Lake, L.W., Carroll Jr., H.B., 1986. Reservoir Characterization. Academic Press, INC, Orlando, FL.
Lake, L.W., Hartmann, D., Hewett, T.A., et al., 1989. SPE Reprint Series No. 27 Reservoir Characterization-1. Society of Petroleum Engineers, Richardson, TX, Printed in U.S.A. 1-55563-022-7.
Lanxiong, Y., 2011. Reservoir changing in process of SAGD development and its influences in Ng petroleum pool in Liaohe Basin. Special Oil and Gas Reservoirs 18 (6), 86–88.
Lars, H., Petter, M., Bjørn, F.N., et al., 2003. Stochastic structural modeling. Mathematical Geology 35 (8), 899–914.
Le Ravalec-Dupin, M., Hu, L.Y., Roggero, F., 2008. Reconstruction of existing reservoir model for fitting production dynamic data. Earth Science Frontiers 15 (1), 176–186.
Lei, G., Lianbo, Z., Fengbin, M., et al., 2012. Application of fractal geometry method in complex fracture system description. Journal of Hunan University of Science and Technology (Natural Science Edition) 27 (4), 6–10.
Leite, E.P., Vidal, A.C., 2011. 3D porosity prediction from seismic inversion and neural networks. Computers &Geosciences 37, 1174–1180.
Lemiski, R.T., Hovikoski Dr., J., Pemberton Dr., S.G., et al., 2011. Sedimentological, ichnological and reservoir characteristics of the low-permeability, gas-charged Alderson Member (Hatton gas field, southwest Saskatchewan): implications for resource development. Bulletin of Canadian Petroleum Geology 59 (1), 27–53.
Leroy, P., Revil, A., Altmann, S., et al., 2007. Modeling the composition of the pore water in a clay-rock geological formation (Callovo-Oxfordian, France). Geochimica et Cosmochimica Acta 71, 1087–1097.
Leyuan, F., Xiaomin, Z., Kun, S., et al., 2005. High-resolution sequence-stratigraphic framework and its control on the reservoir heterogeneities of Bdagang Tectonic Belt in Huanghua Sag. Journal of Stratigraphy 29 (4), 355–361,367.
Li, L., Wenyan, X., Lijuan, J., et al., 2003. An approach to genesis of Cenozoic volcanic fractures in eastern sag of Liaohe basin. Special Oil and Gas Reservoirs 10 (1), 18–21.
Li, Z., Jianmeng, S., Zhiqiang, S., et al., 2012. Application of multiple-point geostatistics in 3D pore structure model reconstruction. Journal of China University of Petroleum 36 (2), 105–109.
Li, L., Caers, J., Sava, P., 2015. Assessing seismic uncertainty via geostatistical velocity-model perturbation and image registration: an application to subsalt imaging. The Leading Edge 9, 1064–1070.
Liao, M., Cai, Z., 2000. Founding of distribution forecasted model of reservoir pore throat volume. Journal of Southwest Petroleum Institute 22 (2), 8–10, 53.
Lianbo, Z., 2008. Formation and Distribution of Fractures in Low Permeability Sandstone Reservoirs. Science Press, Beijing.
Lianbo, Z., Jiyong, Z., Shengju, Z., et al., 2008. Study on the effect of rock heterogeneity on fracture development. Progress in Natural Science 18 (2), 216–220.
Lianbo, Z., Shizhen, K., Yang, L., 2010. compiled Research Methods of Low Permeability Reservoir Fracture. Petroleum Industry Press, Beijing.
Lichun, K., Zhongchun, S., Min, O., et al., 2013. Complication lithology logging identification of the Lucaogou Tight Oil reservoir in Jimusaer Depression. Well Logging Technology 37 (6), 638–642.

Lideng, G., Caineng, Z., Fengchang, Y., et al., 2010. Practical Four – Dimensional Seismic Monitoring Technology. Petroleum Industry Press, Beijing.
Lideng, G., Xiaofeng, D., Xin, Z., et al., 2012. Key technologies for the seismic reservoir characterization of high water-cut oilfields. Petroleum Exploration and Development 39 (3), 365–377.
Lifeng, W., Guangke, M., Jiaxiong, Z., et al., 2009. Reservoir heterogeneity description of Dongfang gas field. Natural Gas Industry 29 (1), 38–40.
Lihong, Z., Qinglong, D., Xueyan, J., et al., 2015. Characteristics and strategies of three major contradictions for continental facies multi-layered sandstone reservoir at ultra-high water cut stage. Acta Petrolei Sinica 36 (2), 210–216.
Lihua, C., Jiahua, W., Ying, L., 2000. Researches Technique of Petroleum Reservoir. Petroleum Industry Press, pp. 198–206.
Lihui, G., Jiagen, H., Yupeng, L., et al., 2015. Application of DS-MPS algorithm in the modeling of reservoir sedimentary facies. Petroleum Geology and Oilfield Development in Daqing 34 (1), 24–29.
Lijuan, N., 2014. Effect of pressure sensitivity on the elastic productivity of low permeability reservoir. Science Technology and Engineering 14 (3), 137–140.
Lijun, W., Xinjun, C., 2003. Influence of reservoir heterogeneity on the distribution of remaining oil. Geological Science and Technology Information 22 (2), 71–73.
Likun, Z., 2013. Application of PNN Logging in well completion data missing evaluation. Petroleum Instruments 27 (1), 52–54.
Lili, L., Zhongping, Z., Liang, L., et al., 2008. Application of the variable scale fractal technique in fracture prediction and reservoir evaluation. Oil & Gas Geology 29 (1), 31–37.
Lillah, M., Boisvert, J.B., 2013. Stochastic distance based geological boundary modeling with curvilinear features. Mathematical Geosciences 45, 651–665.
Lin, Z., Ximin, Z., Shiyan, H., 2013. Research on the Core Issues of Low Permeability Reservoir Description. Petroleum Industry Press, Beijing.
Lingyun, Z., Binggao, X., 2009. Quantitative evaluation of tight reservoir in Baise Basin by analytic hierarchy process. Foreign Logging Technology 46–48. Total issue 172.
Linyu, L., Qing, C., Yiqun, L., et al., 2006. Chang 81 sandstone diagenesis and the lnfluence on Chang 81 reservoir of south area of Baima. Ordos Basin. Acta Geologica Sinica 80 (5), 712–717.
Linyu, L., Zhenliang, W., Yiqun, L., et al., 2008. Micro-anisotropic experimental analysis on Chang-8 sandstone in Xifeng Area of Ordos Basin. Rock and Mineral Analysis 27 (1), 29–32,36.
Linyu, L., Jinzhou, C., Zhenliang, W., et al., 2009. The microanisotropy study of Chang3 reservoirs in Baibao area. Journal of Northwest University (Natural Science Edition) 39 (2), 269–272.
Liqiang, S., 2002. Logging Geology Application Technology. Petroleum Industry Press, Beijing.
Liqiang, Z., Youliang, J., Wenjie, M., et al., 1998. Fractal geometry characteristics and reservoir evaluation of sandstone pore structure in piedmont zone of Bogda. Journal of Petroleum University (Natural Science Edition) 22 (5), 31–33.
Lixin, L., 2006. The technology of reservoir geological model and the meaning role in reservoir description. Journal of Northwest University (Natural Science Edition) 36 (6), 988–991.
Longxin, M., Ailin, J., 2003. Study on Fan Delta Sedimentary Reservoir Model and Prediction Method. Petroleum Industry Press, Beijing.
Longxin, M., Yinan, Q., 1999. Description of Reservoir in Different Development Stages. Petroleum Industry Press., Beijing.
Longxin, M., Ailin, J., Liang, C., et al., 2000. Fine Reservoir Research Methods - Outcrop Reservoir and Modern Sedimentary and Fine Geological Modeling at Home and Abroad. Petroleum Industry Press, Beijing, pp. 51–73.

Longxin, M., Liqing, Z., Xiaowu, Z., et al., 2006. Fine Reservoir Description and Integration Technology. Petroleum Industry Press, Beijing.
Lopez Jimenez, B.A., Aguilera, R., 2016. Flow units in shale condensate reservoirs. SPE Reservoir Evaluation & Engineering 19, 1–16.
Luo, Z., Cui, B., Huang, S., et al., 1991. The theoretical approach and the case application of the control of reservoir evaluation for the clastic reservoir by means of clay minerals. Journal of Chengdu College of Geology 18 (3), 1–12.
Ma, Q., Junjie, E., Li, W., et al., 2000. Reservoir evaluation of deep-seated igneous rocks in Beipu Region, Huanghua Depression. Oil & Gas Geology 21 (4), 337–340,344.
MacBeth, C., Stammeijer, J., Omerod, M., 2006. Seismic monitoring of pressure depletion evaluated for a United Kingdom continental-shelf gas reservoir. Geophysical Prospecting 54, 29–47.
Maesano, F.E., Toscani, G., Burrato, P., et al., 2013. Deriving thrust fault slip rates from geological modeling: examples from the Marche coastal and offshore contraction belt, Northern Apennines, Italy. Marine and Petroleum Geology 42, 122–134.
Mahjour, S.K., Al-Askari, M.K.G., Masihi, M., 2016. Flow-units verification, using statistical zonation and application of Stratigraphic Modified Lorenz Plot in Tabnak gas field. Egyptian Journal of Petroleum 25, 215–220.
Mahsanam, M., Diganta, B.D., 2007. Dynamic effects in capillary pressure-saturations relationships for two-phase flowin 3Dporous media: implications of micro-heterogeneities. Chemical Engineering Science 62, 1927–1947.
Mallon, A.J., Swarbrick, R.E., 2008. Diagenetic characteristics of low permeability, non-reservoir chalks from the Central North Sea. Marine and Petroleum Geology 25, 1097–1108.
Manichand, R., Mogollón, J.L., et al. 2010. Preliminary assessment of tambaredjo heavy oilfield polymer flooding pilot test. In: Proceedings of the SPE Latin American and Caribbean Petroleum Engineering Conference, Lima, Peru, December 2010, Paper Number: SPE-138728-MS,:1-3.
Maolei, C., Yunhong, D., Chengguo, X., et al., 2013. Percolation law of large-scale natural sandstone physical model with ultra-low permeability. Journal of Central South University (Science and Technology) 44 (2), 695–700.
Mariethoz, G., Caers, J. 2015. Multiple-point geostatistics-stochastic modeling with training images. Printed and bound in Malaysia by Vivar Printing Sdn Bhd, New Delhi, India.
Masoud, N., Aminzadeh, F., 2001. Past, present and future intelligent reservoir characterization trends. Journal of Petroleum Science and Engineering 31, 67–79.
Mastalerz, M., Schimmelmann, A., Drobniak, A., et al., 2013. Porosity of Devonian and Mississippian New Albany Shale across a maturation gradient: Insights from organic petrology, gas adsorption, and mercury intrusion. AAPG Bulletin 97 (10), 1621–1643.
Matthew, J.P., Colette, B.H., David, A.B., 2005. Scales of lateral perophysical heterogeneity in dolomite lithofacies as determined from outcrop analogs: implications for 3-D reservoir modeling. AAPG Bulletin 89 (5), 645–662.
Matthew, J.D., Nicole, D.R., Peter, S.M., et al., 2006. The effect of carbonate cementation on permeability heterogeneity in fluvial aquifers: an outcrop analog study. Sedimentary Geology 184, 267–280.
Maximoff, M. K., 1980. Translated by Z. Wei and Q. He. *Fundamentals of oilfield development geology* (second ed.): 120–140.
McCaffrey, W., Kneller, B., 2001. Process controls on the development of stratigraphic trap potential on the margins of confined turbidite systems and aids to reservoir evaluation. AAPG Bulletin 85 (6), 971–988.
McHargue, T., Pyrcz, M.J., Sullivan, M.D., et al., 2011. Architecture of turbidite channel systems on the continental slope: patterns and predictions. Marine and Petroleum Geology 28, 728–743.

McLaurin, B.T., Steel, R.J., 2007. Architecture and origin of an amalgamated fluvial sheet sand, lower Castlegate Formation, Book Cliffs, Utah. Sedimentary Geology 197, 291–311.

Meina, W., Jihong, L., Zhaojie, G., et al., 2004. The Influence of water-flooding development on reservoir properties of the No. 2 Member of Shahejie Formation of Tuo 30 Fault Block in Shengtuo Oilfield. Acta Scientiarum Naturalium Universitatis Pekinensis 40 (6), 855–863.

Meysam, N., Behzad, R., Maryam, K., 2015. Effect of heterogeneity on the productivity of vertical, deviated and horizontal wells in water drive gas reservoirs. Journal of Natural Gas Science and Engineering 23, 481–491.

Miall, A.D., 1996. The Geology of Fluvial Deposits. Sedimentary Facies, Basin Analysis, and Petroleum Geology. Springer-Verlag, Berlin Heidelberg, Printed in Italy.

Mikes, D., 2006. Sampling procedure for small-scale heterogeneities (crossbedding) for reservoir modelling. Marine and Petroleum Geology 23, 961–977.

Mikes, D., Geel, C.R., 2006. Standard facies models to incorporate all heterogeneity levels in a reservoir model. Marine and Petroleum Geology 23, 943–959.

Min, W., Yuehua, Z., 2002. A discussion on the special reservoir geological features of Sumu Structural Belt in Baolang Oilfield. Journal of Chengdu University of Technology 29 (1), 31–36.

Min, X., Qintian, W., 2003. Study on pore structure and oil displacing in Panhe fault region. Oil & Gas Geology 24 (1), 42–44.

Ming, Z., Haining, Z., 2002. Application of two new logging methods for evaluating remaining oil saturation. Acta Petrolei Sinica 23 (5), 73–77, 82.

Ming, W., Li, D., Dianbin, G., et al., 2013. Laboratory experimental research on waterflooding sweep eficiency of large-scale and two-dimensional inter-layer heterogeneous physical model. Petroleum Geology & Experiment 35 (6), 698–701.

Minggang, L., Jiafu, Q., Hengmao, T., et al., 2010. Cenozoic fault structure and hydrocarbon accumulation in Western Sag, Liaohe Depression. Petroleum Exploration and Development 37 (3), 281–288.

Mingguo, L., Xiaolan, T., 2007. Characteristics of pore structures and oil production potential of sandstone reservoir of Paleogene Xingouzui Formation in Jianghan Basin. Petroleum Geology and Oilfield Development in Daqing 26 (4), 31–34.

Mingjun, C., Jiagen, H., 2009. Residual Oil Distribution of High Water Cut Reservoir by Composite Flooding: A Case Study of Yangsanmu Oilfield in Dagang oil area. Petroleum industry press, Beijing.

Mingli, G., 2003. Several methods for characterizing residual oil distribution in complex fault block oilfield by numerical simulation. Fault-block Oil & Gas Field 10 (2), 48–50.

Minglu, W., Jun, Y., Xiyu, L., et al., 2009. Research on remaining oil distribution using streamline numerical well testing method. Petroleum Drilling Techniques 37 (3), 95–98.

Mingyi, H., Shixiang, L., Guoqi, W., et al., 2006. Reservoir apprecisal of tight sandatones of upper triassic Xujiahe Formation in the Western Sichuan Foreland Basin. Natural Gas Geoscience 17 (4), 456–458,462.

Mørk, M.B.E., 2013. Diagenesis and quartz cement distribution of low-permeability Upper Triassic–Middle Jurassic reservoir sandstones, Long year by en $CO_2$ lab well site in Svalbard, Norway. AAPG Bulletin 97 (4), 577–596.

Morrow, D.W., 2001. Distribution of porosity and permeability in platform dolomites: insight from the Permian of west Texas: discussion. AAPG Bulletin 85 (3), 525–529.

Nabawy, B.S., Géraud, Y., Rochette, P., et al., 2009. Pore-throat characterization in highly porous and permeable sandstones. AAPG Bulletin 93 (6), 719–739.

Naiju, W., et al., 1999. Summary of Reservoir Development Model in China. Petroleum Industry Press, Beijing.

Narr, W., Schechter, D.W., Thompson, L.B., 2006. Naturally Fractured Reservoir Characterization. Sosiety of Petroleum Engineers, USA.
Nelson, P.H., 2009. Pore-throat sizes in sandstones, tight sandstones, and shales. AAPG Bulletin 93 (3), 329–340.
Nengqiang, L., 2008. compiled the aboriginal Practical Modern Well Test Interpretation Method (Fifth edition). Petroleum Industry Press, Beijing.
Nguyen, B.T.T., Jones, S.J., Goulty, N.R., et al., 2013. The role of fluid pressure and diagenetic cements for porosity preservation in Triassic fluvial reservoirs of the Central Graben, North Sea. AAPG Bulletin 97 (8), 1273–1302.
Nima, R., Ioannis, C., 2011. Characterization of heterogeneities in porous media using constant rate air injection porosimetry. Journal of Petroleum Science and Engineering 79, 113–124.
Ningning, M., Huaishan, L., Jinliang, Z., et al., 2016. Heterogeneity and its impact on remaining oil distribution of low permeability reservoirs in the Hei46 Block. Marine Geology & Quaternary Geology 36 (1), 143–150.
Odunowo, T.O., Moridis, G.J., Blasingame, T.A., et al., 2013. Evaluation of well performance for the slot-drill completion in low-and ultra low-permeability oil and gas reservoirs. SPE Journal 19, 748–760.
Olariu, M.I., Aiken, C.L.V., Bhattacharya, J.P., 2011. Interpretation of channelized architecture using three-dimensional photo real models, Pennsylvanian deep-water deposits at Big Rock Quarry, Arkansasq. Marine and Petroleum Geology 28, 1157–1170.
Oliveira, G.P., Roque, W.L., Araújo, E.A., et al., 2016. Competitive placement of oil perforation zones in hydraulic flow units from centrality measures. Journal of Petroleum Science and Engineering 147, 282–291.
Ouahed, A.K.E., Tiab, D., Mazouzi, A., 2005. Application of artificial intelligence to characterize naturally fractured zones in Hassi Messaoud Oil Field. Algeria 49, 122–141.
Pan, D., Mengjin, C., Zherong, G., et al., 2002. Log response and explanation of structural fractures in volcanic rock reservoir. Acta Petrolei Sinica 23 (6), 32–36.
Panja, P., Conner, T., Deo, M., 2013. Grid sensitivity studies in hydraulically fractured low permeability reservoirs. Journal of Petroleum Science and Engineering 112, 78–87.
Park, H., Han, J., Sung, W., 2015. Effect of polymer concentration on the polymer adsorption-induced permeability reduction in low permeability reservoirs. Energy 84, 666–671.
Peigang, R., Cunyin, X., Yuan, L., et al., 2010. Application of self-organizing neural network to logging reservoir evaluation. Geological Science and Technology Information 29 (3), 114–118.
Peihua, D.X., 2003a. Water Flooded Layer Logging Technology for Oilfield Development. Petroleum Industry Publishing House, Beijing.
Peihua, Z., 2003b. Logging Interpretation of Water Flooded Layer in Oilfield Development. Petroleum Industry Press, Beijing.
Peijs-van Hilten, M., Good, T.R., Zaitlin, B.A., 1998. Heterogeneity modeling and geopseudo upscaling applied to waterflood performance prediction of an incised valley reservoir: countess YY Pool, Southern Alberta, Canada. AAPG Bulletin 82 (12), 2220–2245.
Peijun, D., 2013. Logging curve response characteristics of the oil-field water flooded layer in high water-cut stage in Shuanghe Oilfield. Oil and Gas Field Surface Engineering 32 (5), 19–20.
Peter, G.T., David, A., Asbjorn, G., 2006. Quantitative analysis of porosity heterogeneity: application of geostatistics to norehole images. Mathematical Geology 38 (2), 155–174.

Peter, A.F., Eloisa, S.B., Cristina, S., et al., 2012. Estimating reservoir heterogeneities from pulse testing. Journal of Petroleum Science and Engineering 86-87, 15–26.

Peter, J.R.F., Mike, A.L., Sarah, J.D., et al., 2015. An integrated and quantitative approach to petrophysical heterogeneity. Marine and Petroleum Geology 63, 82–96.

Petrochina Exploration and development company, 2009. Evaluation Technology and Application of Volcanic Reservoir Logging. Petroleum Industry Press, Beijing.

Philippe, L., Jean, B., Jean-Pierre, M., et al., 2012. Relation between stratigraphic architecture and multi-scale heterogeneities in carbonate platforms: the Barremian-lower Aptian of the Monts de Vaucluse, SE France. Sedimentary Geology 265, 87–109.

Ping, G., Xinquan, R., Yanmei, X., Weigang, H., 2004. Remaining Oil Distribution Study Method. Petroleum Industry Press, Beijing.

Ping, C., Yongchao, L., Xuebin, D., et al., 2010. Dual characteristic of sequence structures and forming mechanism in compressional Abdominal Area of Junggar Basin. Chinese Journal of Geology 45 (4), 1078–1087.

Pingping, S., Xinwei, L., 2009. Carbon Dioxide Geological Storage and Enhanced Oil Recovery Technology. Petroleum Industry Press, Beijing.

Pingping, S., Xinmin, S., Honggui, C., 2003. New Methods of Modern Reservoir Description. Petroleum Industry Press, Beijing.

Pranter, M.J., Ellison, A.I., Cole, R.D., et al., 2007. Analysis and modeling of intermediate-scale reservoir heterogeneity based on a fluvial point-bar outcrop analog, Williams Fork Formation, Piceance Basin, Colorado. AAPG Bulletin 91 (7), 1025–1051.

Pranter, M.J., Hirstius, C.B., Budd, D.A., 2005. Scales of lateral petrophysical heterogeneity in dolomite lithofacies as determined from outcrop analogs: implications for 3-D reservoir modeling. AAPG Bulletin 89 (5), 645–662.

Prather, B.E., 2003. Controls on reservoir distribution, architecture and stratigraphic trapping in slope settings. Marine and Petroleum Geology 20, 529–545.

Prem, B., Jiamin, W., Yongman, K., et al., 2016. Influence of wettability and permeability heterogeneity on miscible $CO_2$ flooding efficiency. Fuel 166, 219–226.

Pujun, W., Zhiqiang, F., et al., 2008. Basin Volcanic Lithology, Lithofacies, Reservoir, Gas Reservoir, Exploration. Science Press, Beijing.

Pyrcz, M.J., Deutsch, C.V., 2014. Geostatistical Reservoir Modeling, second ed. Oxford University Press, New York.

Pyrcz, M.J., Boisvert, J.B., Deutsch, C.V., 2009. ALLUVSIM: a program for event-based stochastic modeling of fluvial depositional systems. Computers & Geosciences 35, 1671–1685.

Qazvini Firouz, A., Jamaloei, B.Y., Torabi, F., et al., 2012. The relationship between the productivity index and the diffusivity coefficient and its application in reservoir characterization. Petroleum Science and Technology 30, 1789–1801.

Qiang, B., Juanru, W., 2009. Applied of fractal geometry on researches of reservoir micro heterogeneities. Petroleum Geology and Engineering 23 (3), 122–124.

Qiaolin, H., Zengwu, G., Xiaojun, Z., et al., 2017. Waterflooding experiment and micro distribution of remaining oil in JIV reservoir in Lunnan-2 Oilfield. Xinjiang Petroleum Geology 38 (1), 81–84.

Qifeng, D., Yuefeng, S., Charlotte, S., 2011. Rock-physics-based carbonate pore type characterization and reservoir permeability heterogeneity evaluation, Upper San Andres reservoir, Permian Basin, west Texas. Journal of Applied Geophysics 74, 8–18.

Qigu, C., Gongyang, C., 2010. Fine Description of Low Permeability Sandstone Reservoirs and Development Evaluation Techniques. Petroleum Industry Press, Beijing.

Qihong, F., Cunliang, C., Sen, W., 2013. Research methods of dynamic connectivity in low permeability reservoirs. Special Oil and Gas Reservoirs 20 (5), 100–102.

Qin, Z., Xiaomin, Z., Guiqiu, L., 2007. Heterogeneity of clastic reservoir in Dongying Sag and of the Shahejie Formation of Paleogene its hydrocarbon significance. Journal of palaeogeography 9 (6), 661–668.
Qingchang, L., Meng, W., Lichun, Z., et al., 1997. Development of Conglomerate Oilfield. Petroleum Industry Press, Beijing.
Qinglong, X., Zhanghong, S., et al., 2010. Fine Reservoir Description Technology of Heavy Oil Field. Petroleum Industry Press, Beijing.
Qingsheng, J., 2009. Microscopic physical simulation experiment of remaining oil constrained by flow unit. Petroleum Geology and Recovery Efficiency 16 (3), 90–91.
Qinlin, G., 1999. New Technology of Improving Water Flooding Effect in High Water Cut Oil Field (Part 1). Petroleum Industry Press, Beijing.
Qiquan, R., Yongjun, W., Yuanhui, S., et al., 2011. Gas Reservoir Characterization Technology of Volcanic Rock. Petroleum Industry Press, Beijing.
Qitai, Y., 1997. Study on residual oil. Petroleum Exploration and Development 24 (2), 46–50.
Qiuqiang, S., Zhansong, Z., Chong, Z., et al., 2013. The application of logging facies – facies analysis in complex lithology. Journal of Oil and Gas Technology 35 (7), 78–81.
Rajesh, J.P., Edwin, B.E., Earl, M.W., 2001. Geostatistical characterization of the Carpinteria Field, California. Journal of Petroleum Science and Engineering 31, 175–192.
Raju, K.U., Al-Buali, M.H., Al-Saadoun, T.F., et al. 2010. Scale inhibitor squeeze treatments based on modelling studies. In: Proceedings of the SPE International Conference on Oilfield Scale, Aberdeen, UK., SPE130280:26-27.
Ramstad, T., Øren, P.-E., Bakke, S., 2010. Simulation of two-phase flow in reservoir rocks using a Lattice Boltzmann Method. SPE Journal 15, 923–933.
Ray, S., Chakraborty, T., 2002. Lower Gondwana fluvial succession of the Pench-Kanhan Valley, India: stratigraphic architecture and depositional controls. Sedimentary Geology 151, 243–271.
Reed, A.J., et al., 1992. Application of resistivity logging in oil and gas exploration and development in Xiluo Oilfield. Foreign Oil and Gas Exploration 4 (5), 76–87.
Remeysen, K., Swennen, R., 2008. Application of microfocus computed tomography in carbonate reservoir characterization: possibilities and limitations. Marine and Petroleum Geology 25, 486–499.
Renyuan, S., Zichao, M., Jianshan, Z., et al., 2013. The influence of steam flooding on the core wettability of low permeability heavy oil reservoirs. Special Oil and Gas Reservoirs 20 (6), 69–71.
Rezvandehy, M., Aghababaei, H., Tabatabaee Raissi, S.H., 2011. Integrating seismic attributes in the accurate modeling of geological structures and determining the storage of the gas reservoir in Gorgan Plain (North of Iran). Journal of Applied Geophysics 73, 187–195.
Richard, A.S., Jordan, J.F.J., 1999. Reservoir Characterization-Recent Advances. AAPG Memoir 71. The American Association of Petroleum Geologists, Tulsa, Oklahoma.
Rongcai, Z., 2010. Wenhuaguo. In: Fengjie, L. (Ed.), High reSolution Sequence Stratigraphy. Geological Publishing House, Beijing.
Rongcai, Z., Chaorong, W., Zuowei, R., et al., 1999. Sequence stratigraphy and source-reservoir-cap assemblage of the deep Shahejie Formation in the western sag of Liaohe Depression. Compound Oil and Gas Fields 4, 48–53.
Rongcai, Z., Qi, Z., Hua, W., et al., 2009. The sequence architecture and sandbody predicition of the Second Member of Shanxi Formation in Changbei Gas Field, Ordos Basin. Geological Journal of China Universities 15 (1), 69–79.
Rongqiang, H., Di, M., Shizhong, M., et al., 2016. Identification of flow units and distribution of remaining oil controlled by the architectural structure of point bar. Journal of China University of Mining & Technology 45 (1), 133–140,156.

Ruili, N., Jinzhuang, X., Hongjuan, L., et al., 2004. Application of casing resistivity technique in residual oil saturation evaluation of Daqing Oilfield. Journal of Daqing Petroleum Institute 28 (5), 16–18,27.

Rusciadelli, G., Simone, S.D., 2007. Differential compaction as a control on depositional architectures across the Maiella carbonate platform margin. Sedimentary Geology 196, 133–155.

Saemi, M., Ahmadi, M., Varjani, A.Y., 2007. Design of neural networks using genetic algorithm for the permeability estimation of the reservoir. Journal of Petroleum Science and Engineering 59, 97–105.

Sagan, J.A., Hart, B.S., 2006. Three-dimensional seismic-based definition of fault-related porosity development: trenton-Black River interval, Saybrook, Ohio. AAPG Bulletin 90 (11), 1763–1785.

Samantha, T., Gary, J.H., Matthew, D.J., 2010. High-resolution stratigraphic architecture and lithological heterogeneity within marginal aeolian reservoir analogues. Sedimentology 57, 1246–1279.

Shafiei, A., Dusseault, M.B., Zendehboudi, S., et al., 2013. A new screening tool for evaluation of steamflooding performance in naturally fractured carbonate reservoirs. Fuel 108, 502–514.

Shanley, K.W., Cluff, R.M., 2015. The evolution of pore-scale fluid-saturation in low permeability sandstone reservoirs. AAPG Bulletin 99 (10), 1957–1990.

Shanwen, Z., Yongshi, W., Chuansheng, P., et al., 2003. Description of Deep Heavy Oil Reservoir in Glutenite Rock in Luojia-Kengxi Area of Jiyang Depression. Petroleum Industry Press, Beijing.

Shaochun, Y., Ruili, W., Gaiyun, W., et al., 2006. The variation characteristics of reservoir plane heterogeneity in diferent development stages of oilfield: take the third m ember of Dongying formation in the second block of Shengtuo Oilfield as an example. Geological Journal of China Universities 12 (4), 493–499.

Shaochun, Y., Yaru, W., Yuanyuan, L., et al., 2016. Characterization on clastic reservoir heterogeneity based on data envelopment analysis method. Journal of Central South University (Science and Technology) 47 (1), 218–224.

Shaohua, L., Changmin, Z., Taiju, Y., et al., 2007a. Geographic information system aided reservoir flow unit division. Acta Petrolei Sinica 28 (5), 114–117.

Shaohua, L., Yanshu, Y., Chang, Z., 2007b. native Reservoir Stochastic Modeling Technology. Petroleum Industry Press, Beijing.

Shaohua, L., Changmin, Z., Youbin, H., et al., 2009. Simulation of internal physical distribution trend of channel sand body. Journal of Oil and Gas Technology 31 (1), 23–25.

Shaohua, L., Changmin, Z., Yanshu, Y., 2012. indigenous Reservoir Modeling Algorithm Analysis. Petroleum Industry Press, Beijing.

Shaoxian, F., 1998. Hou Fanghao Compiled. Reservoir Geology of Oil and Gas. Petroleum University Press.

Shaoyang, Y., Zhansong, Z., Quan, L., et al., 2016. Porosity evaluation based on lithology indentification using bayesian classifier. Well Logging Technology 40 (3), 281–285.

Shaoying, C., Yongfeng, Z., Peng, C., et al., 2017. Application of geology-engineering integration in high-eficiency remaining oil potential tapping and its results: a case study on YM32 dolomite oil reservoirs in Tarim Basin. China Petroleum Exploation 22 (1), 46–52.

Shedid, S.A., 2007. A novel technique for determining microscopic pore size distribution of heterogeneous reservoir rocks. Petroleum Science and Technology 25, 899–914.

Shehadeh, K.M., 2002. Studying the effect of wettability heterogeneity on the capillary pressure curves using the centrifuge technique. Journal of Petroleum Science and Engineering 33, 29–38.

Shenghe, W., 2010. Reservoir Characterization and Modeling. Petroleum Industry Press, Beijing, pp. 136–174.
Shenghe, W., Kewen, L., 2005. Multiple-point geostatistics: theory, application and perspective. Journal of Palaeogeography 7 (1), 137–144.
Shenghe, W., Qihua, X., 1998. compiled Hydrocarbon Reservoir Geology. Petroleum Industry Press, Beijing.
Shenghe, W., Zhenkui, J., Cangdian, H., et al., 1999. Reservoir Modeling. Petroleum Industry Press, Beijing.
Shenghe, W., Zheng, F., Changfu, X., et al., 2012. Internal architecture of alluvial fan in the Triassic Lower Karamay Formation in Karamay Oilfield, Xinjiang. Journal of Palaeogeography 14 (3), 331–340.
Shengli, L., Xinghe, Y., Xingjun, G., et al., 2003. A new method for studying remaining oil distribution- Grey association analysis. Oil & Gas Geology 24 (2), 175–179.
Shengli, X., Rongxi, L., Deming, Z., et al., 2013. Formation of low perm eable reservoir Chang 4 + 5 in Jiyuan region, Ordos Basin. Chinese Journal of Geology 48 (4), 1164–1176.
Shibin, Z., Jun, C., Yongxian, Z., et al., 2012. Analysis of characteristics of low-permeable reservoir with micro-fracture and their origins of the Neogene Youshashan Formation in Nanyishan Oilfield, Qaidam Basin. Journal of Palaeogeography 14 (1), 133–141.
Shiguang, Z., Shuangfang, L., Yan, Z., et al., 2009. Application of high-resolution sequence stratigraphy in the study of the reservoir macro-heterogeneity: taking the Fuyu oil layer of Chao 1-Chaoqi 3 blocks in Chaoyanggou oil reservoir of Songliao Basin as an example. Acta Sedimentologica Sinca 27 (3), 458–469.
Shijun, Z., Zhimin, D., Zhaoguo, S., et al., 2005. Quantitative Reservoir Modeling. Petroleum Industry Press, Beijing.
Shiqing, C., Engao, T., Linfeng, X., 2007. Research on rational development of complex lithologic multi-bottom water fault block reservoir. Special Oil and Gas Reservoirs 14 (3), 62–65.
Shiwen, X., Wei, Z., Qingming, L., et al., 2015. Potential tapping of remaining oil in the late stage of offshore oilfield development based on multidisciplinary integration concept: a case study of H4C thin reservoir in X3 oilfield pearl river mouth basin. China Offshore Oil and Gas 27 (5), 68–75.
Shiyi, Y., Xinmin, S., Qiquan, R., 2004. Development Technology of Fractured Reservoir. Petroleum Industry Press, Beijing.
Shizhong, M., Guiyou, L., Baiquan, Y., et al., 2008. Research on three-dim ensional heterogeneous model of channel sandbody controlled by architecture. Earth Science Frontiers 15 (1), 57–64.
Shouqing, W., 1993. Fan Delta Model. Petroleum Industry Press, Beijing.
Shouyu, X., 2005. Oil Reservoir Description Method Principle. Petroleum Industry Press, Beijing.
Shujuan, Z., Yanbin, W., Xuemei, G., et al., 2011. New description technology for carbonate buried-hill oil reservoirs at high water-cut stage. Petroleum Geology and Oilfield Development in Daqing 30 (6), 71–74.
Shuxin, H., Hongmei, S., Cheng, W., et al., 2007. Application of microcosmic testing techniques for conglomerate reservoir in study about Xujiaweizi Gas Reservoir. Petroleum Geology and Oilfield Development in Daqing 26 (4), 43–46,50.
Shuyuan, S., Hui, Q., Xuan, Z., et al., 2014. A method making paleokarst training image based on GoogleEarth. Complex Hydrocarbon Reservoir 7 (4), 6–10.
Shuyuan, S., Yanshu, Y., Jingyang, H., et al., 2011. A random walk-based multiple-point statistics modeling methods. Geological Science and Technology Information 30 (5), 127–131,138.

Sijing, H., Yonglin, Y., Yuming, S., et al., 2000. Effect of waterflooding on pore structure insandstone reservoir. China Offshore Oil and Gas (Geology) 14 (2), 122–128.
Siyu, Y., Shaohua, L., Youbin, H., et al., 2016. Multiple-point geostatistics algorithm based on pattern scale-down cluster. Acta Petrolei Sinica 37 (11), 1403–1409.
Skelly, R.L., Bristow, C.S., Ethridge, F.G., 2003. Architecture of channel-belt deposits in an aggrading shallow sandbed braided river: the lower Niobrara River, northeast Nebraska. Sedimentary Geology 158, 249–270.
Songjiang, D., Jiaxi, Z., 2003. Remaining oil distribution in complicated faulted blocks and its accesspory potential exploitation measures. Petroleum Exploration and Development 30 (5), 90–93.
Songjiang, D., Qingkui, W., Jinzhong, N., et al., 2008. Research on flow units of thick sandstone reservoir in Fault Block Guan 142 of Dagang Oilfield. Modern Geology 22 (1), 76–80,142.
Sruoga, P., Rubinstein, N., 2007. Processes controlling porosity and permeability in volcanic reservoirs from the Austral and Neuquen basins, Argentina. AAPG Bulletin 91 (1), 115–129.
Stoudt, E.L., Paul, M.H., 1995. Hydrocarbon Reservoir Characterization- Geologic Framework and Flow Unit Modeling. SEPM (Society for Sedimentary Geology), Tulsa, Oklahoma.
Sweet, M.L., Blewden, C.J., Carter, A.M., et al., 1996. Modeling heterogeneity in a low-permeability gas reservoir using geostatistical techniques, Hyde Field, Southern North Sea. AAPG Bulletin 80 (11), 1719–1735.
Sylvester, Z., Pirmez, C., Cantelli, A., 2011. A model of submarine channel-levee evolution based on channel trajectories: implications for stratigraphic architecture. Marine and Petroleum Geology 28, 716–727.
Taghavi, A.A., Mørk, A., Kazemzadeh, E., 2007. Flow unit classification for geological modelling of a heterogeneous carbonate reservoir: cretaceous sarvak formation, dehluran field, SW Iran. Journal of Petroleum Geology 30 (2), 129–146.
Taiju, Y., Changmin, Z., Hongjing, Z., et al., 2004. Prediction of residual oil distribution in complex fault-block districts during high water-bearing periods. Petroleum Geology & Experiment 26 (3), 267–272.
Taiju, Y., Changmin, Z., Hongjung, Z., 2006. Remaining oil prediction based on comprehensive geological analysis methods. Advances in Earth Science 21 (5), 539–544.
Takahashi, S., Kovscek, A.R., 2010. Wettability estimation of low-permeability, Siliceous Shale using surface forces. Journal of Petroleum Science and Engineering 75, 33–43.
Tao, Z., Chuanying, M., Wenyuan, H., et al., 2015. Impact of low oil prices on the geophysical prospecting industry. International Petroleum Economics 23 (7), 60–65, 97.
Taware, S., Friedel, T., Datta-Gupta, A. 2011. A practical approach for assisted history matching using grid coarsening and streamline-based inversion: Experiences in a giant carbonate reservoir. In: Proceedings of the SPE Reservoir Simulation Symposium, The Woodlands, Texas, USA, 21-23.
Tiantai, W., Ming, Z., Jinsheng, Z., et al., 2005. Reservoir characteristic of gas accumulation and analyzed of harm elements in Keyi tectonic belt. Drilling & Production Technology 28 (6), 55–58.
Tidwell, V.C., Wilson, J.L., 2000. Heterogeneity, permeability patterns, and permeability upscaling: physical characterization of a block of massillon sandstone exhibiting nested scales of heterogeneity. SPE Reservoir Evaluation & Engineering 3 (4), 283–291.
Timothy, T.E., 2006. On the importance of geological heterogeneity for flow simulation. Sedimentary Geology 184, 187–201.
Vera, L.B., Ariffin, T., Trabelsi, A., et al., 2010. Assessing fluid migration and quantifying remaining-oil saturation in a mature carbonate reservoir. Journal of Petroleum Technology 62, 46–48.

Wanchao, S., 2003. High Water Cut Oilfield Development Technology and Method. Geological Publishing House, Beijing.

Wangshui, H., Qiuyan, L., Songze, L., et al., 2012. Macroscopic heterogeneity of reservoir and its influence on the distribution of remaining oil-By taking Tengger formation of Xinlinhaolai area in the Baiyinchagan sag as an example. Petroleum Geology and Engineering 26 (6), 1–4.

Weber, J., Ricken, W., 2005. Quartz cementation and related sedimentary architecture of the Triassic Solling Formation, Reinhardswald Basin, Germany. Sedimentary Geology 175, 459–477.

Weger, R.J., Eberli, G.P., Baechle, G.T., et al., 2009. Quantification of pore structureand its effect on sonic velocity and permeability in carbonates. AAPG Bulletin 93 (10), 1297–1317.

Wei, Z., Ningsheng, Z., 2000. Network modeling of the migration of solid particles in fractal porous media. Journal of Xi'an Petroleum Iustltute (Natural Science Edition) 15 (2), 15–17.

Wei, Y., Renzi, L., Xiaohui, J., et al., 2001. Researches of fluid heterogeneity of Qian12 block in Liaohe Oilfield. Journal of Jianghan Petroleum Institute 23 (1), 14–16.

Wei, Z., Haiwen, W., Yanmei, L., et al., 2007. Application of Boon-Neutron life logging-An example of Xiahezhuang Field. Xinjiang Petroleum Geology 28 (3), 372–374.

Wei, T., Renqi, T., Xijun, B., 2008a. Application of fractal dimension theory to reservoir physics. Acta Petrolei Sinica 29 (1), 93–96.

Wei, Z., Chengyan, L., Chunmei, D., 2008b. Application of multiple-point geostatistics in geological modeling of D Oilfield in Peru. Journal of China University of Petroleum 32 (4), 24–28.

Wei, L., Zhongxiang, X., Siyu, Y., et al., 2009. Comparison researches of reservoir pore evaluated methods which used NMR logging data. Oil Geophysical Prospecting 44 (6), 773–778.

Wei, Z., Shaoqiu, G., Zijian, C., et al., 2013. Fuzzy mathematics-based reservoir evaluation in the southeastern Pre-Caspian Basin. Oil & Gas Geology 34 (3), 357–362.

Weidong, D., Changmin, Z., Taiju, Y., et al., 2007. High resolution sequence stratigraphic correlation of Penglaizhen Formation, Upper Jurassic of Baimamiao gas field, Western Sichuan. Acta Sedimentologica Sinica 25 (5), 708–715.

Weilin, Y., Jiandong, Z., Puwang, Z., 2012. Identifying method of the fluids in the complex-lithology reservoirs of Tamutsag Basin. Petroleum Geology and Oilfield Development in Daqing 31 (1), 158–162.

Weirong, X., Zhanfeng, Z., Shiqing, C. (Eds.), 1999. Geology of Oil and Gas Field Development. Petroleum Industry Press, Beijing, pp. 10–27.

Weiwei, Z., Shuangfang, L., Shiguang, Z., et al., 2010. The physical properties of igneous reservoir and its influencing factors – taking Longshen 1 well area in Yingtai fault depression in southern Songliao Basin as an example. Acta Sedimentologica Sinca 28 (3), 563–571.

Weizhi, S., Qiaofeng, L., Yujiao, D., et al., 2004. On evaluation of waterflooded zones with NMR logging and its application. Well Logging Technology 28 (1), 34–38,44.

Weizhi, S., Yujiao, D., Ya, L., et al., 2009. The application of NMR log data in evaluation of reservoir pore structure. Well Logging Technology 33 (1), 52–56.

Wenbiao, Z., Taizhong, D., Lei, Z., et al., 2015. Generation and application of three-dimensional MPS training images based on shallow seismic data. Oil & Gas Geology 36 (6), 1030–1037.

Wenjie, F., Shenghe, W., Changfu, X., et al., 2015a. Water flooding channel of alluvial fan reservoir and its controlling distribution pattern of remaining oil: a case study of Triassic Lower Karamay Formation, Yizhong area, Karamay oilfield, NW China. Acta Petrolei Sinica 36 (7), 858–870.

Wenjie, F., Shenghe, W., Xia, Q., et al., 2015b. Micro-facies modeling of alluvial fan reservoir based on geological vector information: a case study on the Triassic Lower Karamay Formation, Yizhong Area, Karamay Oilfield, NW China. Geological Journal of China Universities 21 (3), 449–460.

Wenlin, L., 2010. Rock physics for remaining oil distribution prediction in water flood field. Lithologic Reservoirs 22 (3), 91–94,123.

Wenling, L., Daxing, W., Xihang, X., et al., 2016. Isochronous stratigraphic correlation of logging-seismic combination for complex reservoirs: a case study from Wangxuzhuang oilfield. Geophysical Prospecting for Petroleum 55 (4), 540–549.

Wenqi, J., Xiaojun, W., Fengpeng, H., et al., 2010. Characteristics of Chang-6 oil bearing Formation changes after long term water flood development in Ansai oilfield. Journal of Earth Sciences and Environment 32 (3), 239–244.

Wenqi, J., Shuheng, D., Xiangwei, L., et al., 2016. Revelation on remaining oil excavation from stress interlayer division in special low permeability reservoir: take Yanchang Formation of Baiyushan Reservoir in Ordos Basin as an example. Acta Scientiarum Naturalium Universitatis Pekinensi 52 (6), 1034–1040.

Wenquan, Y., Li, L., Tao, F., et al., 2010. Quantitative research for porosity evolution in low permeability deep gas reservoir of rift-subsidence basin. Natural Gas Geoscience 21 (3), 397–405.

Wenrui, H., 2009. Introduction to Low Permeability Oil Field (The First Volume). Petroleum Industry Press, Beijing.

Wenxiang, H., Le, Y., Chaoya, M., et al., 2011. Effect of micro-pore structure parameter on seepage characteristics in Ultra-Low permeability reservoir: a case from Chang6 Reservoir of Ordos Basin. Natural Gas Geoscience 22 (3), 477–481,517.

Wilkes, H., Vieth, A., Elias, R., 2008. Constraints on the quantitative assessment of in-reservoir biodegradation using compound-specific stable carbon isotopes. Organic Geochemistry 39, 1215–1221.

Williams, B.J., Robertd, H., 2004. Variable alluvial sandstone architecture within the Lower Old Red Sandstone, Southwest Wales. Geological Journal 39, 257–275.

Wu, X., 2004. Geological research method of residual oil distribution in fault block reservoir. Petroleum Geology and Recovery Efficiency 11 (1), 58–60.

Wuguang, L., Shenglai, Y., Xianjie, S., et al., 2011. Application of variogram in reservoir evaluation and development: taking Yangjiaba Oilfield for example. Geological Science and Technology Information 30 (4), 83–87.

Wuren, X., Wei, Y., Guang, Y., et al., 2010. Pore structure features of sandstone reservoirs in the Upper Triassic Xujiahe Formation in the central part of Sichuan Basin. Natural Gas Geoscience 21 (3), 435–440.

Xiang' an, Y., Youfu, W., Keliang, W., 2007. were compiled Basis for Enhancing Oil Recovery. Petroleum Industry Press, Beijing.

Xiangping, Z., 2010. Application of reservoir structure research in the fine exploitation of oilfields. Petroleum Exploration and Development 37 (4), 483–489.

Xianguo, Z., Tao, Z., Chengyan, L., et al., 2015. Lithology recognition with well logs in complex lithologic low permeability reservoirs in Tanan Depression. Journal of Southwest Petroleum University (Science & Technology Edition) 37 (1), 85–90.

Xiantai, L., Jun, L., Jun, W., et al., 2013. Study on the technologies of identification and delicate description of low level faults. Special Oil and Gas Reservoirs 20 (1), 44–47.

Xiaobo, G., Zhilong, H., Xiaoxian, T., et al., 2013. Identification and application of complex lithology of Lucaogou tight reservoir in Malang Sag, Santanghu Basin. Xinjiang Petroleum Geology 34 (6), 649–652.

Xiaodong, W., Yanqing, Z., Lili, C., et al., 2011. Application of amplitude spectrum gradient attribute of seismic data in reservoir evaluation of WC area. Geophysical Petroleum Exploration 46 (2), 281–284.
Xiaofeng, M., Qi, W., Jian, S., et al., 2012. Lithologic and lithofacies characteristics of the Carboniferous – Permian volcanic rocks and their control actions on the reservoirs in Luxi area, Junggar Basin. Special Oil and Gas Reservoirs 19 (1), 54–57.
Xiaojie, Z., Honggang, L., Dana, L., et al., 2015. Remaining oil distribution of low amplitude with great bottom water reservoir in high water cut stage in Tahe. Science Technology and Engineering 15 (5), 86–90.
Xiaojun, W., Xiaomei, L., Dan, X., et al., 2015. Application of multiple-point geostatistics method to structure modeling of alluvial fan. Lithologic Reservoirs 27 (5), 87–91.
Xiaoli, Z., Zhengwen, X., 2006. Reservoir characteristics of Chang8 of Yanchang Formation (Triassic) in Longdong Area, Ordos Basin. Journal of Mineralogical and Petrological 26 (4), 83–88. Total 106.
Xiaoming, Z., 2012. Application of natural gamma ray spectrometry logging in carbonate reservoir evaluation in Yubei area. World Well Logging Technology 190 (4), 35–36, 39.
Xiaoqin, F., Zhenhua, Y., Chunhua, X., et al., 2008. Application of RMT in water-flooded conglomerate reservoir with low-permeability of Karamay Oilfield. Well Logging Technology 32 (2), 180–184.
Xiaoqing, H., Tingen, F., Hui, W., et al., 2015. Fine characterization of thick reservoirs with complex lithologies and its influence on oil recovery – A case from reservoirs of the 1st and 2nd Members of Shahejie Formation in A oilfield, Shijiutuo uplift in Bohai Bay Basin. Oil & Gas Geology 36 (5), 835–841.
Xiaoxia, G., Wanping, L., Hongliang, T., et al., 2015. The impact of low oil prices on deepwater oil and gas exploration and development. International Petroleum Economics 4, 63–67.
Xiaoyan, T., Xingyuan, W., Yonghong, Z., et al., 2009. Use synthetic probability method to evaluate the fracture development of volcanic rocks. Natural Gas Exploration and Development 32 (1), 26–27, 38.
Xiaoyong, X., Hongming, T., Yingfeng, M., et al., 2009. Applied of gas bubble pressure methods in reservoir pore structure tested. Journal of Southwest Petroleum University (Science& Technology Edition) 31 (5), 17–20.
Xiaoyu, H., Qiang, B., 1995. Application of logging hole comprehensive probability method in Sinian reservoir evaluation in Ziyang area. Drilling and Production Technology 18 (4), 44–46.
Xiarong, J., Shan, J., Yong, Y., et al., 2009. Quantitative calculation of residual oil saturation by reservoir engineering method. Special Oil and Gas Reservoirs 16 (4), 48–50.
Xijin, Z., Zongquan, H., Xiaomin, Z., 2002. Jurassic reservoir of the Northwest edge in Junggar Basin. Journal of Palaeography 4 (1), 90–96.
Xinbian, L., Shimin, W., 2003. Application of fractai techniques in heterogeneous carbonate reservoir. Geophysical Prospecting for Petroleum 42 (3), 309–312.
Xing, L., Gongshe, Z., Qin, C., et al., 2008. Study on distribution of residual oil saturation in water-drive reservoir based on information of well test. Oil and Gas Well Testing 17 (2), 1–4.
Xingguo, L., 1994. Application of micro-structure and reservoir sedimentary microfacies to study remaining oil distribution. Oil and Gas Recovery Technology 1 (1), 68–80.
Xinghe, Y., 2004. Geological Model and Hierarchical Interface Analysis of Braided River Reservoir. Petroleum Industry Press, Beijing.
Xinghe, Y., 2009. Petroleum Reservoir Geology Foundation. Petroleum Industry Press, Beijing.
Xingli, L., Yunlong, L., Yi, Q., 2015. Permeability evaluation of complex lithology reservoir using ECS log data. Well Logging Technology 39 (6), 720–723,728.

Xingxin, W., Shuxin, Z., 1992. compiled Sandstone Reservoir Clay Mineral and Oil Layer Protection. Geology Press, Beijing.

Xinlong, D., Yili, K., Lijun, Y., et al., 2013. Review of micro flow mechanism and application in low-permeability reservoirs. Geological Science and Technology Information 32 (2), 91–96.

XinTao, Y., Pingping, S., 2007. Continental strata correlation of high-resolution sequence in reservoir development phase. Acta Petrolei Sinica 28 (6), 87–91.

Xiongyan, L., Jinyu, Z., Hongqi, L., Shaohua, Z., Yihan, C., 2012. Computational intelligent methods for predicting complex lithologies and multiphase fluids. Petroleum Exploration and Development 39 (2), 243–248.

Xiugang, P., Yongping, W., Jiansheng, Z., et al., 2005. Fractal structure and physical properties evaluated new parameters of reservoir pore of low-permeability reservoir. Natural Gas Industry 25 (12), 37–39.

Xiurong, G., Shoutian, C., Xing, L., 2001. Classification and comparison of small layers in reservoir description - a case study of K71 fault block dongying group in kenxi oilfield. Geological Science and Technology Information 20 (2), 55–58.

Xiwen, L., Rongcai, Z., Tao, Z., et al., 2006. Accurate sequence stratigraphy and reservoir assessment of point reefs and shoals in the Late Permian Changxing Formation of the Jiannan structure, Lichuan, Hubei, China. Journal of Chengdu University of Technology (Science & Technology Edition) 33 (4), 407–413.

Xudong, D., Qihui, Z., Guohui, N., et al., 2009. Reservoir evaluation and oil-gas distribution of Yurubcheno oil Field in Russia. Journal of Jilin University (Earth Science Edition) 39 (6), 968–975.

Xueru, Z., Heping, C., Jichang, Z., et al., 1999. Structural Fracture Research Technique of Metamorphic Rock Reservoir. Petroleum Industry Press, Beijing.

Xuesong, L., Baoquan, S., Yan, J., et al., 2015. Optimized design of high-efficiency adjusting wells near the faults of high-watercut mature oilfield. Petroleum Geology and Oilfield Development in Daqing 34 (1), 56–58.

Xupeng, M., 2010. Relationship between reservoir properties parameters and its micro pore structure. Progress in Exploration Geophysics 33 (3), 216–219.

Yamada, T., Okano, Y., 2007. A volcanic reservoir: integrated facies-distribution modeling and history matching of a complex pressure system. SPE Reservoir Evaluation & Engineering 10, 77–85.

Yan, H., Jun, Y., Niansheng, W., 2001. Geostatistical method for reservoir heterogeneity description. Journal of Southwest Petroleum Institute 23 (3), 13–15.

Yan, H., Yinlai, Z., Niansheng, W., 2005. The application study of the chlorine logging in Weicheng Oil-field. Journal of Southwest Petroleum Institute 27 (1), 5–7.

Yan, L., Xiaoqi, D., Xuewei, L., 2013. Application of high resolution sequence stratigraphy on remaining oil distribution of continental reservoirs: case of Karamay formation of east Karamay oilfield. Petroleum Geology and Recovery Efficiency 20 (2), 15–20.

Yanbin, B., Chengdou, M., Hongping, S., et al., 2003. Application of variational function in describing reservoir plane heterogeneity. Xinjiang Petroleum Geology 24 (3), 251–253.

Yang, L., 2003. Flow unit mode and remaining oil distribution in reservoir. Acta Petrolei Sinica 24 (3), 52–55.

Yang, L., 2011. Characterization of Remaining Oil Enrichment Region of Terrestrial Reservoir with Water Drive. Petroleum Industry Press, Beijing.

Yang, L., Jianmin, L., 2005. Principles and Methods of Mobile Unit Research. Geology Press, Beijing.

Yang, L., Jianmin, L., 2007. Geology of Reservoir Development. Petroleum Industry Press, Beijing, pp. 12–39.

Yang, L., Yanchao, Z., 2008. Application of multiple-point geostatistics in fluvial reservoir stochastic modeling. Geological Science and Technology Information 27 (3), 68–72.

Yang, L., Zhi, F., Jie, L., et al., 2012. Application of real-time micro-seismic wave monitoring technology for artificial cracks to low permeability reservoir reformation. Xinjiang Petroleum Geology 33 (4), 500–501.

Yangquan, J., Sitian, L., Zhen, L., et al., 1995. Analysis of the Sedimentary System and the Structure of the Typical Skeleton Sand Body in Qu River and Lake Delta. China University of Geosciences Press, Hubei Wuhan.

Yangquan, J., Sitian, L., Zhen, L., et al., 1998. Heterogeneity of porosity and permeability in clastic bock reservoirs. Oil & Gas Geology 19 (2), 89–92,98.

Yangquan, J., Liqun, W., Hui, R., 2015. Coal Accumulating Basin Sedimentology. China University of Geosciences Press, Hubei Province Wu Han.

Yanhui, Z., Xuemei, Z., Yingbiao, W., et al., 2011. Numerical simulation of polymer flooding in Class III reservoir of Daqing Oilfield. Fault-block Oil & Gas field 18 (2), 232–234.

Yanjie, S., Jianfeng, Z., Weilin, Y., et al., 2007. The method of complex lithologic logging identification based on support vector machine. Journal of Daqing Petroleum Institute 31 (5), 18–20,46.

Yanshu, Y., Shenghe, W., Rui, Z., et al., 2008. The forecasting of fluvial facies reservoir distribution by using simpat. Joumal of Southwest Petroleum University (Science & Technology Edition) 30 (2), 19–22.

Yanshu, Y., Changmin, Z., Shuyuan, S., et al., 2011. A new method for modeling fluvial facies by integrating random walk process and multi-point statistics. Journal of Oil and Gas Technology 3398, 44–47.

Yanshu, Y., Changmin, Z., Shaohua, L., et al., 2014. A pattern-based multiple point geostatistics method. Geological Review 60 (1), 217–221.

Yanwu, J., Guilin, Y., Yuqi, K., et al., 2001. The application of artificial nerve network system for reservoir elvaluation. Mud Logging Technology 12 (2), 8–12.

Yanzhong, W., Yingchang, C., Guoqi, S., et al., 2008. Application of production test effective reservoir in deep data to evaluation of the clastic of Bonan Sag. Acta Petrolei Sinica 29 (5), 701–706, 710.

Yao, M., Wenhou, L., Zhe, L.X., 2016. Characteristics of microscopic pore structure in the low permeability sandstone reservoir: a case study of Chang 9 of Yanchang Formation in Zhijing-Ansai area of Ordos Basin. Geological Bulletin of China 35 (2/3), 398–405.

Ye, Y., Changmin, Z., Shangfeng, Z., et al., 2011. Sedimentary facies recognition using lithological statistics – Example from delta deposits of Pearl River M outh Basin. Journal of Palaeogeography 13 (3), 271–277.

Yefang, L., Sijing, H., Jie, L., et al., 2011. Influences of authlgenic chlorite on pore structure in sandstone reservoir: a case study from Upper Triassic Yanchang Formation in Ordos basin, China. Geological Bulletin of China 30 (1), 134–140.

Yefei, C., Shimi, P., Guiru, S., et al., 2003. Inter- well prediction of flow units and remaining oil distribution. Acta Petrolei Sinica 24 (3), 74–77.

Yi, X., Hongliang, W., Hongwen, D., et al., 2008. Sedimentary evolution in the framework of high resolution sequences of Dongying Formation and prediction of favorable zones in Qingshui sag Liaohe. Acta Sedimentologica Sinica 26 (6), 1014–1020.

Yi, H., Qirong, Q., Cunhui, F., et al., 2012a. Application of weight evaluation method in volcanic reservoir evaluation: a case study of Carboniferous volcanic reservoir in Wuba area, Zhongguai Area. Petroleum Geology and Engineering 26 (3), 25–27,31.

Yi, T., Chuanli, X., Chao, L., et al., 2012b. Application of grey correlation analysis method in reservoir evaluation of Qingdong Sag. Natural Gas Geoscience 23 (2), 381–386.

Yichao, W., Huaizhong, W., Lianmin, L., et al., 2011. Application of constant velocity mercury injection technique to study microscopic pore structures on various reservoir flow unit in Kongnan Area. Natural Gas Geoscience 22 (2), 335–339.
Yikun, L., Shuang, L., Chunsen, Z., et al., 2015. Water-oil percolation law of complex fault block in middle and high water-cut stage. Science Technology and Engineering 15 (17), 141–144.
Yiling, H., Zifang, Z., Mingde, Z., et al., 2004. The reservoir evaluation method for calculating the ratio of porosity to permeability with log data. Well Logging Technology 28 (2), 135–137.
Yin, Y., Feng, S., Yin, T., 2012. Comparison study on modeling method of meandering river reservoir. Fault-block Oil & Gas Field 19 (1), 44–46, 64.
Yinan, Q., Ziqi, C., 1996. compiled Reservoir Description. Petroleum Industry Press, Beijing.
Yinan, Q., Jixin, C., Liangmiao, Y., et al., 1993. Foreign Reservoir Description Technology. CNPC, Beijing.
Yinan, Q., Shuhao, X., et al., 1997a. Oil and Gas Reservoir Evaluation Technology (revised edition). Petroleum Industry Press, Beijing, pp. 249–254.
Yinan, Q., Shuhao, X., Fengxiang, Y., 1997b. Continental Hydrocarbon Reservoirs in China. Petroleum Industry Press, Beijing.
Yingchang, C., Zaixing, J., 2003. Sedimentological Experimental Methods and Techniques. Petroleum Industry Press, Beijing.
Yingli, Z., 2008. Study on interpretation method of heavy oil flooded layer logging. World Well Logging Technology 23 (4), 26–31.
Yingli, W., Jianhui, Z., Xinguo, Z., 2011. Characteristics and evaluations of reservoirs in Carboniferous- Permian, Northeast China. Petroleum Geology & Experiment 33 (5), 499–504.
Yingnan, W., Yuqing, X., 2009. Study on the characteristics of volcanic lithology, facies, and pore structure at the First Member of Yingcheng Formation in Xingcheng Area, Songliao Basin. China Petroleum Exploration 1, 24–29.
Yining, W., Xiaojun, H., Lin, G., et al., 2015. Application of high precision structural model in tight well pattern reservoir prediction and remaining oil tapping. Journal of Xi'an Shiyou University (Natural Science Edition) 30 (6), 17–21.
Yiwei, Z., Qihua, X., Zhizhang, W., et al., 1997. Terrigenous Clastic Depositional Reservoir Description. Petroleum Industry Press, Beijing.
Yong, W., Zhidong, B., Hu, L., 2007. Researches math methods used in reservoir evaluated of low-porosity and low-permeability reservoir. Journal of Southwest Petroleum University 29 (5), 8–12.
Yongguang, L., 2015. Study on fluid identification method of complex lithologic reservoir. Journal of Yangtze University (Natural Science Edition) 12 (29), 36–39.
Yongjun, W., Lin, Y., Qiquan, R., et al., 2007a. Lithologic identification of deep volcanic rock gas reservoir in Xingcheng gas field. Journal of Southwest Petroleum University 29 (2), 78–81.
Yongjun, W., Yongle, H., Qiquan, R., et al., 2007b. Fracture development evaluation for deep volcanic gas reservoir. Natural Gas Industry 27 (8), 31–34.
Yongli, G., Zhiguo, Z., 2011. Valuation on different of pore throat structure of low permeability sandstone by constsp mercury penetration technique. Geological Science and Technology Information 30 (4), 73–76.
Yongmei, T., Zehong, Y., Xiangyang, H., et al., 2010. The complex lithology and hydrocarbon reservoir identification in the May 8 region of junggar basin. Journal of Oil and Gas Technology 32 (1), 257–260.
Yongmin, S., Jin, H., Yuguang, Z., 2004. Description of Fine Oil Reservoir in the Middle and Late Period of Continental Oil Field Development. Petroleum Industry Press, Beijing.

Yongsheng, C., 1993. Oilfield Heterogeneity of Countermeasure Theory. Petroleum Industry Press, Beijing.
Youguo, S., Jingqun, W., Shaohua, W., et al., 2003. Application of comprehensive evaluation of reservoir logging in residual oil distribution. World Well Logging Technology 18 (2), 26–28.
Youjing, W., Xinmin, S., Changbing, T., et al., 2015. Dynamic fractures are an emerging new development geological attribute in water-flooding development of ultra-low permeability reservoirs. Petroleum Exploration and Development 42 (2), 222–228.
Youliang, J., Liqiang, Z., 2007. Subsurface Geology of Oil and Gas Fields. Tongji University Press, Shanghai, pp. 56–85.
Youliang, J., Shenghe, W., Rui, Z., 2012. Recognition of auto-cycle and allo-cycle and its role in strata correlation of reservoirs. Journal of China University of Petroleum 36 (4), 1–6.
Youxi, L., Caijing, W., 2004. Prediction of remaining oil distribution using fractal condition simulation and streamling simulator. Natural Gas Geoscience 15 (1), 42–46.
Yu, Z., 2016. Development countermeasure analysis of 3 blocks in Xinmin Oilfield at the high water-cut stage. Journal of Yangtze University (Natural Science Edition) 13 (20), 56–61.
Yu, K., Yongmin, S., Boren, L., et al., 2010. The complex lithology rock-electricity features of volcanic rocks in Kelameili gas field. Acta Petrologica Sinica 291–301.
Yuan, Y., Jianping, L., Shengbin, F., et al., 2015. Heterogeneity of the massive tight sandstones and its inspirations for the exploration and development of the tight oil. Petroleum Geology and Oilfield Development in Daqing 34 (4), 168–174.
Yuanyan, W., Shenghe, W., Zhengqi, C., 2005. compiled Petroleum Mineral Geology. Petroleum Industry Press, Beijing.
Yubao, L., Jiang, Z., Xiangui, L., et al., 2008. Pore structure features of reservoirs at late high water-cut stage, Lamadian Oiifieid, Daqing. Petroleum Exploration and Development 35 (2), 215–219.
Yubao, L., Zhongwei, J., Zhanjie, H., et al., 2014. Micro-flow characteristics of the oil and water at the late stage of high water cut. Petroleum Geology and Oilfield Development in Daqing 33 (1), 70–74.
Yudan, H., Zhiqiang, M., Lizhi, X., et al., 2005. An improved method of using NMR T2 distribution to evaluate pore size distribution. Chinese Journal of Geophysics 48 (2), 373–378.
Yue, W., Shiyue, C., 2016. Meandering river sand body architecture and heterogeneity: A case study of Permian meandering river outcrop in Palougou, Baode, Shanxi province. Petroleum Exploration and Development 43 (2), 209–218.
Yueming, L., Weixin, L., Xuequn, T., et al., 2007. Diagenesis and reservoir evaluation on the Upper Palaeozonic reservoir sandstones in Daniudi Gas Field, the Ordos Basin. Petroleum Geology & Experiment 29 (4), 384–390.
Yuhong, S., Jiangong, W., Shumei, G., et al., 2006. The application of dielectric log in Erlian Basin. Well Logging Technology 30 (2), 158–160.
Yujiao, D., Shuqin, W., Qinghe, L., et al., 2000. Dielectric logging data correction and its qualitative evaluation of oil, water bed. Well Logging Technology 24 (2), 130–134.
Yukun, Z., Hui, W., Xiaoqing, H., et al., 2016. Reservoir modeling of complex lithologies with sparse wells: a case from A oilfield in Shijiutuo uplift. Bohai Bay Basin. Oil & Gas geology 37 (3), 451–456.
Yumei, S., Wenlin, L., Fengchang, Y., et al., 2003. Feasibility of seismic monitoring for remaining oil in alternate thin-beds reservoir with water-flooding recovery. Acta Petrolei Sinica 24 (5), 52–56.
Yuming, L., Can, L., 2014. Nuclear magnetic resonance log analysis of complex lithologic reservoir. Journal of Oil and Gas Technology 36 (6), 75–78.
Yuming, L., Jiagen, H., 2016. Three-Dimensional Geological Modeling of Fractured-Vuggy Carbonate Reservoirs - A Case Study of Ordovician Reservoirs in Tahe Oilfield. Petroleum Industry Press, Beijing.

Yuncheng, W., Yonggui, Z., Zong, H., 2003. Reservoir Development Earthquake. Sichuan Science and Technology Press, Xinjiang Science and Technology Health Press (K), Chengdu, Sichuan.
Yupu, W., Yikun, L., Qingjun, D., et al., 2014. Development present situation and the countermeasures of the Chinese continental sandstone oilfield. Journal of Northeast Petroleum University 38 (1), 1–9.
Yuqing, Y., Fuxi, P., Hong, T., et al., 2010. Characteristics and classification and evaluation of low porosity and permeability reservoir in Shahejie Formation of BZ25-1 Oilfield. Geoscience 24 (4), 685–693.
Yusheng, Z., Wanming, S., Zhihong, Y., et al., 2015. Evaluation on remaining oil with hydrologic model in braidriver deltas front subfacies: a case of swept layer in Wenxisan Block. Journal of Northwest University (Natural Science Edition) 45 (3), 453–459.
Zafari, M., Reynolds, A.C., 2007. Assessing the uncertainty in reservoir description and performance predictions with the ensemble Kalman Filter. SPE Journal 12, 382–391.
Zaixing, J., 2003. Sedimentary. Petroleum Industry Press, Beijing, pp. 270–282.
Zekri, A.R.Y., Shedid, S.A., Almehaideb, R.A., 2013. Experimental investigations of variations in petrophysical rock properties due to carbon dioxide flooding in oil heterogeneous low permeability carbonate reservoirs. Journal of Petroleum Exploration and Production Technology 3, 265–277.
Zerong, L., Quanlin, X., Weifeng, W., et al., 1993. Oil Reservoir Description Principle and Method Technology. Petroleum Industry Press, Beijing.
Zeyu, Z., Tangqian, Z., Jirui, H., et al., 2016. Visible research on remaining oil after nitrogen flooding in fractured-cavity carbonate reservoir. Petroleum Geology and Recovery Efficiency 23 (2), 93–97.
Zhan, L., Kuchuk, F., Al-Shahri, A.M., et al., 2010a. Characterization of reservoir heterogeneity through fluid movement monitoring with deep electromagnetic and pressure measurements. SPE Reservoir Evaluation & Engineering 13, 509–522.
Zhan, Z., Shenghe, W., Changfu, X., et al., 2010b. Lithofacies and reservoirs of allluvial fan in the Lower Keramay Formation in the Block6 of Karamay Oilfield, the Junggar Basin. Oil & Gas Geology 31 (4), 463–471.
Zhang, T., 2008. Integrating geological conceptual model and its interpretation using multi-point geostatistics in reservoir modeling. Geoscience Frontiers 15 (1), 26–35.
Zhang, Q., Bao, Z., Song, X., et al., 2008. Division and genesis analysis of single sand body in Fuqian Oilfield. Petroleum Exploration and Development 35 (2), 157–163.
Zhaolun, Hongwei, L., Xiangzhong, Z., et al., 2016. Relationship between sandstone architecture – a case of the Kumkol South oilfield in and remaining oil distribution pattern: South Turgay Basin, Kazakstan. Petroleum Exploration and Development 43 (3), 433–441.
Zhaowei, Y., Xiaolong, Q., Shichen, G., et al., 2017. Modeling of different sedimentary facies and assessments for features of spatial structures in the Sulige Gasfield. Special Oil and Gas Reservoirs 24 (1), 32–37.
Zhen, L., Jianchun, D., Wanxuan, Z., 1995. Reservoir evaluation with revised DIVA method in structure LD. Geophysical Prospecting for Petroleum 34 (4), 53–58.
Zhengming, Y., Hanqiao, J., Guangya, Z., et al., 2008. Research on reservoir evaluation index for low-permeability water-bearing gas reservoir. Acta Petrolei Sinica 29 (2), 252–255.
Zhengming, Y., Lingjing, H., Yapu, Z., et al., 2010. Mechanism of non-liner gas seepage flow in water-bearing volcanic gas reservoir. Natural Gas Geoscience 21 (3), 371–374.
Zhengyan, X., 2003. Study on the geological model of reservoir heterogeneity in low-permeability sandstone oil pools. Petroleum Geology & Experiment 25 (5), 505–507, 512.

Zhenliang, G., Congjiao, X., Hu, D., et al., 2009. 3D Imaging and visualization technology of micro pore structure in porous media. Geological Science and Technology Information 28 (2), 115–121.
Zhenlin, Y., Shenghe, W., Du, Q., et al., 2010. An accurate anatomizing method for structure of reservoir of alluvial fan: a case study on Lower Karamay Formation, Liuzhong Area, Karamay Oilfield. Journal of Jilin University: Earth Science Edition 40 (4), 939–946.
Zhenqi, D., Zerong, L., Shaochun, Y., 1998. Study on micro structure of Shengtuo Oilfield. Journal of the University of Petroleum, China 22 (6), 31–33.
Zhenquan, L., Jian, H., Xulong, C., et al., 2005. Microscopic simulation for influence of microscopic reservoir parameters on remaining oil distribution. Acta Petrolei Sinica 26 (6), 69–73.
Zhenyu, P., Wei, S., Jinbu, L., et al., 2013. Microscopic pore structure and seepage characteristics of low-permeability and tight sandstone gas reservoir: taking blocks Su 48 and Su 120 in Sulige Gas Field as an example. Geological Science and Technology Information 32 (4), 133–138.
Zhetan, L., Yuncheng, W., 1986. compiled Pore Structure of Petroleum Reservoir Stratum. Science Press, Beijing.
Zhidi, L., Hongqi, L., Shihua, D., et al., 2008. Study on volcanic rock fracture quantitative identification method using logging data. Petroleum Geology and Oilfield Development in Daqing 27 (5), 132–134.
Zhigao, W., Huaimin, X., Lidong, D., et al., 2004. The analysis of forming-distributing mode and its controlling factors of the remains in the heavy oil reservoirs — an example from the Dalinhe Heavy Oil Reservoir of the Shu II District in the Liaohe Oilfield. Journal of Anhui University of Science and Technology (Natural Science) 24 (3), 19–23.
Zhigui, X., Xuejun, J., Xiaohong, X., 1998. Application of trace elements in the metamorphic stratigraphic correlation. Special Oil and Gas Reservoir 5 (1), 10–13.
Zhihao, Q., Zhuangjing, Y., Shengliang, Z., et al., 1994. Description of Volcanic Reservoir. Northwest University Press, Xi'an.
Zhihong, T., Chunhua, G., Wenming, W., 2007. Technique of dynamic descriptions to the crack and cave carbonate rock reservoir in the Tahe oil field, Xinjiang, China. Journal of Chengdu University of Technology (Science & Technology Edition) 34 (2), 143–146.
Zhijie, H., Yiwei, Z., Qihua, X., et al., 2008. Development of four-dimensional model for remaining oil distribution under control of reservoir sedimentary facies. Acta Petrolei Sinica 29 (4), 562–566.
Zhiliang, G., 2011. Digital oil Field in China. Science Press, Beijing, pp. 6–30.
Zhilin, D., 2000. Study on Reservoir Flow Unit. Petroleum Industry Press, Beijing.
Zhilin, D., 2014. Description and reserves calculation of fractured-vuggy carbonate reservoirs. Petroleum Geology & Experiment 36 (1), 9–15.
Zhipeng, L., Chengyan, L., Bo, D., et al., 2012. Improvement measules and influencing factors of the water flood development of low permeability reservoirs. Earth Science Frontiers 19 (2), 171–175.
Zhiyong, W., Jianhua, G., Dongsheng, W., et al., 2000. Assessment method of carbonate reservoir based on well log data: an example of Shuu-103Block buried hill in west Liaohe Depression. Oil & Gas Geology 21 (2), 157–160.
Zhizhang, W., Jinhua, Y., 1999. Description and Prediction of Low Amplitude Structural Reservoirs. Petroleum Industry Press, Beijing.
Zhizhang, W., et al., 1999a. Description and Prediction of Fractured Reservoirs. Petroleum Industry Press, Beijing.

Zhizhang, W., Yi, C., Lei, Y., 1999b. Changing Law and Mechanism of Reservoir Parameters in Mid-Later Develpoment Period. Petroleum Industry Press, Beijing.
Zhizhang, W., Haiying, H., Yuetian, L., et al., 2010. The complex fractured reservoir: staged numerical simulation and remaining oil distribution prediction. Xinjiang Petroleum Geology 31 (6), 604–606.
Zhizheng, L., 1984. Quality Control Methods and Management. Defense Industry Press, Beijing.
Zhong, C., 2000. Study on relationship between pore structure and displacing oil efficiency. Petroleum Exploration and Development 27 (6), 45–46,49.
Zhong, H., Huaqing, L., Mingjun, S., Haiqing, N. 2012. Application of Seismic Sedimentology in the Area with Complex Lithologies: A Case from the Second Member of Shahejie Formation in Qibei Sag. 24(4): 40-44.
Zhongfeng, L., Shunli, H., Chengquan, M., et al., 2005. Experimental study on heterogeneity 3D model water driving remaining oil. Oil Drilling & Production Technology 27 (4), 41–44.
Zhonghong, Z., Zhengming, Y., Xiangui, L., et al., 2012. A grading evaluation method for low-permeability reservoirs and its application. Acta Petrolei Sinica 33 (3), 437–441.
Zhongran, L., Yanshan, N., Fengling, X., et al., 2004. Application log constraining reservoir invension technology to low permeability reservoir description. Journal of Daqing petroleum institute 28 (6), 89–91.
Zhongshan, S., Xuejing, M., Jiahua, W., et al., 2013. Application of multi-point geostatistical modeling in reserves calculation of Daqing Mijing pattern oilfield. Journal of Xi an Shiyou University (Natural Science Edition) 28 (4), 64–68.
Zhongsheng, H., Zhigang, C., Guanghui, C., et al., 2016. Evaluating residual oil saturation and water-flood degree with hydrocarbon ratio logging. Journal of Southwest Petroleum University (Science & Technology Edition) 38 (5), 41–49.
Zi, R., Cheng, C., Jun, L., 2011. Characterization of reservoir configuration and residual oil distribution in Block X10-2. Special Oil and Gas Reservoirs 18 (6), 40–43.
Zifei, F., Kongchou, L., Jianxin, L., et al., 2014. A study on remaining oil distribution in a carbonate oil reservoir based on reservoir flow units. Petroleum Exploration and Develpopment 41 (5), 578–584.
Ziqi, S., Lilei, Y., Hong, W., et al., 2007. Comprehensive evaluation method of reservoir flow unit with grey system. Petroleum Geology and Oilfield Development in Daqing 26 (3), 76–81.
Zitao, W., Chengyan, L., Shizhen, C., et al., 2017. Sensitivity analyses of parameters for multiple-point geostatistics modeling. Journal of Xi'an Shiyou University (Natural Science Edition) 32 (1), 44–51.
Zongbao, L., Li, Y., Fei, G., et al., 2014. Remaining oil enrichment regularity and extract method of oilfield in high water-cut stage fault edge Taking Putaohua reservoir of North Fault Block in Xingnan Oilfield in Songliao Basin as an example. Journal of Northeast Petroleum University 38 (4), 52–58.
Zongxiong, T., 1994. Development Classification and Description of Reservoir. Petroleum Industry Press, Beijing.
Zubing, L., Minggao, Y.Q.L., 2007. Reservoir heterogeneity indication applied in researching for heterogeneity of the grail oil: an example of oil group Below V in Shuanghe Oilfield. Geological Science and Technology Information 26 (6), 83–87.

# Index

*Note*: Page numbers followed by "*f*" and "*t*" refer to figures and tables, respectively.

## N



## O

## X



## Y



## Z

FSC
www.fsc.org
MIX
Papier | Fördert
gute Waldnutzung
FSC® C083411

Kolibri 360 GmbH
Ferdinand-Jühlke-Straße 7
99095 Erfurt, Deutschland
produktsicherheit@kolibri360.de